EIB

Installation Bus System

Edited by Thilo Sauter,
Dietmar Dietrich,
Wolfgang Kastner

PUBLICIS

Die Deutsche Bibliothek – CIP-Cataloguing-in-Publication-Data

A catalogue record for this publication is available from Die Deutsche Bibliothek

Originally published in German language
by Hüthig GmbH & Co. KG, Heidelberg
under the title
Dietrich, Kastner, Sauter: EIB. Gebäudesystem. 1st ed. 2000
Copyright by Hüthig GmbH & Co. KG, Heidelberg

This book was carefully produced. Nevertheless, authors and publisher do not warrant the information contained therein to be free of errors. Neither the authors nor the publisher can assume any liability or legal responsibility for omissions or errors. Terms reproduced in this book may be registered trademarks, the use of which by third parties for their own purposes my violate the rights of the owners of those trademarks.

http://www.publicis-komm.de/books

ISBN 3-89578-175-4

Editor: Siemens Aktiengesellschaft, Berlin and Munich
Publisher: Publicis Corporate Publishing, Erlangen
© 2001 by Publicis KommunikationsAgentur GmbH GWA, Munich
This publication and all parts thereof are protected by copyright. All rights reserved. Any use of it outside the strict provisions of the copyright law without the consent of the publisher is forbidden and will incur penalties. This applies particularly to reproduction, translation, microfilming or other processing, and to storage or processing in electronic systems. It also applies to the use of extracts from the text.

Printed in Germany

Thilo Sauter / Dietmar Dietrich / Wolfgang Kastner (Eds.)   EIB

The authors of this book:
Professor Dietmar Dietrich, Vienna University of Technology, Austria
Dr. Uli Doberer, Bosch, Germany
Professor Peter Fischer, FH Dortmund, Germany
Dr. Horst Gerlach, Siemens, Germany
Mr. Marc Goossens, EIBA, Belgium
Mr. Mikhail Gordeev, Vienna University of Technology, Austria
Mr. Markus Haag, Vienna University of Technology, Austria
Dr. Christian Heite, Busch-Jaeger, Germany
Dr. Wolfgang Kastner, Vienna University of Technology, Austria
Mr. Martin Knizak, Vienna University of Technology, Austria
Mr. Hans R. Kranz, Siemens, Germany
Dr. Martin Manninger, Austria Card, Austria
Mr. Peter Palensky, Vienna University of Technology, Austria
Mr. Thomas Rauscher, Vienna University of Technology, Austria
Dr. Alexander Redlein, Vienna University of Technology, Austria
Dr. Heinrich Reiter, EIBA, Belgium
Mr. Norbert Reiter, Vienna University of Technology, Austria
Dr. Thilo Sauter, Vienna University of Technology, Austria
Dr. Erich Schäfer, FH Wiener Neustadt, Austria
Mr. Norbert Stampfl, Vienna University of Technology, Austria
Mr. Hanns-Karl Tronnier, EIBA, Belgium

Translation:
Sharon Tenniswood

Foreword

There are numerous books on the subject of EIB aimed at installers and system integrators that guarantee the application of this standard technology across all disciplines. There is already enough introductory literature on the market covering the practical application of the EIB system. For those of a more technical/scientific background, i.e. students, product manufacturers, software developers or experienced system integrators, the EIB manual published by the European Installation Bus Association (EIBA) has long been the book of choice for concrete, technical background information on EIB. In its current issue, version 3.0, this book contains the entire specification of the EIB system including all relevant details. However, the shear scope of the EIB manual (around 2000 pages) indicates that gaining a comprehensive knowledge of the fundamentals of EIB would be a very time consuming business indeed. The idea of writing a scientifically based reference book was born.

The pure application of the EIB system was not to be the main focus – too great is the current scope of this area. An exact description of the EIB protocol as well as the necessary technical background information on system components were given top priority; the 21-strong team of authors also wanted to examine peripheral areas such as PC software and the Internet. The result of these efforts, under the watchful eye of Professor Dietrich of the Institute for Computer Technology of the University of Vienna, is the book before you, which not only meets the technological and scientific needs but also outlines the complex relationships between the many participant technologies. The rapid development of the Internet is sure to provide the basis for many more EIB products. The merging of Internet technologies with the EIB is just as significant as the current convergence process, which brings the EIB into contact with the Batibus and EHS systems. With that, EIB is expanding its sphere of influence in different directions. We would like to take this opportunity to thank all those authors who took part in the writing of this book.

The various established technologies associated with the Internet, the rise in the use of mobile radio as well as a number of newly developed protocols for PC peripherals and entertainment electronics mark the transition into an age of universal networking. These areas are constantly subjected to radical change. This is a totally natural process, but when it comes to the field of home and building electronic systems, sustainability is an important consideration. The system user must be able to assume that his investment is protected in the long term. For home and building electronic systems it is therefore necessary to create standards by using open and extendable technologies that are not only capable of meeting future requirements but also take into account characteristics such as compatibility and non-depreciation of value. In the future, we will see a whole range of new EIB products that can be used in connection with other networks. This book can only provide a preview; a detailed explanation covering all complex network links would far exceed the scope of this book. With that, this book can also be viewed as a basis for future work. It is a text book, a reference book and specialist "bible" all rolled into one.

Foreword

My last ten years as President of EIBA have been shaped by growth and success. With EIB, we have achieved an expression that is and will continue to be associated with uncompromising interoperability, complete openness and standardized training. This book underlines such aspects by linking some of the underlying technical principles with relevant strategic focuses. Such a book greatly enhances the world of EIB. In the last few years in particular, there has been a rising demand for scientific literature from fields outside academia. I am looking forward to the publishing of this book in English and Russian with great expectations. I am convinced that this global circulation will make a significant contribution to the greater and sustained success of EIB.

Günter G. Seip
President of EIBA

Preface

To most people, the word "networking" brings to mind the Internet, not many associate this term with local networks in offices or operations. Contrary to this, the term fieldbus has not yet gained the status that it deserves despite the fact that many different fieldbus systems are now used in so many areas such that we can no longer imagine life without them. There are many examples of their usage in fields such as industrial automation, automobile construction, air and space travel. Many of the processes in these automation systems are no longer imaginable or even feasible without the use of a corresponding fieldbus. In addition to the obvious advantages in terms of performance, the use of fieldbus technology has meant a dramatic reduction in both installation and maintenance costs.

More recently, the fieldbus has also made an impact on the field of building automation and control. Admittedly, traditional installation technology still has a firm hold in this area, but there has been a definite upwards trend in recent years. More and more buildings, in the private sector too, are being networked, and greater numbers of companies are developing products for the private end user. Even entertainment electronics and white goods are being adapted for this future-oriented technology. In addition to installation and maintenance costs, the area of energy saving should also be viewed as a significant cost advantage. If we look further into the future, the concept of an "intelligent house" can scarce be imagined without the fieldbus.

One representative of a fieldbus system tailor-made for buildings is the European Installation Bus (EIB). It represents a clear, well thought out concept. Whilst many fieldbuses demand a clean break from the technology previously used within their areas, it has been possible to integrate EIB into the existing installation. With that, it both facilitates a smooth transition from traditional technology to modern networking whilst simultaneously offering all the advantages of this new technology. This means that devices must be certified before they can be awarded the EIB logo. This also guarantees however, that they can then be integrated into corresponding systems by suitably trained personnel without fear of compatibility problems. Even the training itself is certified, ensuring the highest quality on all levels.

EIB has been certified by CEN TC 247 within ENV 14154-2. It is an open system, which means that companies are able to develop their own products. More than 110 companies are now producing EIB devices and more and more universities are examining this bus concept with the aim of finding new design possibilities and application areas. EIB can be used equally well in small systems or large systems with more than 10,000 nodes. New challenges include the various possibilities of linking these networks to the Internet or the adaptation to building backbones such as for example, Ethernet, BACnet and FireWire (IEEE 1394). Much of this work, together with further standardization activities, already provides a clear indication of the direction that EIB is taking.

Preface

There is no doubt among the experts that this technology will eventually replace the traditional installation technology. There is also no doubt that we are about to experience radical change in the field of installation technology and building networks. There is enormous potential for greater energy savings, improved safety measures, increased quality but equally for greater comfort right through to an economically feasible, automated facility management even in the private sector. Here there is much to do and much to be expected. EIB has already become established and there is now a market out there. However, we are still waiting for the real upturn in the field of EIB components, associated applications and extended research areas. This phenomenon can be seen repeatedly in the field of computer technology: Totally new concepts that are based on computer technology need time before the necessary level of acceptance is attained. Once it has been, the advantages are immense – demand increases production numbers and variety and the prices fall rapidly. A new market is opened up that can push classic technology into the background – not least due to the high piece numbers that are now possible in this climate of globalization.

In the writing of this book, the expectations of the authors were high. It was conceived as an easy-to-understand introduction to the technology and a reinforcement of existing knowledge, but also as an in-depth reference. It has been written by university lecturers and professors, some of whom have been involved with the development of EIB right from the very beginning. The list of authors also includes development and application engineers who in the field of EIB are the true experts. There have also been contributions from a number of other authors who have the necessary knowledge concerning practical application, research and training in their respective areas. However, this book is not simply a collection of articles from various authors. Each chapter has been read and revised by several authors and adapted and coordinated so that the final product is a uniform work. The publishers have made every effort in cooperating with the authors to ensure a comprehensive work that can be used as a reliable reference for anyone working with EIB.

We would like to thank the publishing house for their patience with this project. The book, as any book concerned with the networking of systems, is the result of a wide-ranging, democratic consensus – and this demands time, a great deal of time. Despite any shortcomings, we are convinced that we have succeeded in providing a top class piece of work. We would also like to thank the European Installation Bus Association (EIBA), in particular Mr. Steven deBruyne, who has provided invaluable support on more than one occasion. Without them, this work could not have been achieved. They have also helped in having the book translated into Russian. This will enable us to provide support to more than 100 universities in Russia in the area of fieldbus technology. We believe that a country undergoing such fundamental change will see long-term benefits from the use of this technology.

Finally, the present English edition was also made possible by EIBA. The outline basically follows the German original, however it is not just a simple translation. The chapters on network integration and tools have been updated to include the latest achievements in IP connectivity. This additional work of coordination and proofreading is also reflected in the changed order of editors, which no longer alphabetical like in the German edition.

Vienna, September 2001

Thilo Sauter
Dietmar Dietrich
Wolfgang Kastner

eibbuch@ict.tuwien.ac.at

Contents

1	**Introduction**	15
1.1	The Significance of Communication	15
1.2	The Basics	24
	1.2.1 Hierarchies	24
	1.2.2 Principles of the ISO/OSI Model	27
	1.2.3 OSI Levels	27
	1.2.4 Layer Layout	31
	1.2.5 Topologies	33
	1.2.6 Interaction Time Diagrams and SDL	36
	1.2.7 Tools	39
	1.2.8 EMC, Lightning and Overvoltage Protection	41
1.3	Common Fieldbus Systems – An Overview	42
	1.3.1 Interbus	43
	1.3.2 P-NET	44
	1.3.3 Profibus	47
	1.3.4 CAN	49
	1.3.5 EIB	51
	1.3.6 LonWorks	53
2	**Networks, Disciplines, Applications**	56
2.1	Cabling in Functional Buildings	56
2.2	Cabling in Domestic Buildings	57
2.3	Building Categories	59
2.4	Building Disciplines	61
3	**EIB Protocol**	63
3.1	Introduction	63
	3.1.1 OSI Model	63
	3.1.2 EIB in the OSI Model	65

Contents

		3.1.3	A Brief Overview of EIB	68
3.2	Physical Layer			69
	3.2.1	Topology		70
	3.2.2	Logical Unit		71
	3.2.3	Medium Attachment Unit		76
3.3	Data Link Layer			79
	3.3.1	Logical Link Control		80
		3.3.1.1	Types of Addressing	81
		3.3.1.2	Datagram Service (L_Data)	82
	3.3.2	Medium Access Control		87
	3.3.3	Frame Formats (LPDU)		90
		3.3.3.1	Data Frame	90
		3.3.3.2	Acknowledgement Frame	94
	3.3.4	Error Detection		95
	3.3.5	Bridges and Routers		96
3.4	Network Layer			96
	3.4.1	Unicast (N_Data)		98
	3.4.2	Multicast (N_Group) and Broadcast (N_Broadcast)		99
	3.4.3	Routers		101
		3.4.3.1	Routers and End Device Addressing	101
		3.4.3.2	Routers and Group Addressing	104
		3.4.3.3	Services	105
	3.4.4	Network Protocol Data Unit (NPDU)		108
3.5	Transport Layer			109
	3.5.1	Connection-Oriented Communication		109
		3.5.1.1	Connection Setup (T_Connect)	111
		3.5.1.2	Data Transfer (T_Data)	113
		3.5.1.3	Termination (T_Disconnect)	117
	3.5.2	Connectionless Communication		122
		3.5.2.1	Unicast (T_Data_Unack)	122
		3.5.2.2	Multicast (T_Group)	123
		3.5.2.3	Broadcast (T_Broadcast)	124
	3.5.3	Transport Protocol Data Unit (TPDU)		125
3.6	Application Layer			126
	3.6.1	Communication Objects		126
	3.6.2	Device Configuration		130
	3.6.3	Memory Access		134
	3.6.4	Analog/Digital Converter Access		136
	3.6.5	Property Access		136
	3.6.6	Security		139

		3.6.7	Other Services	141
		3.6.8	Application Layer Protocol Data Unit (APDU)	141

4		**Application Environment and Network Management**		143
4.1	Application Environment			143
	4.1.1	Communication Objects		143
	4.1.2	EIB Objects		148
4.2	Network Management			149
	4.2.1	Introduction		150
	4.2.2	Configuration Tables		151
	4.2.3	Network Configuration		154
	4.2.4	Device Configuration		155

5		**EIB Hardware**			157
5.1	Topology				157
5.2	Media				159
	5.2.1	Twisted Pair			159
		5.2.1.1	Data Transmission Criteria		160
		5.2.1.2	Global Power Supply		160
		5.2.1.3	Connection Possibilities		163
	5.2.2	Radio Medium			165
	5.2.3	Powerline			166
5.3	Communication End Devices				167
	5.3.1	Components of an End Device			168
	5.3.2	Transceivers			169
		5.3.2.1	Twisted Pair Transceiver		169
		5.3.2.2	Powerline Transceiver		174
		5.3.2.3	Radio Transceiver		176
		5.3.2.4	Internal Power Supply		177
	5.3.3	Processing Unit			178
		5.3.3.1	Twisted Pair Implementations		178
		5.3.3.2	PL Implementation (Operating System 1.x)		181
		5.3.3.3	RF Implementation (Operating System 2.x)		181
	5.3.4	Communication/Application Module Interface			181
	5.3.5	Certified Communication Module			185
5.4	Couplers and Interfaces				185
	5.4.1	Twisted Pair Coupler			186
	5.4.2	Powerline Band Stop			187
	5.4.3	Powerline Phase Coupler and Repeater			189

		5.4.4	Media Couplers	189
		5.4.5	Data Interface	191
6		**EIB Software**		**192**
6.1		System Software		194
		6.1.1	Operating System Implementation 1.2	194
		6.1.2	Operating System Implementation 2.x	198
		6.1.3	Operating System Implementation 70.x	202
		6.1.4	PL-Implementation	204
		6.1.5	RF-Implementation	207
6.2		Interface to the Internal Application		207
		6.2.1	Group Addressing with Communication Objects	207
		6.2.2	Physical Addressing with EIB Objects	210
		6.2.3	Physical Addressing with Direct Memory Access	212
6.3		Interface to External Applications		212
		6.3.1	Serial Protocols for PEI Configuration Types 12 and 16	213
		6.3.2	Serial Protocol for PEI Configuration Type 10 (FT1.2)	214
		6.3.3	Message Exchange	214
7		**Integrating Building Networks**		**216**
7.1		Types of Couplings		216
7.2		Using Couplings		218
		7.2.1	Extending the Range	218
		7.2.2	Remote Control	219
		7.2.3	Remote Maintenance and Remote Services	220
7.3		Examples of Internet Access		221
		7.3.1	Java-Oriented Approach	221
		7.3.2	SNMP-Based Approach	222
		7.3.3	SOAP/HTTP-based approach	224
7.4		IP Connectivity Models and Applications for EIB		225
		7.4.1	Device Network Protocol Transfer	227
		7.4.2	Common Datapoint Model	228
		7.4.3	Applications in EIB Controlled Buildings	228
7.5		Security Aspects		228
		7.5.1	Data Security Concepts and Definitions	229
		7.5.2	Attacking Fieldbuses	230
		7.5.3	Secure Architecture Elements	231

		7.5.3.1	Chip Cards and Smart Cards	231
		7.5.3.2	Security Features of the Third Version of SNMP	233
	7.5.4	Construction of Secure System Architecture		233
		7.5.4.1	General Considerations	234
		7.5.4.2	Data Security at the Fieldbus Level	234
		7.5.4.3	Security Concepts for a Gateway and a Client	236

8	Tools		238
8.1	Development and Test Tools		238
	8.1.1	Integrated Development Environment	239
	8.1.2	Bus Monitor	239
	8.1.3	EIB Interworking Test Tool	239
8.2	EIB Tool Software		240
	8.2.1	Application Areas of ETS	241
	8.2.2	Manufacturer Extensions of ETS	242
	8.2.3	ETS2 as the DDE Server	242
	8.2.4	ETS via Internet	243
	8.2.5	EIBnet/IP and iETS Gateway "Server"	243
8.3	EIB Tool Environment Component Architecture		243
	8.3.1	An Overview of Previous EIB Components	244
	8.3.2	Models Behind the New eteC Components	245
	8.3.3	eteC Falcon	246
	8.3.4	eteC Eagle	246
	8.3.5	Other Components	247
8.4	EIB OPC Server		248

9	Facility Management		249
9.1	Definition		249
9.2	Prerequisites		252
	9.2.1	Profitability	253
	9.2.2	Data Acquisition	255
	9.2.3	Maintaining and Updating Data	256
	9.2.4	Facility Manager	257
9.3	EDP Tools for Facility Management		258
	9.3.1	CAIFM Tools	259
	9.3.2	Facility Management and Fieldbus Systems	260
	9.3.3	System Architecture	260
	9.3.4	Facility Management and EIB	261

9.4	Virtual Facility Management	262
10	**Standardization**	264
10.1	Interoperability	264
10.2	EIB Interworking Standard	268
10.3	The Standardization of EIB	270
	10.3.1 The Current Situation	270
	10.3.2 Relevant Standardization Committees	271
10.4	Certification	274
10.5	European Installation Bus Association – EIBA	275
	10.5.1 Development of EIB	275
	10.5.2 The Aims and Tasks of EIBA	276
11	**Performance Aspects**	278
11.1	Measurements on the Bus Level	278
	11.1.1 Latency Time Measurement	278
	11.1.2 Reaction Time Measurement	279
	11.1.3 Response Time Measurement	280
	11.1.4 Data Throughput	280
11.2	Measurements on the User View	284
	11.2.1 Waiting Time Measurement	284
	11.2.2 Reliability Measurement	285
12	**Outlook**	287
12.1	Convergence	287
12.2	BACnet	289
12.3	EIB and Jini	290
12.4	IEEE 1394	295
12.5	Intelligent Software Agents	298
	Abbreviations	301
	Bibliography	304
	Index	311

1 Introduction

Networks such as EIB (European Installation Bus) should be viewed as part of an evolutionary stage that will have longer lasting effects than previously thought.

1.1 The Significance of Communication

The possibilities of networking electrical components and the meaning of communication are more easily understood if we take a look at some examples from a selection of current scientific works. Two such scenarios are represented below. They are essentially based on the work of [Tis 00] and general reports from MIT (Massachusetts Institute of Technology, USA), FeT'95 and FeT'97 [Diet 95, Diet 97].

Vehicle technology

Contrary to home and building automation, the networking of vehicle technology is already well advanced. Despite this, it is not yet as widely used when compared with the airline industry (take the train crash in Eschede for example – this could easily have been avoided if networked sensors had been used to test the sensitive parts of the vehicle chassis online). The basic idea is to record the systems using sensors, manage the information in as decentralized a manner as possible and provide the driver with the most relevant information. This includes simple things such as continuously checking the lights, and not just when switched on (built-in test systems [Stein 99]), but also security systems that render it impossible to steal a motor vehicle by disabling the major systems such as the brakes or fuel injection, or using frequency measurements and subsequent analysis to establish the onset of any cracks in the mechanical parts.

To ensure that this complex technology does not lead to mismanagement in normal operation, it needs to be incorporated not only reliably but within a communication algorithm that allows the system to independently detect any errors and even to remove them or render them ineffective. With regard to maintenance, the error must be identified to such an extent that allows it to be rectified within a simple workshop with as little effort as possible.

Vehicle navigation systems coupled with GPS (Global Positioning System[1]) were introduced into the motor vehicle industry some time ago, but remain relatively expensive. It won't be long however until we see the mass introduction of this technology. Vehicles that have such navigation systems need a PC for the volume of data that needs to be stored. It also continuously receives and processes the current traffic information and is required for other reasons too: Microsoft, DaimlerChrysler, Toshiba, Nokia and Mannesmann among others are cur-

---

[1] Satellite-based location system

1 Introduction

rently working to make motor vehicles complete communication units. Not only because e-mail and fax connections are soon to become standard in vehicles or to allow bored passengers in the back to be able to surf the net whenever they want. The aim is to automate position determination and path-finding mechanisms as far as possible[2], to design the vehicle as a communication center and perhaps more importantly, to be able to carry out essential vehicle maintenance online, wherever it happens to be.

With the e-mail connection, the vehicle can correspond with the depot whenever necessary and in the case of errors, suitable measures can be suggested to the driver. A typical scenario could run as follows: The vehicle addresses the driver with the following message – "The headlight bulb is defective, should I check which depots in the area can deal with the problem?" After a "yes" from the driver, the vehicle sets up an e-mail connection with the control center, which holds a record of all depots, and receives a list stating who can do the job immediately and at what price. The driver need only decide which option to take. Other similar scenarios are also possible.

An important characteristic of communication systems becomes clear – the integration of a communication unit which allows children traveling in the back to surf the net is difficult to justify and analysis systems that check critical components only are hardly ever installed (again we go back to the accident in Eschede – why didn't the train have such alarm systems in all the chassis parts? They probably will now). However, if a network was to be integrated into the vehicle to which all the various units could be easily connected and at low cost, there would be no more arguments. Microsoft, DaimlerChrysler and other firms have understood this principle and are currently working on putting this idea into practice.

The situation is similar for automation systems in the home.

Home automation

Private homes will soon have a set-top box or other home-gateway solution (see Figure 1.1) that opens up various communication paths with the outside world. Standard access paths include ISDN[3] or ISBN, TV[4] and mobile radio (talk at the moment, except in Great Britain, is about broadband transmission via the power supply network (400V powerline [Gois 98])). In the house, this set-top box would be connected to a backbone, which could for example be based on FireWire (IEEE 1394), Fast-Ethernet or Gigabit-Ethernet. It has sufficient bandwidth to simultaneously allow multiple TV channels, telephone conversations, data transmission between PCs as well as the transfer of diverse control information. There will be access to fieldbus systems such as EIB in the individual rooms, which represents networking at the field level.

This will allow the implementation of functions that can currently only be imagined in very limited form. We will list a few of these functions to give you a rough idea of what will be possible:

---

[2] Timely information concerning traffic jams together with the suggestion of alternative routes make such an investment both economical and ecological.

[3] ISDN: Integrated Services Digital Networks, ISBN: Integrated Services Broadband Network

[4] TV: Television channel

1.1 The Significance of Communication

- The room detects the person, identifies him, and attempts to adapt itself accordingly; possible identification criteria are: figure, size, smell, habits or perhaps simply a smart card.
- A personal tracking system is activated, in order to monitor his security, e.g. from a health point of view.
- The room will pose questions and react to unusual behavior.
- Graphic and acoustic representation units (corresponds to the PC/TV, which will only be available in the future as a combined unit) or acoustic/visual input units will represent the room-human-interface, in exactly the same way as microphones and loudspeakers that will be distributed around the rooms.
- Garbage bins will be checked automatically to see how full they are and upon reaching a specific threshold an e-mail will be triggered.
- The heating will be regulated by an intelligent control mechanism that will derive its data from a simple weather recording system in the house as well as via the current weather forecast in the Internet.
- Proper ventilation is achieved via CO_2 and CO sensors, with the later use of dedicated olfactory sensors, promised by scientists in the field micro-mechanics.
- Soon, the regular light switches in use today will be viewed as nothing more than backup switches, as the room knows when the people present need light and when they don't. The intelligent control can also decipher what level of lighting is required – the light needed to read something from a monitor is different to that needed to read a newspaper. The person of the future will also be able to verbally inform the system of the lighting he requires.

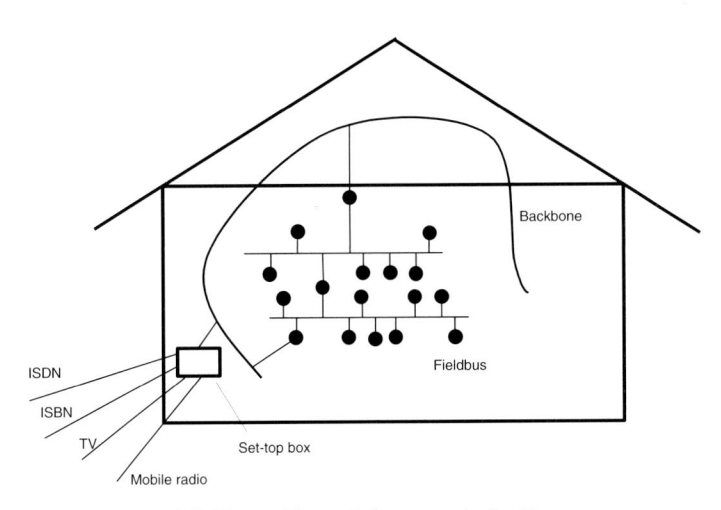

Figure 1.1: Possible Future Network Structure in the Home

The demands placed on today's simple event controls, whether centralized or decentralized, are clearly excessive. However, they will still be used in the future as the basis for AI home

1 Introduction

systems[5]. Fuzzy controllers and neuronal networks or intelligence-based systems will be used on top of these simple process controls.

The following scenario could become commonplace. You are taking an afternoon nap on the sofa in the lounge, after having just hung out some washing to dry. The house AI system has established that it is about to start raining and wakes you with the words "According to Internet reports and the local data it looks like rain, maybe you should bring the washing in". You could then respond with "Show me the weather report!" The room would then respond with the corresponding weather map and ask whether you would like to see more data from houses in the immediate vicinity. The room-human-interface would be represented by a series of spoken dialogues and monitors installed in every room.

The skepticism of many, with regard to the financial feasibility of all this, is easily dispelled. After the current liberalization of the communication and energy markets, the various service providers are now forced to offer their customers advantages. We are already in the midst of this in the communications field, and we are standing on the threshold in the energy market. Pre-payment will be introduced with the advantage of easier settlement. There will be no more personal customers. The anonymous inhabitants of a house will buy their power and water etc. in specific quantities, provided via addressed meter controllers. Energy will be priced differently according to the time of day, helping to reduce peak loads. Companies capable of successfully implementing these concepts will be able to supply their energy at lower prices. The result will be a fall in the cost of water and electricity, etc.

If the house system could inform me via my mobile phone that someone was trying to break in, allowing me to take immediate action such as locking all doors and windows, completely blacking out all lights and sounding an alarm to scare off the intruders and warn the neighbors, which would also be possible via e-mail or fax etc., then I would be prepared to pay for this. Especially when the necessary electronic components are already integrated into the house and "all" that is required is the corresponding additional software.

If through corresponding measures, the systems in the house could be fully integrated with the result of a significant saving in energy [Reit 98], then to a certain extent, the profitability of such systems could already be proven today [Diet 95].

A decisive counterargument is often thrown into the field: Does security, in the sense of the inviolability of man, fall by the wayside? What help is it, if the technology allows a great deal of possibilities, but privacy, data protection and controlled access to information cannot be guaranteed.

[Mann 98] examines this subject and shows ways of solving these problems. Banks are changing to electronic payment methods, mainly for economic reasons. Cash in electronic form only makes sense, if among other things

1. The national bank or central European bank can continue to supervise and safeguard the money reserves
2. The unauthorized production of electronic money is too expensive
3. Usability is simple and flexible

---

[5] AI: Artificial intelligence (understanding natural speech, image recognition, solving problems with expert knowledge, etc.)

1.1 The Significance of Communication

4. The associated costs are decidedly lower than the money circulation costs of today

The measures to be derived from this are clear. Money transfers must continue to be conducted via the central bank, which will continue to play the central, controlling role in the bank-trader-customer triangle (see Figure 1.2). As the high security of electronic units can only be guaranteed within the secure confines of the bank or in the form of smart cards (and all other electronic systems such as PCs and encapsulated computer units are relatively easy to manipulate), the communicative transfer of electronic payments will continue to be based on the use of smart cards. Compared to the level of profit to be gained, corrupting a card would be too costly making smart cards almost immune to vandalism[6].

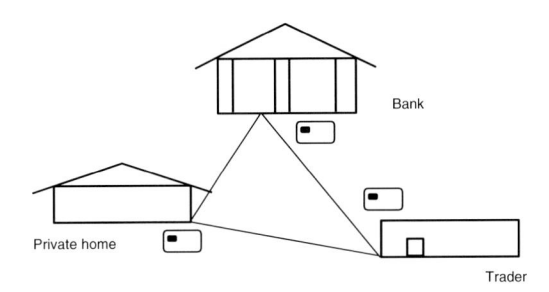

Figure 1.2: The Bank-Trader-Customer Triangle

With that we come to point 3 of the above list – usability and flexibility. The customer does not want to have to travel to a machine to load his card. This should also be possible from the home, where in the opposite direction, he can make anonymous payments via the Internet. This demands the corresponding access possibilities, highlighted in Figure 1.1. At once, the set-top box is given a new, central function. It now forms the central access point to a house and with that needs to be properly secured. It is clear that a number of further (hierarchically ordered) security measures[7] need to be implemented (when reading the meter, specific sensor data, etc.). This can be achieved on the simple basis of bilateral communication between two smart cards (Figure 1.3). If these smart cards each contain a controller chip, in which the coded communication protocol is integrated, they are then well protected against attack, as the decoding would involve a great deal of effort. In their own interests, the banks must provide a guarantee against this.

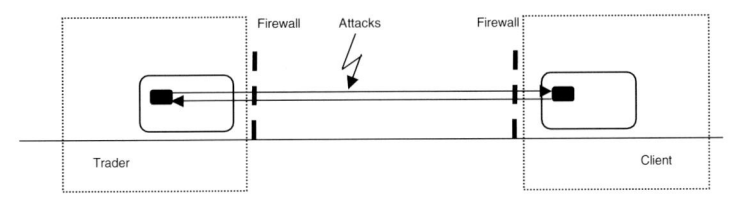

Figure 1.3: Secure Point-to-Point Connection

---

[6] Vandalism will never be eliminated altogether, we can only try to limit the damage caused.

[7] What are meant are security measures against vandalism. The "boundaries" should not be obvious to everyone and such measures should make it difficult to cause damage/disturbance. Their sense is to make the limits obvious to everyone and to make it difficult to annoy others.

The logical conclusion is as follows: For economical reasons the banks need "electronic commerce". This can only be implemented when it becomes possible to pay from home using smart cards, where it is also possible to reload them. The banks then require correspondingly security measures, which must be incorporated into the concept of an "intelligent house". Here, as already mentioned, [Mann 98] suggests a concept that is not only compatible with the networks within a house, but is also part of them. With that, the smart card can be used as the "key" for the intelligent house, offering a very high standard of security. This topic will be examined in more detail later in the book.

Another argument that is often used against fieldbus systems within the home is reliability. It is a well-known fact that the simplest systems offer the greatest reliability, which in more complex systems requires an additional and enormous effort. As long as the lights are switched on and off manually via a 230V switch and as long as the elevators are controlled via simple contactors, then with regular maintenance not much can go wrong. If instead of these switches a computer is responsible for switching on the lights and setting the elevators in motion, there are a multitude of failure possibilities – all of which must be taken into consideration if the system is to offer the same or improved level of reliability. This leads us to the question of how to achieve this.

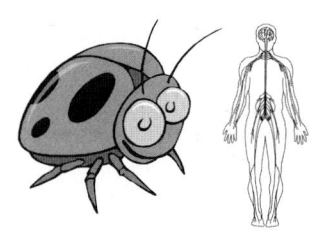

Figure 1.4: Comparing Insects and Vertebrates

We will first look at a comparison in nature. Insects of course have a much simpler information system than vertebrates (Figure 1.4). For a start, they have far fewer nerves[8], and secondly they have a simple, economically optimized structure [Dawk 99]. Beetles for example have ganglion cells on the legs, each of which directly controls a leg, and on the head to control the complex antennae movements etc. The remaining nervous system, which controls the digestive tract and other functions within the body, is kept relatively simple as the entire body is protected by an armored shell. It is quite different for vertebrates! Take the brain for starters, organized on a decentralized basis but placed centrally within the body. Added to this is the enormous quantity of different sensors that are arranged in parallel throughout the body. Two principles immediately jump out – the continuous checking of the body units (in the case of external attacks or failed functioning the corresponding nerves report directly to the brain, where a significant part of this alarm is the pain we feel) and the high level of redundancy.

It is these natural principles that must be incorporated. Continuous checking is relatively easy to achieve, it only requires us to do a bit of rethinking. Every modern digital IC (inte-

---

[8] The ability of creating highly developed social structures in insects for example, is related to the ratio of the weight of the nerves to the total body weight. [Lind 67] has already established the following ratios: cockchafer: 1/3500, worker ant: 1/628 and worker bee: 1/174.

grated circuit[9]) of a fieldbus system requires built-in test units for testing production. These can be expanded quite easily to allow the subsequent use of online tests. The fact that this demands an equally high level of investment in the field of network management is obvious and this is sure to remain a comprehensive research area for many years to come. It won't be long however before every electric razor is able to send an e-mail shortly before it breaks down, giving the manufacturer a comprehensive picture of the quality of its products, and soon every vacuum cleaner will be able to inform the user that the motor has failed and needs replacing.

The question of redundancy is more difficult. The effectiveness of redundancy is only really seen when an error arises. However, we learn the opposite by experience – things are usually good, so why do we need redundancy? Here we must apply integral thinking and probable failures must be taken into consideration. We must learn to assess which errors actually occur, the extent to which they can be accepted and which, for whatever reason, need to be avoided. This is a principle that commonly occurs within the field of security but which unfortunately is seldom applied. All too often, it is only the price that is considered when buying products; the buyer does not often consider the possibility of damage. Who buys a car on the basis of how much it is going to cost over the next 7 years or however long it is owned?

When integrating networks such as the fieldbus system, it is necessary to consider right from the planning stage just how advantageously redundancy can be presented. More and more planners are demanding higher levels of redundancy for exchange systems, elevator controls, lighting systems etc. The main considerations are not just safety related; they are also a question of economics.

The basic idea behind networking via fieldbuses such as EIB

EIB and other fieldbus systems in building automation and control (such as LonWorks) do not yet include the characteristics to satisfy the above-mentioned specifications. They only represent part of the described concept, i.e. networking at the field level.

Previous applications have only really replaced the traditional installations; the functional boundaries remain the same. Integrated heating sensors, for example, are not used for other purposes. The room temperature, which is recorded for the purpose of controlling the heating system, is not used for a fire alarm and the data is not used to simultaneously control the windows. Why should we not create a functional link to prevent the windows being left open in the winter, causing the heating system to freeze up?

Until now it has not been realistically possible to implement these functions, as the systems were not compatible with one another[10]. There were no computers or software able to process the data in an economic fashion. A network demands interoperability between the individual units; it demands that companies stick to interface and protocol agreements.

When all these systems can communicate with one another, it will inspire the creation of additional hardware and software components, the integration of new components, the deri-

---

[9] ICs are also referred to as chips

[10] Until now each company has used its own room temperature controllers. How, therefore, can a uniform display be developed? And there has not been much profit in the manufacture of temperature displays for living rooms.

vation of new scenarios and we will see a situation like that in the automobile industry, where electronic components are starting to far outnumber conventional ones. Twenty years ago, a large motor vehicle may have contained around 50 electrical components, today, you could expect to find in the region of 500. Twenty years ago, electrical parts may have made up 5% of the total costs, today this value has increased to 30%. Similar figures are expected in the home and building automation field. If we look at the development of PCs, the Internet, luxury goods, electronic games, the need for greater protection against fire and break-ins in the home, it all leads us to believe there is a huge market out there for networking within buildings.

Consequences that cross fields and disciplines

Professional organizations pay meticulous attention to ensure that the level and quality of their professions are maintained. For the user, they offer a guarantee that unqualified persons and operations are prevented from taking on orders. With that, the corresponding organizations protect the user against botched jobs on the one hand, and on the other hand they guarantee a certain level of income for their members.

However, more and more people are starting to question whether these social agreements are still justified. In this context, it is necessary to differentiate between economic-political aspects and technical aspects. With regard to the economic-political aspects, the common argument is that such professional organizations prevent the founding of small, modern companies and with that represent economic braking, especially in light of the current globalization of the market. We will not go into a discussion at this point as to how far this aspect is justified. It is however closely linked to the technical aspect, an aspect that we will be examining in more detail. Whilst until now professional organizations have largely been able to act independently of one another, the picture is rapidly changing.

Twenty years ago not many believed that the computer could have such an influence on the various disciplines. Today, computers are taken for granted and operation without them is almost unthinkable. Until now however, these computers worked largely autonomously, interfaces to the outside world took second place. Communication between remote processes was achieved via the human interface – information was sent using post (snail mail), or we reached for the telephone. We have since learnt to automate this communication; the market has been liberalized, the prices of computer technology are falling dramatically with each passing year. And more recently the Internet has shown that a dramatic change is upon us. Technology has been given a new component – electronic communication via networks. The information we need, and which was previously conveyed via the human interface, is now collected via a network, evaluated by computers and passed directly on to control the relevant processes. The "human interface" is slowly being eliminated. And this is not only true for information that is exchanged exclusively between PCs (personal computers), but also for data that is exchanged between sensors, actuators, control units and PCs.

If information needs to flow between lighting and heating systems, electronic window systems and alarm systems and back into the heating systems for example, then this information must first be standardized but the interfaces must also be designed accordingly. With that, communication forms a connecting link that demands more than the definition of interfaces (although this is the first prerequisite for successful communication).

In the field of home and building automation, the various disciplines have until now, as already mentioned, worked independently of one another (Figure 1.5). It was up to the architects and planners to design the compound system and to coordinate the various disciplines.

Now, all the electric components are to be interlinked (Figure 1.6), which means that the systems and components of the individual disciplines exchange data, exert a mutual influence on one another and form a higher-order system. This facilitates new functions, which until now were unthinkable, but it also opens up new possibilities for optimization between the devices and processes. Persons with conventional thought patterns are quickly overwhelmed, the coordination of the disciplines still causes a lot of difficulties.

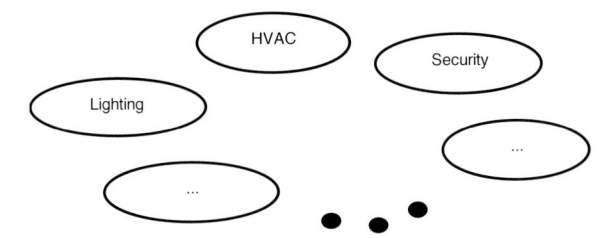

Figure 1.5: Building Disciplines Working Independently

One solution presents itself and that is the formation of a new occupation – the system integrator. The system integrator is responsible for the tasks that cross the various branches and disciplines. This necessitates a technical and organizational fusion that would have previously been unthinkable. If the lighting is to function in a way that depends on the security and HVAC (heating, ventilation, air-conditioning) etc., the company with the greatest chance on the market would be the one with the most know-how. Isolated solutions are gradually disappearing from the market. This has consequences for general training programs, for the disciplines and the way in which they work together.

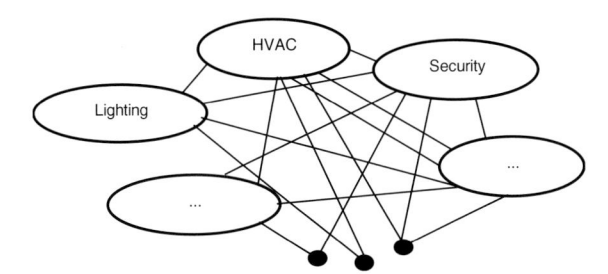

Figure 1.6: The Interlinking of Disciplines

Building automation and control

Whilst the home automation field is concerned with the private home sector, the field of building automation and control deals with all other kinds of buildings. These are typically office blocks, public buildings, banks, factories, etc., i.e. relatively large buildings which can easily contain upwards of 40,000 to 50,000 nodes (network nodes of a fieldbus). The principles outlined in the above section on "home automation" are also valid here, but with building automation and control the dimensions involved are much greater, and the question of economics is somewhat different. Here, comfort plays a lesser role.

In the field of building automation and control the considerations are slightly different. Certain things must be viewed differently and other things represent additional requirements.

Take facility management for example. In the home automation area this is hardly an issue, whereas in building automation and control it is a driving factor for the introduction of fieldbus systems. This chapter, which is meant only as a general introduction to the subject, would become very long-winded were we to now examine all these aspects in detail. The authors have decided to limit an exploration of these aspects to the corresponding chapters.

1.2 The Basics

This section deals with the necessary basics of communication technology, knowledge of which is a prerequisite for an understanding of this book. It not only represents an introduction to this chapter but also a common platform for the terminology used throughout. Protocol technology, networks etc. are a very young field, which with regard to terminology, is a long way from being established. Terms are not only used differently, but also even with contradiction. A standardized definition is therefore essential.

A representation of the basics can only be used as an outline. For a better understanding, it is recommended that you consult other literature, for example [Diet 99, Tane 90].

1.2.1 Hierarchies

A network node is a component that comprises a complete computing unit and an interface to the network. Every EIB unit of a switch, a light etc. represents a network node. Today, there are buildings in which more than 10,000 such nodes are integrated. Such complex systems, consisting of such large numbers of intelligent units, must be

– described by suitable models,

– designed with suitable tools,

– serviced with equally suitable tools and operating units.

Here, suitable means economically justifiable and applicable. A suitable model for describing complex networks reduces the represented information to a workable minimum. Two hierarchic models are most commonly used today: the hierarchic model of the network structure and the hierarchic protocol model. The hierarchic ISO model (International Organization for Standardization) that is currently used in almost all "traditional" protocols, will be dealt with in the next section. Hierarchic network structure models are not as a rule defined in detail. The simple aim is to split the network, that must be viewed in terms of a large computer system, into hierarchic levels, to which various sub-units (computers) are assigned. The individual levels can then be further split into networks and domains (Figure 1.7), until such a level is reached in which the individual computers are listed.

An initial division (listed below) follows the functional properties of the networks[11].

---

[11] Today, a division on the basis of length only is insufficient. FANs have since been developed for monitoring oil fields which stretch over distances exceeding 30 km. On the other hand, WANs are now used in urban areas for shorter distances (in shopping malls for example). The terms therefore should be interpreted on the basis of their functions.

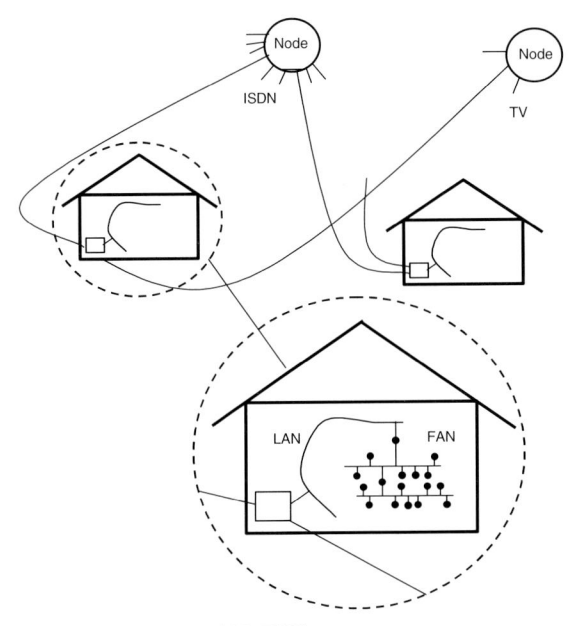

Figure 1.7: WAN, LAN, FAN

GAN (Global Area Network)

Contrary to WANs and LANs this represents a global network that is implemented via satellites and transatlantic cables. It is generally used to connect WANs.[12]

WAN (Wide Area Network)

The term was mainly conceived to distinguish from LANs and is used to represent the public network. After the changes to the European telecommunications laws, this term can now be used to describe any network that is laid on public ground and which is distinguished by the fact that it is not associated with any specific data. With that it is used to connect LANs (and can also be viewed as a backbone for these), but at the same time, the ISDN is also considered to be a WAN.

LAN (Local Area Network)

As the name suggests, this is a network that links computers within a specific site, building or area. LANs are suitable for transferring large amounts of data; the connected computers are usually workplace computers.

FAN (Field Area Network)

The word "field" refers to general processes that use sensors and actuators. Standardization committees mainly deal with industrial, technical or building-specific processes. Fieldbuses

---

[12] The function of a GAN is to globally network systems, this does not contradict the previous footnote

1 Introduction

are therefore networks that connect computer units within the "field". These could be intelligent[13] temperature sensors or even control units.

Nowadays, this simplified differentiation is no longer adequate. The terms are by no means fixed and are often revised as a result of technical advances or legal changes. They do help however to establish a general classification.

If we now turn our attention to concrete networks in fixed areas, it is possible to formulate the terms more uniquely. As we move away from centralized computers, this level sees the use of dedicated networks, i.e. networks to which specific functions are assigned. In the CEN TC247 standardization committee, responsible for the fieldbuses used in building automation and control, three levels have been defined:

– Management level

– Automation level

– Field level

Additional levels are often defined in concrete applications depending on the particular discipline, the philosophy behind it as well as the actual networks used and the number of nodes. Examples are shown in Figure 1.8.

An interesting fact should be pointed out here. Whilst just a few years ago, relatively large numbers of levels were used to describe industrial systems, often 6 or 7, the trend now is for far fewer. There are several reasons for this. The more levels the greater the network overhead. On the other hand, a high number of defined levels reduces the complexity of each individual level making it more transparent. If therefore there are efficient tools for the networks, there is no need to define many levels. A further reason for the reduction in the number of levels however is the increasing performance of modern networks. We will examine this point in more detail, as it will show how network development will progress in the near future.

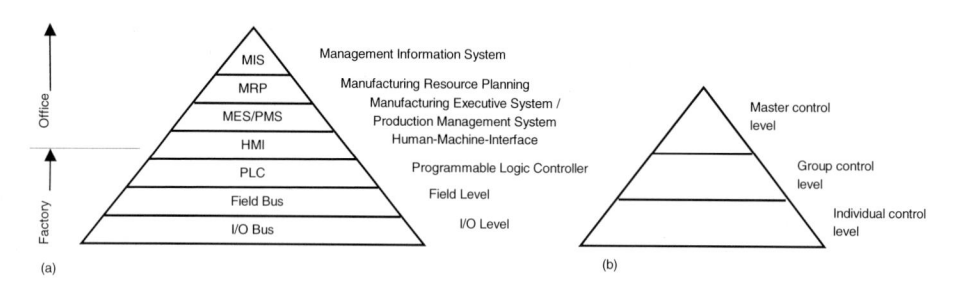

Figure 1.8: Different Network Hierarchies for Various Disciplines: (a) [Müll 98], (b) [Krie 00]

---

[13] Many people have a problem with the term "intelligent". Each field has its own definition of the term, often borrowed from other fields. In communication technology at the field level, this term is used to represent units that contain control and computing elements and which permit pre-definable algorithms or procedural processes. In a network, intelligent nodes must be able to manage the communication protocols.

1.2.2 Principles of the ISO/OSI Model

Before the definition of the ISO/OSI model, transmission systems were described using pure electrical parameters but mostly via the statuses of the units taking part in the communication. Communication systems were mainly viewed in terms of signaling technology, which was not surprising when most developers were pure electrical engineers. Something that has hardly changed in communication systems is the underlying philosophy of a state machine [Patt 94]. This is advantageous as it allows all states of the machine to be uniquely described. The machine can be developed so that even in the case of power failure, it can switch to a state in which an ordered procedure is still possible.

It is reasonable to assume that a description of electrical connections and simple handshake procedures (I send you something, you send me something, I send you something else, etc.) cannot be used to describe more complex systems. This prompted the people at ISO (the then responsible CCITT had largely ignored this task, they continued to see only the electrical aspects of communication as important) to develop a communications model that was to provide developers with a basis for implementing largely uniform communications systems[14]. They called it OSI (Open Systems Interconnection [Tane 95]). It combined all conceivable communication functions and assigned them to seven layers. Only the lowest level described the physical aspects, all the others dealt with functions such as data transmission security, path finding, access protection, etc. The idea was already there of creating a model that covered as many requirements as possible, but at that time experience was limited to the fields of voice communication, telex and mainframes. The significance of networking smaller units in the fieldbus area was not yet understood. No one considered the necessity of integrating real-time systems either. The remote control of processes was not considered a major problem.

The current central problem of network management was also overlooked, as was the possibility of numerous protocol standards or that the tools for integration and maintenance could play such a significant role; the list goes on. It becomes obvious that the ISO/OSI model can only be used as a basis. A lot more is required to create efficient solutions and systems that satisfy the needs of today. It may perhaps still prove necessary to create totally new models that are more flexible and more importantly, more powerful.

1.2.3 OSI Levels

The aim of this section is to highlight the most important functions that have been assigned to the individual layers. The following section then deals with the various processes between the levels.

By definition, the application is placed above the levels (above the communication unit defined by ISO). In a PC (personal computer), that is firstly the operating system, which is then overlaid with the actual applications. The "communication column" can be viewed as part of the operating system as it provides services to the application. Management of the communication unit has been omitted from Figure 1.9.

---

[14] This is important for the overall understanding: The OSI model is not an implementation it is simply a recommendation of how one can develop efficient protocols. The corresponding implementations must be adapted for the boundary conditions.

1 Introduction

Before we go into an explanation of the individual levels, two quick answers to the questions: Why was an underlying hierarchic concept developed instead of a democratic one? Why were seven layers defined?

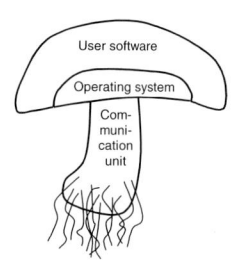

Figure 1.9: Hierarchic Computer System

A hierarchic concept does not have a high overhead, it is distinguished by a capacity for rapid reaction, simple scenarios are possible for cases of error and it is simple to describe as well as to implement. Democratic systems are generally fairer[15] and more flexible. As communication systems at the time of the OSI model definition were relatively easily understood, such a concept was the more obvious one. Today, in this era of "total networking" in which the number of nodes within a network has reached enormous proportions (take buildings for example which now contain anything upwards of 50,000 addressable nodes), and which are no longer so transparent, the issue demands a rethink.

The second question is also easily answered: All the communication functions were counted up and ordered into overlying groups. The functions of one group built on the functions of the underlying group. Opinions differed in the international committees such that it proved impossible to come up with a single logical number. As a compromise they settled on the magic number 7. There couldn't be too many more levels, as this would have created too great an overhead for the communication between them. There couldn't be too few either, as the low level of modularization would have complicated the connections.

Level 7 (Application Layer)

Level 7 is a boundary layer (interface to the application) and with that occupies a special position. It forms the interface between the application and the communication unit. In this level the procedures or protocol processes of various application functions are defined, for calling up data, file transfer etc. The purpose of layer 7 is a transparent representation of the communication. If, for example, a system accesses databases via the communication unit, layer 7 must be designed so that it does not require any knowledge of the individual tasks of the underlying levels. With that, an efficient communications system allows a database that is distributed amongst various different locations to be viewed as a single interconnected database. An intelligent switch that transmits information via an OSI communication unit, such as "turn light on", does not need to know anything about the actual communication

---

[15] In this context "fair" is used to imply that it is not necessary to treat differently classified processes in accordance with strict specifications. Mechanisms are integrated into fair systems that can deviate from strict specifications, in order to create a balance between units within a specific context.

protocol; it simply knows the "name" of the lights as well as the functions "turn light on", "turn light off" or "dim light by 40%", etc.

Level 6 (Presentation Layer)

Level 6 deals with interpreting the incoming data and coding the data to be transmitted. This means that level 6 carries out syntactic and semantic tasks. These include for example the meaning of the sequence of bits of a character, to be interpreted as a letter, interpretation of currency as well as physical units, cryptographic tasks, etc.

Level 5 (Session Layer)

After the communication procedures have been defined in layer 7 and the interpretation of the data in layer 6, the underlying layers must deal with the setup and termination of the communication in a consistent manner. The main task of layer 5 therefore is to bring together several end devices into a session and to synchronize the "conversation". This also involves identification or authentication (e.g. password check). Close co-operation between this layer and the operating system is of vital significance. For this reason, layers 6 and 7 are implemented in fieldbus practice with high transparency for these functions, or a separate channel is created to bypass them.

It is also the task of layer 5 to introduce any necessary synchronization markers, so that it knows when to resume after a breakdown in communication. A prerequisite for this is the constant confirmation of the transmitted frames (see level 4).

Level 4 (Transport Layer)

The transport layer sets up an end-to-end connection. This means that the receiver does not route the data further, but passes it on to layer 5, already prepared. There are various mechanisms available to layer 4. If the data packets to be transmitted are too big, layer 4 can split them up and transmit each piece individually. If the transmission times are long or there are a number of possible transmission paths it is useful to number the split packets. The receiver station must then recombine the individual packets in the right order.

If several sessions are activated, the transport layer can manage them via the same channel. If, on the other hand, there are several channels available, the transport layer can transmit the individual split packets via the different paths.

If any data packets are lost, layer 4 must ensure they are reliably transmitted again. Layer 5 relies on secure data transmission, otherwise it must terminate the connection via layer 4.

The most common example of a layer 4 protocol today is TCP (Transmission Control Protocol), which was basically the foundation for the Internet.

Level 3 (Network Layer)

If there are nodes between the end points of an end-to-end connection, packets must be routed. In layer 3, the paths between origin and destination are established via the specified target addresses. This is easy if the corresponding path lists are available in the nodes. It becomes more complicated when the paths are to be optimized on the basis of various criteria: fees, quality, load, delay times, etc. or if the path conditions change during a transmission, packets need to take different paths due to bandwidth considerations or the packet size is unsuitable for certain paths, etc.

The task of layer 3 is by no means a trivial one, especially when there are various physical transmission media within the network with different transmission speeds. It is also necessary to ensure that jams do not occur anywhere along the paths, which would cause the maximum delay times to be exceeded.

A differentiation is made between connectionless and connection-oriented services. In the case of a connectionless service (datagram service) there is no allocation of fixed channels; every transmitted package must include the complete address and is sent as an independent unit.

The best-known connectionless protocol is IP (Internet Protocol), also used to handle the Internet addresses (version 4: 4 numbers between 0 and 255, separated by dots ".").

With a connection-oriented service (virtual circuit service) a virtual channel is made available, which from the point of view of the user offers the advantage that the data packets need not include any addresses. One of the first protocols of this type to be implemented and which is still in use today is the X.25 (ISO 8473).

In line bus systems, containing neither routers[16] nor gateways, layer 3 can be omitted.

Level 2 (Data Link Layer)

This layer is a pure point-to-point connection with the task of guaranteeing transmission between two stations. This firstly involves the setup of the frame; which is why we talk of formation of the data frame (Figure 1.10). In the underlying layer, all that remains is to supply the individual bits of the created frame. There is no recognition of allocation to individual blocks. The user data contained within (simply referred to as "data" in Figure 1.10) is only useful from the point of view of layer 2. In reality, it contains additional information from every layer. Each layer adds its information to the end of the "user data packet" received from above.

The second task of the data link layer is the coding and checking of the frame (e.g. via CRC: cyclic redundancy check[17]), to allow any transmission errors to be detected or even eliminated (correction mechanism). This also includes the checking of timeouts, or verification that the corresponding responses and confirmations are received from the opposite party. This means that it is the task of the data link layer to provide the following service to the above lying layer 3: Setup of a logic channel to an opposite end device which cannot include any intermediate nodes, or the transmission of a data frame between two end points.

If at implementation the layers are too abundant, then it is recommended that further modularization is carried out. This is often done for layer 2. In this situation a differentiation is made between the LLC (Logical Link Control, see IEEE 802.2), which sets up the connection to layer 3 (to which the error detection mechanism is assigned) and the MAC (Medium Access Control) to link to layer 1 (this generally controls who is able to transmit when).

The best-known example of a layer 2 protocol is the HDLC protocol (High-Level Data Link Control) that was first implemented by the SDLC (Synchronous Data Link Control) IBM protocol for the ISDN, but can now also be found in other standards such as Ethernet.

---

[16] Routers and gateways etc. will be examined in more detail in later chapters.

[17] CRC was integrated into SDLC (Synchronous Data Link Control) by IBM and with that has become very well known.

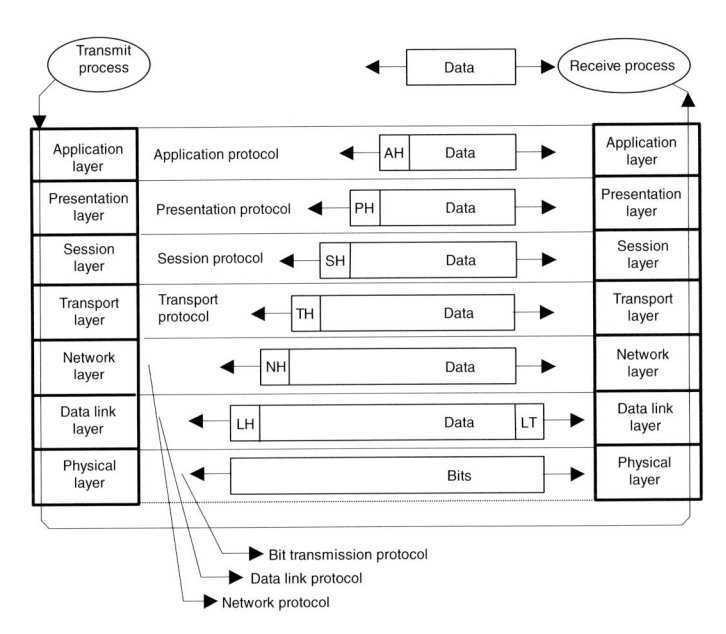

Figure 1.10: Data Frame Formation

Level 1 (Physical Layer)

As the name suggests, this layer describes all mechanical, physical, optical, electrical and logical properties. This includes, for example, definition of the connector type, impedances, transmission frequencies, line temperature properties, line lengths, line types, the type of coding with regard to transmission properties (such as NRZI: Non Return to Zero Inverted), etc. The best-known standard of layer 1 in fieldbus technology is the RS 485 interface, which naturally only forms a small part of a complete layer 1 definition.

The X.21 standard has found widespread use in WANs.

1.2.4 Layer Layout

The OSI model not only describes the allocation of the communication functions to the individual layers, it also proposes communication procedures between the layers. A differentiation is made between vertical and horizontal communication (Figure 1.11), whereby it is only opposite-lying layers that are able to communicate directly, layers lying above one another exchange data via services.

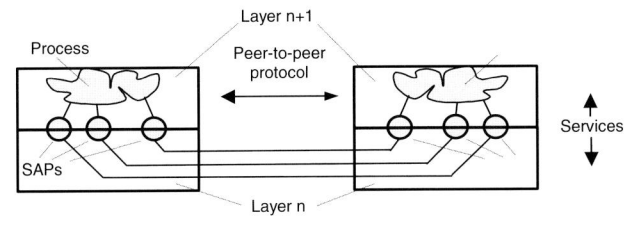

Figure 1.11: Connection Possibilities between Entities

1 Introduction

The vertical communication is described via service relationships, whilst the horizontal communication (peer-to-peer communication) is defined via protocols. Two opposite lying entities (layers), also referred to as service users (SU), exchange data via their protocols, making use of the service provider (SP, Figure 1.12).

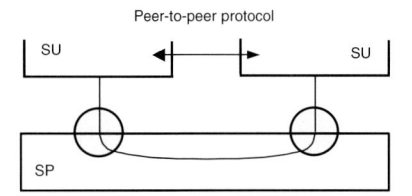

Figure 1.12: Service Users (SU) Using the Service Provider (SP)

Or put differently: A service provider offers the overlying service user a range of services enabling it to fulfill its tasks.

The exact procedure is highlighted in Figure 1.13. The information, which layer n of a communications column wants to transmit to an opposite-lying layer, is called the service data unit (SDU). The interface process of layer n places an interface control information (ICI) onto the message header (the SDU header), whereby the ICI and SDU together form the interface data unit (IDU). The interface data unit is transferred via a service access point (SAP). In the underlying layer, the interface control information (ICI) is now decoded and a corresponding process is initiated. The following process (execution of a protocol) should be viewed as a logic process, except in the lowest layer, the physical layer. In the procedure initiated by the interface control information (ICI) layer n-1 packs the SDU into a protocol data unit (PDU). This involves adding a protocol control information (PCI) to the service data unit (SDU). The protocol data unit (PDU) is now transmitted which means that layer n-1 uses the lower layers and itself now behaves as a service user. Only layer 1 actually physically transmits the information.

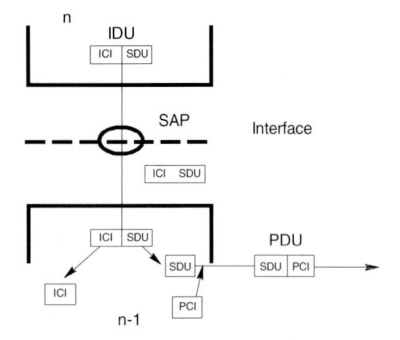

Figure 1.13: Interface between Two Overlying Layers

This results in a staggering of the data packets (layers 3 to 7) and the data frame (layer 2) in accordance with Figure 1.10.

1.2.5 Topologies

Describing networks using topological forms should help to provide a clearer abstract view. There are essentially 6 different forms (Figure 1.14), all of which are distinguished by certain characteristic properties. This generalization however should be used with caution, as in principle, the topological form need not necessarily refer to layer 1 of the OSI model (physical layer). There are networks such as the Profibus (EN 50170) for example, which has a line structure with regard to layer 1, but with regard to layer 2 has a ring structure.

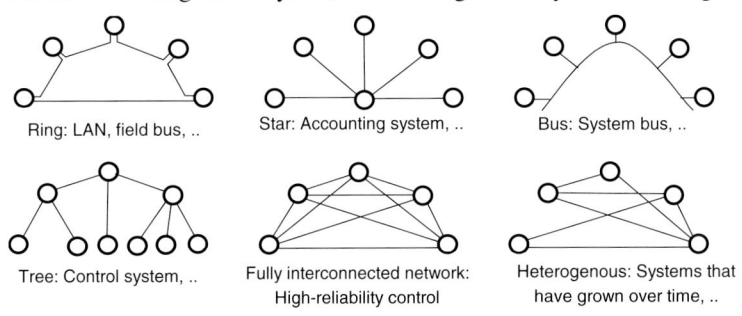

Ring: LAN, field bus, ..   Star: Accounting system, ..   Bus: System bus, ..

Tree: Control system, ..   Fully interconnected network: High-reliability control   Heterogenous: Systems that have grown over time, ..

Figure 1.14: Topological Network Structures

The *star* commonly occurs in three versions. Typical telephone exchanges, regardless of whether they handle analog channels or ISDN, are organized centrally. The advantage is obvious – connections between end devices (telephone terminals, PCs, faxes) and the control center are pure point-to-point connections. Large transmission paths and optimum capacities are relatively easy to implement. There are no interfering stubs, which are often permitted in a line structure and can cause interference in the form of reflections. Economically speaking, the star formation is interesting when it is possible to set up the control center so that the units of the individual partners can be implemented at low cost, or if the maintenance of the system is simpler on a central basis. The performance (mailbox systems, change-over strategies, path finding mechanisms) and data (databases, partner lists, loadable test functions,...) are then based on central network nodes.

The second area in which the star formation is often used is control technology. The PLC (programmable logic control), often thought to have been assigned to the scrap heap by the fieldbus, can still be of great value. It represents a compact solution, for which there are excellent tools and which is easy to handle. Should modifications become necessary, it is generally only necessary to change the PLC program, no correlation lists in the various modules. Interoperability problems, such as those seen in line systems, are not a matter for concern. They have been on the market a long time and have proven their worth. It would take a lot to convince that the classic PLC is not capable of meeting the requirements on a long-term basis.

The third form is found in LANs, which electrically and logically are designed as ring or line networks, but for reasons of security and ease of laying[18] are cabled in a star formation. In the star, the line is only looped through so that any faulty station can be easily disconnected

---

[18] In new buildings, the ring cabling often presents problems. Point-to-point cabling is much easier for the installation engineer.

from the line or removed from the ring. In many new buildings it is more economical and easier to ask the cable-laying contractor to lay a separate line for every computer in the control center, than to link up each computer one after another according to a pre-specified principle.

Another simple topological form is the *ring*. It is characterized by two independent access points (usually an input and a separate, independent output)[19]. There are two main principles. Control technology is often concerned with real-time requirements and demands equally high data throughput for all stations. The nodes are arranged one after another in the form of a chain and a shift register is implemented in every node. In its entirety the ring can be viewed as one shift register, in which it is the task of a central unit (master) to transfer the information to the remaining nodes (slaves) or in the opposite direction retrieve their data. If the master contains a reproduction of all slave shift registers, this process can be achieved with a single shift process (see chapter 1.3.1). As there is no need to address the slaves and all slaves are operated simultaneously, it is the fastest method when all slaves need to be polled equally.

The first rings to find widespread use in computer technology were not however master-slave systems but master-master systems, known by the name "token-rings". They are still of important economic value in the LAN worlds of today. They work on the principle that one of the stations has the token at any one time and with that the authorization to transmit. This data then travels once through the circle and is directly manipulated by the receiver station in such a way that the sender knows it has been correctly received. This is an extremely complex procedure, but because the sent data frames are usually much larger in the LAN world than in control technology, the overhead is not such a major consideration.

Rings with outward and return channels in one line were implemented early on, but have only realized their potential with newer, modern concepts such as FDDI.

The *line*, often referred to as simply *bus*, although this term is not unique, is the most commonly used network topology. It is possible to implement simple access algorithms, the connection or removal of units is simple and the cabling is generally low cost. For this reason, we will take a closer look at this technology.

All nodes on the line receive transmitted messages. The respective receiver independently decides whether the message relates to it and if not rejects it. This procedure however is not to be compared with broadcasting. With a broadcast transmission the message is aimed at all addressed end devices. From this we can deduce the most important task for the line network – how does one define bus access, so-called *arbitration*? There are numerous possible solutions, especially in fieldbus technology. Each of these methods has advantages and disadvantages, which are so serious that it is not so easy to place one above the other. We will take a look at various examples of different bus systems which clearly show that many of these methods will also have long-term appeal. Bus access methods are distinguished by the cost of implementation, access delays, the assigning of priorities, fairness with respect to bus specification and the treatment of faulty nodes, etc. It is impossible to highlight every single possibility here, we will limit our discussion to a few of the most important. More details are then given when presenting the individual fieldbus systems or can be read in the reference literature.

---

[19] P-NET is often referred to as a ring. However, in the electrical sense it is not a ring, only in terms of the cable arrangement and the logical ordering of the stations.

In general, a differentiation is made between direct and indirect arbitration. Today, direct arbitration usually occurs on a decentralized basis. This involves a *token* being assigned to one of the end devices which is then able to transmit. The token is then allocated according to specific algorithms. In principle, this procedure is very simple, but in terms of implementation is costly, as it requires every possible error to be taken into consideration.

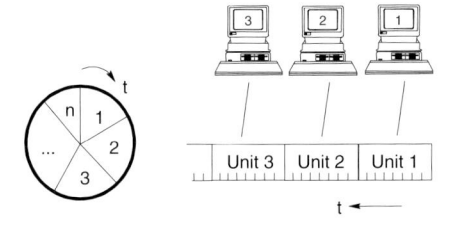

Figure 1.15: Principle of the Time Slot Method

The *virtual token passing* method is a technically interesting variant. It is implemented in the P-NET, which is looked at in more detail below.

Special *time slot methods* (Figure 1.15) have been developed for real-time applications, in which each end device is assigned (guaranteed) a time slot. In the case of safety-related applications this is achieved on a static basis which makes it easy to guarantee that each station can send or receive its data within an assigned time slot. Within telecommunications technology, such systems are referred to as *synchronous bus systems*. Each end device is assigned a time window in which for example the correspondingly digitized voice data can be transmitted. Arbitration is often centralized in these cases.

In the LAN field the use of indirect arbitration is more common, and in particular the CSMA method (Carrier Sense Multiple Access). It also plays an important role in fieldbus technology and is available in a number of different variations, even in systems designed for "semi" time critical applications. The term "random access" is also applied, which implies that the end devices are free to access the bus whenever they want although there are specific rules to be obeyed. The first and most important of which is that before every access it is necessary to check that no other end device is active on the bus. It is only possible to transmit when the bus is free. This can result in collisions, a situation that needs to be resolved. Here we differentiate between the various methods. The most common method is the CSMA/CD (CD: collision detection), in which by simultaneously listening in to the bus the transmitter can detect whether its data is being interfered with by another transmitter. If so, it immediately withdraws and tries again later. The time at which a second attempt is made is determined by the specific algorithm of the implemented method [Tane 90].

With regard to real time, there are no access methods for lines that provide ideal solutions, as the contradiction lies in the principle. Random access on the one hand and guaranteed transmission, with regard to time, which can only be achieved via a compromise. The different methods therefore are attempts at optimization related to specific applications. The question arises of whether the desired real-time requirements are really nothing more than marketing tools and not actually needed for technical reasons. It is nonsensical to place strict real-time requirements (that need to be 100% satisfied) on a device with limited reliability. In this situation, 'softer' real-time requirements would suffice, i.e. access occurs with sufficient probability and is economically justifiable[20].

The *tree structure* (Figure 1.14) is characterized by one or more substations being dependent on a root node. Each substation can also represent a root node. In many cases the connections between the stations are regular point-to-point connections or lines. In terms of communication, the tree structure does not represent anything special.

In automation technology, tree structures are very significant when it comes to network hierarchies (Figure 1.14), as they allow the construction of relatively complex network systems [Ben 92]. In accordance with the hierarchic allocation, both the application and communication processes can be perfectly adapted to the various areas, and with that efficiently implemented in terms of hardware and software. If the root nodes are implemented as gateways and switches, this allows management of complex, heterogeneous systems.

With the *fully interconnected network structure* every node is connected to every other node (Figure 1.14). This holds a real attraction for the development engineer, but there is also the disadvantage of the need to manage an enormous amount of communication. These systems naturally offer a great deal of flexibility and reliability, especially with the level of diversity in the stations[21]. However, it is also necessary to consider the costs, which is why they do not yet play a role in building systems control. As mechanical systems are increasingly being replaced by electronic systems in safety-related applications (control units for elevators, garages, motor vehicles, etc.), redundant, fully interconnected systems are no longer considered such an extravagance within control technology as they once were a few years ago.

Heterogeneous systems are composed of different basic structures. Heterogeneous networks therefore can be broken down into structures such as rings, stars etc. They usually represent systems that have grown over time or special applications.

1.2.6 Interaction Time Diagrams and SDL

There is no doubt that within the field of software technology, the subject of "documentation" is just as problematic, associated with extremely high costs and just as hotly discussed as it was thirty years ago. Even though the methods have improved, they have by no means kept pace with the increasing complexity of the software packets. Significant advances have been made in the field of communications technology. The OSI model, with its relatively simple layout, and the philosophy of basing the model on a simple state machine, allows communication tools to be defined that in turn facilitate good documentation. Complex status diagrams and lists, endless verbal descriptions as found in older protocols, are being increasingly resigned to the past.

The following section describes two description methods (tools), which originally developed for digital telecommunication are now enjoying increasing popularity in the area of fieldbuses. As the economic viability of a bus system essentially depends on the maintenance costs (implementation improvements, extensions, etc.), we will briefly cover this subject and

---

[20] Here we are concerned with the overall reliability, i.e. that a system functions at a specific point in time.

[21] A system that involves diversity consists of parallel units that to the outside world behave exactly the same but internally consist of completely different hardware and software.

make reference to corresponding literature. It is not our aim to give a comprehensive introduction to SDL as this would exceed the scope of this book.

Interaction diagrams

The interaction diagram was introduced to represent simple communication relationships (ISO, Z.120). It is very clear and contains only the most important information (Figure 1.16):

– Communication units (entities)

– All possible system statuses (usually demands a state machine)

– Transitions (action – reaction)

– Timer

This is the main reason why interaction diagrams are used as an initial description of a system. They provide a general overview, stressing the main principles. In Figure 1.16, entity 1 in status S_m sends a message, whilst entity 2 is in status S_n. A timer is started at the same time, which triggers an alarm if the opposite station does not react.

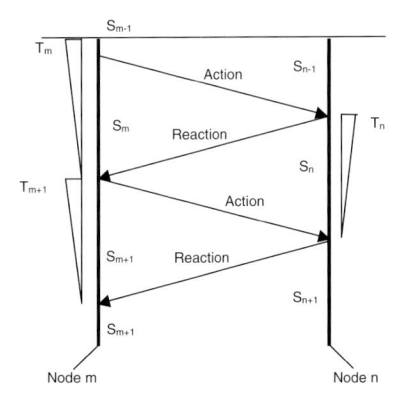

Figure 1.16: Interaction Diagram (S: Status, T: Timer)

It is obvious that this type of representation is very simple and clear, but it does not permit the illustration of problem cases. How could we represent the situation in which 2 or more entities receive messages at virtually the same time? What happens and in what order? How do we describe a complete communication system? Interaction diagrams are not enough to provide answers to these questions. For an exact description of communication on the basis of the state machine, it is necessary to use Lotos (Language of Temporal Ordering Specification), Estelle or SDL (see next section).

SDL (Specification and Description Language)

Compared with other description methods such as Lotos[22] and Estelle, SDL has found a much more widespread use, which is why we would like to describe it here in more detail. It was developed by CCITT for the field of telecommunication and is now managed by ITU. SDL is strictly formalized, which helps to avoid errors. The developer is forced to stick to

---

[22] Lotos: Language of Temporal Ordering Specification

1 Introduction

the rules of communication technology. Some companies have freed themselves from the ties of these strict rules and have developed their own simplified versions of SDL. From experience, we would strongly advise against this, maintenance and upkeep can in the long term be much more expensive. There is also a tendency to drift into the tried and tested flow diagram, but to work with new symbols.

Two syntactic language forms are defined: SDL/PR (PR, phrase representation) and SDL/GR (GR, graphical representation). As a high level language, SDL/PR is formulated in prose. The greatest clarity is achieved with the graphic model. Conversion between the two forms is 1:1 and occurs automatically in tools. As we only wish to offer a brief overview of the subject, it is enough to concentrate on SDL/GR.

It is based on *extended finite automata* in the form of processes which communicate with one another on an asynchronous basis via data paths. Data is stored and manipulated in the processes. A process has either assumed a rest status and is waiting for an input, or an input has arrived and causes a transition to another status in which specific actions are triggered (set timer, manipulate data,...). There are various symbols to describe the various activities of a process and the actual communication.

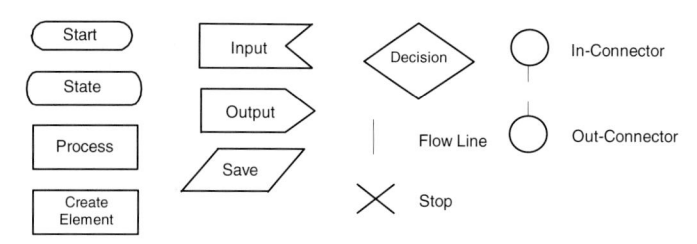

Figure 1.17: Typical SDL Symbols in accordance with Z.100 (ITU)

The 12 basic symbols of SDL/GR are shown in Figure 1.17. They are established in the Z.100 recommendation of the ITU, which has since been significantly expanded. The start symbol signifies the beginning of a process. This is either followed by an initiation (task) or the process reverts to a defined state. A state can only be quit via an *input* or a *save*.

In order to understand this we must first explain the underlying management process which can be viewed as the operating system of the communication unit. Such a management process is assigned to one or more communication processes. One of the most important tasks of this management process is the management of a queue that contains all arriving inputs of the processes. It is organized according to the FIFO principle (first in, first out), which means that incoming messages are buffered one after the other and made available according to the requirements of the process (Figure 1.18).

A typical procedure can be illustrated using the example of Figure 1.19. A process is in status S_0. With that, the process demands the next input from the queue, which in the case of input A generates a process sequence with an output that causes the system to change to status S_1. If the system is in status S_0 and input F occurs, this is rejected and the next input is requested. If input E occurs, it is stored and the next input is processed. Input E, stored in the interim, is then returned to the queue output.

As shown in Figure 1.19, processes always occur between two statuses which means they are easily described uniquely and completely.

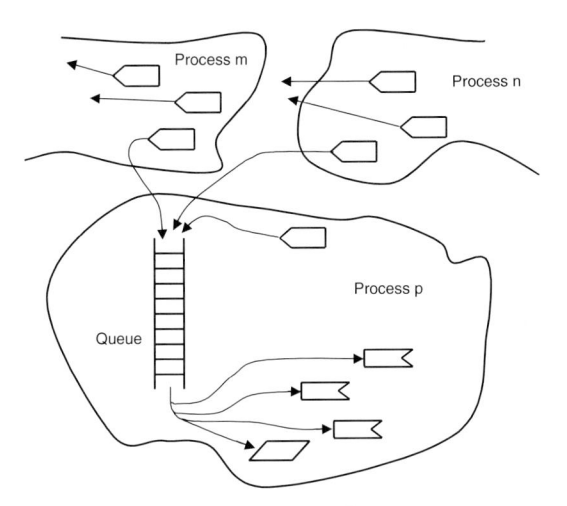

Figure 1.18: Process Communication

Create element generates a new process, which in principle can only end itself (via the *stop* element). *Decisions*, which have more then 2 outputs, are also possible within a sequence. In order to represent the diagrams neatly on A4 pages, so-called *connectors* are used. Connectors do not have any effect on the communication.

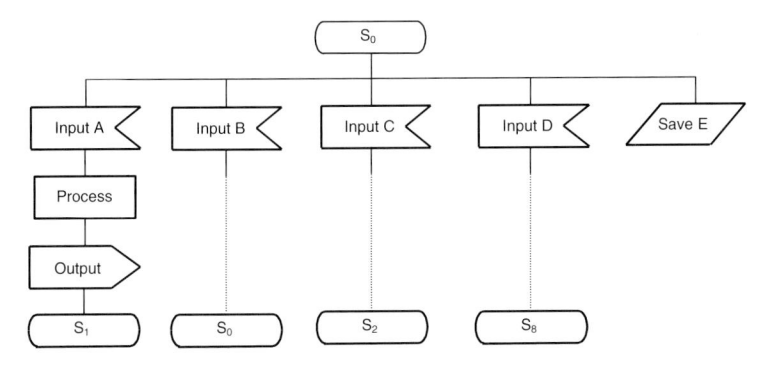

Figure 1.19: Typical SDL Sequence

There are now compilers for SDL which means we can expect the description of communication systems to become even easier and increasingly automated. As SDL is so formally defined, it should not be too difficult to make a conversion in a compiler for a specific bus system [Schm 98].

1.2.7 Tools

Tools for fieldbus systems can be categorized in a number of different ways:

- Tools can be assigned to the respective life cycle sections that characterize a fieldbus installation – from the planning through project engineering right up to commissioning, maintenance and visualization.

- It is also possible to organize tools according to performance characteristics; some tools can only change tables in the fieldbus nodes, others load applications into the device memories and visualization tools offer access to runtime functionality.
- The best method, however, has proved to be the assigning of tools to the respective users. Product developers, application developers, planners, commissioners and many others are assigned to various occupational groups and are representative for the wealth of tools that denote a fieldbus system.

If we look at the tools required for the pure development of fieldbus components, then there is a natural split between hardware and software tools. Depending on the processors used in the individual fieldbus nodes, there is firstly the need for hardware with the corresponding interfaces in order to be able to actually create individual fieldbus nodes. Many fieldbus systems are based on common microprocessors and standardized interfaces. The common PC with its variety of interfaces is sufficient as a development platform for a whole host of fieldbus systems. Additional devices such as EPROM burners may also be required.

As the software tool, a commercially available C-compiler or assembler is often sufficient for many fieldbus systems. Some fieldbus systems require translation programs that are tailored to the machine code in the fieldbus nodes. There are also tool environments that are based on other C-compilers or assemblers, which cleverly integrate the fieldbus specific tasks.

Clearly, the work is not over with the simple development of the fieldbus node. Even when a node has been provided with an application (and this has been compiled without error), we are still a long way from the node working in total compliance with the protocol. Here, there are tools that test the interoperability of a fieldbus node within a network. Such tools differ completely from system to system. For some bus systems the temporal behavior is of particular importance, with others it is more the correct interpretation of the data types. In general we can say that such tools are essential for the product or application developer as they facilitate the fast and efficient testing of new nodes.

Tools that are used for planning fieldbus systems are to be assigned to another category. Here, the level of the applications within the fieldbus nodes should be left out of consideration; it is more important to concentrate on the ability of comfortably designing large networks. A visual user interface and intuitive operation (e.g. drag & drop) characterize modern planning tools. The PC has been established as the universal platform for planning tools for fieldbus systems. An important feature of such tools is the capacity to handle other programs or file types. For example, when planning electrical installations in large buildings, it is useful to be able to directly import CAD files. Of course, the exporting of data is equally important. A typical example is the piece list generator, which after the successful planning of a network lists all necessary devices in a text file so that this can then be further processed in spreadsheet programs.

Tools for commissioning fieldbus systems combine the data model, which underlies the planning tools, with the information that is important within developer tools. Commissioning firstly deals with the entire network but also with the individual fieldbus nodes and their internal structures. A commissioning tool therefore must be able to see not only the entire network but also be able to control every memory cell in every fieldbus node.

In the field of building automation and control in particular, it is often useful during the commissioning stage to allow short-term changes in the planning. The fast switching between commissioning and planning is very useful here. The idea of merging these two categories is an obvious one. And with the performance levels of today's PCs is easily possible.

One category of tools that is becoming ever more significant is the so-called end-user tool. This denotes visualizations, remote control applications, gateway solutions and many more. An important difference with respect to the tools mentioned further above is the user group. Take the example of visualizations for building automation and control – it should not be necessary for the users to have any special knowledge or experience with the respective fieldbus system.

In general, it is true to say that tools have a decisive influence on the cost of a network, and this influence is likely to increase. There are several reasons for this. The cost of developing a tool is immeasurably higher than that of developing a fieldbus component. There are even greater differences in the maintenance costs. There is a constant demand for additional functions when it comes to tools, with shorter and shorter development times. Whilst fieldbus nodes are largely smaller, complete units in comparison to tools, the scope of the software tools is on the increase. In addition, the piece numbers of fieldbus components are growing rapidly; the number of sold tools on the other hand is growing at a much slower rate as the per-head installation[23] increases.

This difficulty is also obvious if we look at the efforts of user groups, company groups and individual companies to define, develop, produce and sell new tools. The current lack of tools on the market is not in the main due to the incapacity of getting such tools off the ground, but of developing powerful systems in the current climate of keeping costs down.

Despite the great difficulties, it has been possible to establish various tools. The direction of the further development of tools, in the sense of top-down design, remains to be explained.

Fieldbuses in building automation and control are characterized by the high number of nodes (20,000 in one building is no longer out of the ordinary and this figure is expected to rise by a further 10% in the next 5 to 10 years), the high number of different sub-networks, the simplicity of the individual nodes, their low price and the necessity of being able to install them quickly and easily without specialized knowledge. There is an ever-greater need for modern tools to satisfy these demands. It must also be assumed that maintenance is only seldom carried out. The tools therefore must be much more user-friendly. The life of the components must be high which demands a long downwards compatibility of the tools.

A further indispensable premise for the long-term development of tools is that they must be applicable to integration right through to maintenance. An integrative module or at the very least tools that are perfectly matched to one another, i.e. have common interfaces, are highly recommended for the purpose of reducing integration and maintenance costs alone.

This concludes the outline of the necessary scope of fieldbus tools.

1.2.8 EMC, Lightning and Overvoltage Protection

Electromagnetic compatibility and overvoltage protection are essential properties of a communication system determining the reliability, availability and usability. By taking appropri-

---

[23] This represents the number of fieldbus nodes installed by one integrator within a year.

ate action when designing systems it is possible to guarantee that the communication and connected devices function properly even in a disturbed environment and do not exert any unwanted influence on the surrounding environment (e.g. TV reception). The generic EN 50082-1 and EN 50082-2 standards describe the general requirements and EN 50090 2-2 the specific requirements of building systems controls, including the tests to be made.

As it is virtually impossible in practice to test a compound system with all possible components or to test the fully installed system, it is necessary to test each individual component in a minimum configuration taking into account the error criteria specified by the manufacturer. In order to be able to assess the resistance of a device or system in detail, it is not enough to simply know whether and to what extent the given thresholds are reached, but to know the permitted behavior of the devices or systems.

The hardware [Pigl 90] and software designs of the device both affect its resistance to jamming. An error-tolerant software implementation can go a long way in avoiding unnecessary efforts in correcting hardware-related problems.

The actual lines used and their arrangement also play an important role. A twisted pair line (TP) together with a symmetric device design represents an excellent solution with regard to non-susceptibility to interference as well as the low level of interference on the environment. In order to avoid interference, it is recommended that a large separation is maintained from the 230 V power supply circuits. This is true only in connection with capacitive couplings. With inductive couplings, particularly those caused by lightning, loops may arise due to the separation from potential equalization lines or power supply lines, which lead to voltage drops and which destroy the isolation between communication and power supply systems. EIB systems therefore are designed with regard to non-susceptibility to interference so that it is recommended when installing, that the bus line is laid in close proximity to the 230 V supply network.

If the EIB system is used in buildings where there is a great risk of overvoltage caused by lightning, then additional measures (both internal and external lightning protection) are required for the entire electrical installation or building in general [Grun 97].

1.3 Common Fieldbus Systems – An Overview

Fieldbuses have always been defined from a particular viewpoint. Their dedicated nature means that comparisons are virtually impossible, which is why we do not use any type of "fieldbus matrix" in this book. It would always distort the facts no matter how objective its design. The following section outlines a few well-known fieldbuses but this is only to show the variety of possibilities. The systems that have been chosen differ widely from one another. It is also necessary to consider, when selecting a fieldbus, the area of application for which it is intended. The demands of industrial automation are completely different from those in building systems control or the automobile industry. However, systems are often taken from different areas as their economic viability is enormous. Take CAN for instance, originally developed for the motor vehicle industry but which is now widely established in many other areas including medicine.

1.3.1 Interbus

The Interbus was introduced by Phoenix Contact in 1987 under the name of Interbus-S but has since been standardized within the CENELEC standard EN 50254. The aim was to develop a sensor/actuator bus, which was optimized for the exchange of process data in the lowest level of the automation hierarchy. An important characteristic of this process data is that it occurs on a cyclic basis and must be processed within a specific time window. This requirement is reproduced in the properties of the Interbus.

Contrary to most other fieldbuses, Interbus uses a true ring topology on the basis of the RS485 interface. The ring topology has certain advantages when compared with the more common bus topology. To start with there is only one sender and receiver per cable segment, which eliminates the problems of access control. At the same time, the electrical isolation of the individual segments means that the overall system can stretch over far greater distances (with Interbus up to 13 km) and it is much easier to locate errors in critical situations. The disadvantage that the breakdown of one end device causes the breakdown of the entire fieldbus is only a minor concern for the main field of application of Interbus, i.e. automation technology, as subdivisions of a system are only seldom meaningful and the breakdown of one end device usually necessitates a total system shutdown anyway.

In order to simplify cabling, the outward and return lines are laid in a common cable, giving the impression of a line structure (Figure 1.20). A differentiation is made between the remote bus, which serves as a type of backbone, and the peripheral bus, which is intended for local cabling (e.g. in control cabinets) and uses somewhat simpler transmission technology. What looks like a stub cable, is in fact part of the ring that is closed via so-called bus terminal modules (BT). In addition to the structuring of the network into subsystems, the bus terminal module also offers the advantage of error localization. A defective peripheral bus can easily be removed from the main line.

The *Interbus Loop* represents a relatively new extension, developed for linking simple sensors and actuators. In accordance with the requirements on simple cable laying, it uses a twisted pair for the simultaneous transmission of supply energy and data, which is modulated as current impulses on the supply. Contrary to the remote bus, the outward and return lines are not laid in the same cable in the Interbus Loop; the ring structure can also be seen here in the cabling layout. Connection to the remote bus or peripheral bus also occurs via bus terminal modules.

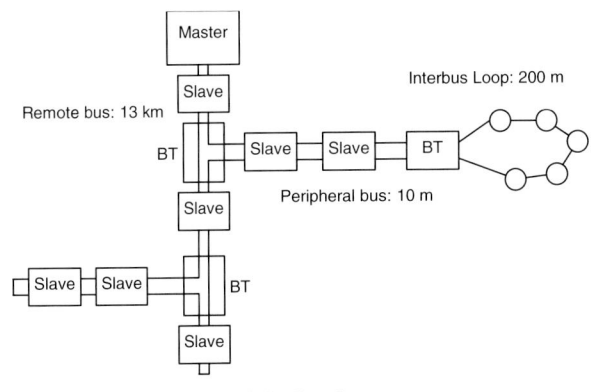

Figure 1.20: Topology of the Interbus

The ring structure means that the data must pass all end devices. It is not necessary for the end devices to have separate addresses, their position in the ring is sufficient identification. The protocol makes the most of these properties. Contrary to many other fieldbuses, the Interbus uses a master-slave protocol with a summation frame (Figure 1.21). Each slave has a predefined share of the data frame that is selected in accordance with its function (or number of data points). The input and output registers of the nodes form a large, distributed shift register and when the entire data frame has moved through the network, each node buffer contains the relevant data. Transmission of the sensor data to the master functions in a similar way. One data frame is sufficient for the cyclic updating of *all* end devices. The overhead that is required for the synchronization (loop check), control and error protection (frame check sequence, FCS) can be kept to a minimum for a complete cycle which in turn yields high protocol efficiency. By virtue of this high efficiency the data rate at 500 kbit/s can be kept relatively low.

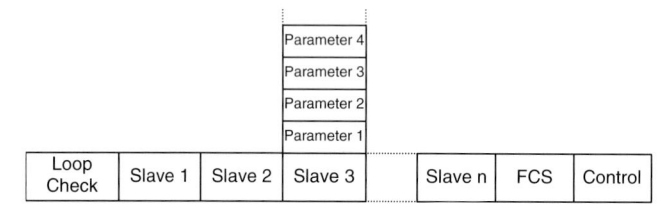

| Loop Check | Slave 1 | Slave 2 | Slave 3 | ... | Slave n | FCS | Control |

Figure 1.21: Summation Frame in the Interbus

The summation frame method is best suited to typical process data that only comprises a few bits but needs to be transmitted in equidistant time periods and in short intervals. For the transmission of larger volumes of data, such as those required for the parameterization of devices, it is possible to define additional parameters. The data is then transferred in small packets without affecting the exchange of process data. The comparatively long duration of transmission is of lesser importance as parameter data is not viewed as time critical data.

1.3.2 P-NET

P-NET is part of the CENELEC standard EN 50170 (1996). It was developed by the company Proces-Data (Silkeborg) in Denmark in the early eighties and has since been integrated into thousands of applications worldwide. Important application areas include process automation and the foods industry. The bus can also be found however in agriculture and building automation.

This fieldbus is a multi-master system into which slaves can be incorporated. Up to 32 masters are permitted with a maximum of 125 end devices (nodes) per bus, whereby contrary to the Profibus, masters can also function as slaves. With that, communication relationships between two master units are simple. Simplicity is the underlying philosophy of P-NET and also one of the main reasons for its success. The cost of high flexibility in hardware and software is often too high. P-NET can be regarded as a successful compromise between simplicity and sufficient flexibility.

This is reflected, for example, in the fixed transmission rate of 76.8 kbit/s, which results in extensive simplifications. The fixed data transmission rate establishes many settings and

1.3 Common Fieldbus Systems – An Overview

parameters that cannot be confused in practice: filter type, register initialization, bus length (max. 1200 m), stub length etc.

Variation in transmission media is also restricted. The underlying standard is RS485, whereby a shielded twisted pair cable should be used. There are no terminating resistors as defined in the standard. Instead, the two ends of the cable are connected together. For this reason, P-NET is often incorrectly referred to as a ring system. This however, is only true from a mechanical point of view; electrically it is a line system.

If point-to-point connections only are required, it is possible to use the RS232 interface. Here however, transmission rates are restricted to between 1,200 and 38,400 bit/s.

In order to keep the overhead of the volume of information that needs to be transferred on the bus to the absolute minimum, an adapted token ring method is employed called "virtual token passing". With this method, the token is simulated by two counters, included in every master, and is not actually passed around the bus. The first is called the "Idle Bus Bit Period Counter" (IC), which is incremented every 13 µs (the bit duration at 76.8 kbit/s) when the bus is free and with that transmits a 1. The IC is reset as soon as a 0 reappears on the bus (signaling renewed bus activities). The second is the "Access Counter" (AC), which is incremented whenever the IC (Idle Bus Bit Period Counter) reaches the values 40, 50, 60, ... 360. If the AC status matches the address status in a master, this means that the node has the token and is therefore able to access the bus. There are however clear restrictions – it can only transmit or receive a single data frame.

If the master does not take up its transmission authorization, it further increments the IC and after a further 10 increments, the AC is also incremented. Once the AC reaches the number of masters contained within the bus system, up to a maximum of 32, it is reset to the value 1. This means that the "virtual token" is "passed on" to the master with address 1.

There are of course other rules that must be observed with this method, but we shall not go into them here. The most important thing to note is that no information is exchanged between the nodes, everything is handled via internal counters whose behavior is subject to strict rules.

Master access to the slaves occurs in the normal way via polling. This means that the temporal behavior for a maximum delay through access by the master is clearly defined, which is why the bus is termed deterministic (real-time capability).

In accordance with the OSI/ISO model, protocol layers 1 to 4 and 7 are implemented in P-NET. An outstanding characteristic of P-NET is its capacity for multi-network structures (Figure 1.22), something that is not offered by any other fieldbus system. Layer 3 facilitates the connection of almost any type of bus. This means that it is possible to integrate up to 32,000 end devices into a single network. Multi-port masters, wrongly referred to as "gateways", are connected between the bus systems, each one separating two bus systems from one another meaning that each bus system operates independently of the other. On the other hand, they can also be set up as redundant units[24], which represents a powerful feature in terms of process automation. It is also abundantly clear that a frame can be assigned any path via various multi-port masters allowing every master to communicate with every other master within the entire network.

---

[24] Here, redundant means that there can be several possible paths between two networks

1 Introduction

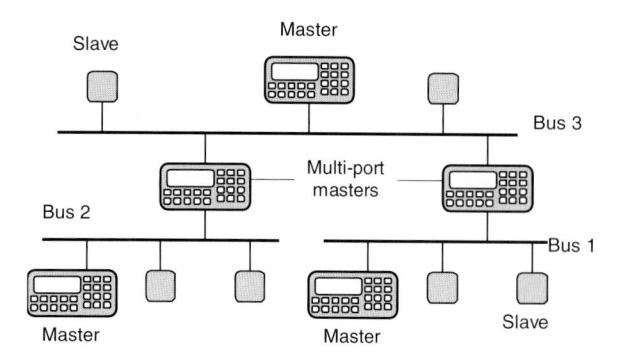

Figure 1.22: P-NET Multi-Network Structure

The data model is one of the defining characteristics of fieldbuses. P-NET links every process signal, e.g. data point, with additional information independent of the current status or value. This information unit contains specific parameters of the configuration, data conversion, scaling, filtering, parameters for error handling etc. It represents a "process object", refers to a variable or function and is called a "channel". The associated data set contains 16 fields (in P-NET terminology – registers). Each of these fields has its own logical address, referred to as the "softwire number" (SWNo). It can be of any data type; the following are standard – digital I/O, analog in, analog out, service etc.

node adr	cont/stat	L_info	info	error

Figure 1.23: Data Frame of the P-NET Protocol

The transmission frame is highlighted in Figure 1.23. It begins with the transmitter address (max. 12 bytes), followed by the receiver address (max. 12 bytes) and a control or status byte, which can contain commands right up to the error information of the slaves. Then comes the length of the information followed by the information itself which can comprise a maximum of 62 bytes and the frame is completed with a check sum of 1 or 2 bytes.

The frame is transmitted through individual UART (Universal Asynchronous Receiver Transmitter) characters with 11 bits, as is familiar from processor technology.

Whether or not a fieldbus gains acceptance is largely dependent on the associated tools that are available to the user. For this, Proces-Data developed a PC-based system called VIGO that is based on the MS-Windows platform. In a very user-friendly way, it allows the user to describe and manage an operational setup on the basis of data structures. VIGO contains various software modules such as a real-time capable communication core, a management information base (MIB) and a server based on OLE2. The OLE2 server offers simple interfaces to MS-Windows applications and software development tools such as C++, Delphi, Visual Basic etc. VIGO also offers various network management and configuration tools. Of particular interest to academic institutions is the fact that most of these tools are available free of charge. The International P-NET User Organization (IPUO) was founded in 1990.

1.3.3 Profibus

The Profibus (PROcess FIeldBUS) is universally applicable at both cell and field level. This is achieved by virtue of its three variations: Profibus-FMS (Fieldbus Message Specification), Profibus-DP (Decentralized Peripherals) and Profibus-PA (Process Automation), which in technical terms cover a wide field of application. It has been internationally standardized via CENELEC (EN50170) and with that represents a solid basis for producers and users independent of individual companies.

Although these three variants have been developed for different application areas, they are clearly coordinated with one another and contain unique interfaces and transitions. This has been achieved on the basis of many similar characteristics. Profibus-FMS and DP are built on the same lowest two levels of the OSI model. With Profibus-DP and Profibus-PA layers 3 to 7 have been omitted and with Profibus-FMS 3 to 6. On the basis of the standards of process engineering, Profibus-PA has a special layer 1, which is integrated into a Profibus-DP system via a connection element (segment coupler).

Profibus-FMS is a multi-master system. A fiber-optic cable or RS485 is used as the transmission medium. RS485 transmission is the most common. This method permits a maximum of 32 end devices in one segment and no more than three repeaters between any two end devices. With a cleverly selected arrangement of repeaters, it is possible to achieve the maximum number of end devices permitted for a Profibus, i.e. 127. The 128th address is reserved for broadcast messages. The bus access method is a token ring between the master stations. When a master has the token it can carry out master-master or master-slave communication. Slave stations can only become active at the request of the master.

Profibus-FMS is the only variant to include layer 7 on which the user-specific programming is based (but which is not a part of it). The services provided here are used to set up and terminate connections, for communication, configuration and for status messages. All these services must be explicitly called up in the user program.

Profibus-DP is in principle a single master system and in the lowest two levels corresponds to Profibus-FMS. There is however no layer 7, which means there are only basic DP functions. In addition to the cyclic exchange of data this includes functions for diagnosis and commissioning and in the extended DP version there is also the function of acyclic data traffic. Profibus-DP must be configured before commissioning. This means that using a configuration tool it is necessary to insert all planned end devices, to assign them with addresses and to allocate the inputs and outputs to an address area of the master station. It is also possible to configure the individual slaves to a certain extent. Certain default settings can be defined, such as what happens in the case of bus failure, i.e. which "safe" status is to be assumed by the outputs. Allocations for the synchronization of slaves or slave groups are also defined in this phase. The command for synchronizing the inputs is called freeze and causes all inputs to retain the valid status at the time of this command until the freeze is lifted or until the next freeze command. The sync command serves to synchronize the outputs and is analogous to the freeze command.

The data that describes the DP network is loaded into the master station. Under normal conditions, i.e. with cyclic data traffic, access only occurs to this assigned memory area of the master. From the point of view of the application, communication between master and slave occurs in the background, thereby relieving the workload. Only diagnostic requests and acyclic data traffic need to be started in the master by calling up the function thereby interrupting the application procedure directly.

1 Introduction

An interesting fact is the ability of operating several DP master systems on one physical bus. In this situation, as with the Profibus-FMS, a token is exchanged among the masters. Each master still has specific assigned slaves to which only it and no other can write; communication between two DP masters is not possible here.

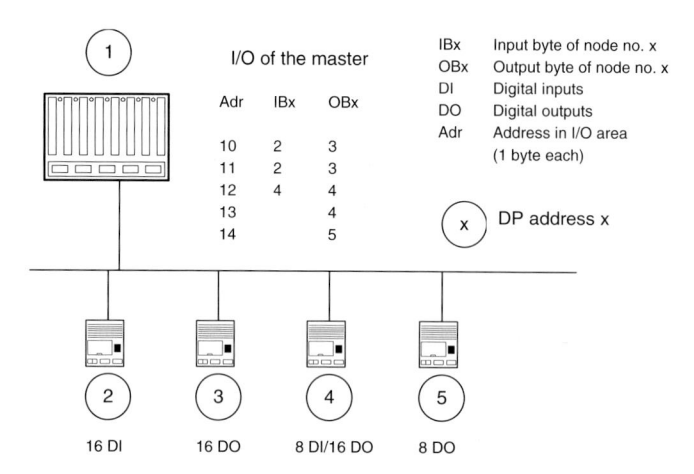

Figure 1.24: Typical Profibus-DP Configuration

The extended DP functions also allow, as mentioned, acyclic data traffic. This is especially advantageous for "intelligent" slaves[25], as it is possible to load new parameters during operation.

In addition to the usual masters, there are also so-called "second class masters". These are used for programming, project engineering and as operating and visualization units. They can be integrated into a Profibus-DP in addition to the other masters. With regard to application functionality however, they do not have any role to play and are also unable to access the communication program.

Profibus-PA has been specially developed for process automation. It is suitable for use in hazardous areas. In its configuration and programming, it corresponds to Profibus-DP. Layer 1 only is formed according to IEC 61158-2. There are two possible ways of adding Profibus-PA devices to a Profibus-DP system: either using a DP/PA link or a DP/PA coupler. The DP/PA coupler transparently connects the DP network with the PA network which means that the application does not detect any difference between the two networks. The coupler does not have a separate address and every PA device is seen from the DP side. The disadvantage lies in the speed. Profibus-PA has a low transfer rate of 31.25 kbit/s. Higher data rates are not possible despite more powerful couplers (e.g. 93.75 kbit/s with Pepperl + Fuchs or 45.45 kbit/s with Siemens).

The DP/PA link appears in the DP part as a slave with a separate address. In the PA part it is a master. The data is buffered in the link, allowing a high data transmission rate in the DP network. A further advantage arises from the fact that it is possible to assign separate

---

[25] In accordance with Profibus terminology, nodes are intelligent when they permit configuration changes. This always involves a controllable processor.

addresses in each PA segment, which significantly increases the number of stations that can be connected to the Profibus-DP. There is also a major difference when it comes to the cabling: with Profibus-PA it is possible to use the same line for data and power supply. Using a DP/PA link therefore provides greater flexibility.

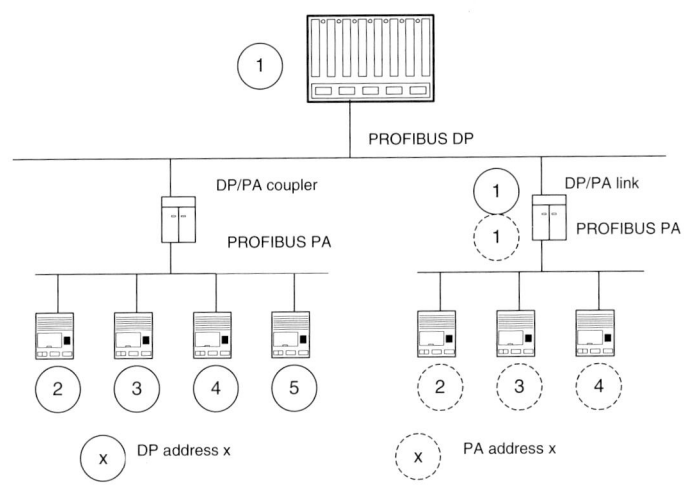

Figure 1.25: Coupling Profibus-DP with Profibus-PA

Summarized:

- FMS is a multi-master system for mid-sized volumes of data at the cell level for which real time is required but no extremely short reaction times.
- DP is a single master system for small volumes of data with short reaction times of between 2 and 5 ms. DP is simple to program and efficient to use.
- PA is a system specially designed for intrinsically safe areas for small volumes of data, is simple to program and ideal in common operation with DP.

Further areas of consideration are conformity and interoperability. For this there are a series of labs specially set up and certified in which Profibus devices are tested for their adherence to the standard. Devices that complete these tests successfully are correspondingly certified. In order however to facilitate the interchangeability of devices in addition to fulfilling the criteria of the standard, there are also a set of profiles in which extra restrictions are formulated.

1.3.4 CAN

The Controller Area Network (CAN) was presented by Bosch in 1982 and was primarily intended for use in motor vehicles. This original application area is also reflected in the properties of the bus system.

Contrary to most other known fieldbuses of the time, CAN uses a message-based form of addressing as opposed to a partner-based method. A message is characterized by an identifier, which in the basic version has 11 bits and in the extended version has 29 bits. On the basis of its configuration, each node knows which of these objects it can transmit and which

1 Introduction

it can receive. This significantly eases any extension to the network. The new nodes no longer need to know with *whom* they can exchange which information, it is enough to know *what* is relevant to each of them – assuming that the allocation of identifiers to messages is already known.

Message-based addressing implies that CAN should be classified as a multi-master system and even as event-controlled. This must be taken into account by a suitable bus access mechanism. In order to prevent the unavoidable delays caused by collisions in the usual random arbitration methods, a non-destructive, bit-wise form of the collision resolution is selected, the CSMA/CA method (Carrier Sense Multiple Access with Collision Avoidance). The bus line is designed as an open collector bus so that the low level is *dominant* and the high level remains *recessive*. This means that a '1' sent from an end device can be overwritten by a '0'. Arbitration ensures that after synchronization, the end devices write the identifier of their message to the bus bit by bit and at the same time observe the current bus level. If two end devices simultaneously access the bus, the identifiers inevitably differ and after the comparison of bits sent and read back, the remaining end device is the one whose message has the lowest identifier (Figure 1.26). This ultimately means that the highest priority object can practically block the bus and messages with lower priorities seldom get through. This negative effect is moderated in the upper protocol layers by a cyclic exchanging of the identifiers within specific priority classes and a restriction in access frequency.

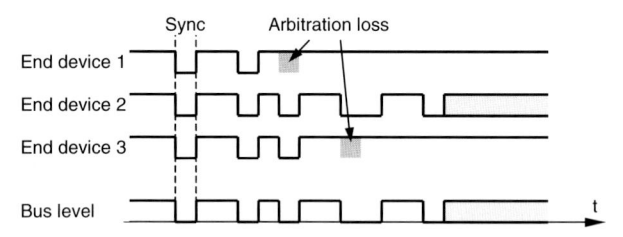

Figure 1.26: Arbitration Method with CAN

The bit-wise arbitration method brings with it a major disadvantage: The propagation time of the signals on the line must be short compared to the bit time to yield quasi-simultaneity for all nodes. With the highest bit rate of 1 Mbit/s this means a maximum bus length of only 40 m. In motor vehicles this restriction is of lesser importance but in industrial automation technology it can lead to a reduction in the data transfer rate.

In the development of CAN particular attention was paid to the detection and secure handling of transmission errors. The safety requirements in vehicles were fully taken into account. In addition to the usual protection of the data frame by means of a check sum, common to other fieldbuses, other mechanisms were also provided to uncover any transmission errors. There is an acknowledge bit in the data frame (Figure 1.27) which must be set by every end device that has correctly received a message. The formal correctness of the data frame is also verified. If a node detects an error, it destroys the telegram by overwriting it with a series of dominant bits thereby violating the CAN bit stuffing rule (there must be no more than 5 consecutive bits that are the same).

The ability of nodes to destroy faulty telegrams can result in faulty nodes blocking the bus. For this reason, the nodes contain error counters which are incremented whenever a transmit or receive error is detected and decremented after successful transmit or receive procedures.

The counter status allows conclusions to be drawn about the reliability of the node and depending on its size the associated node can actively take part in bus traffic or take part with certain restrictions. In the extreme case, it is completely excluded from the bus. This property in particular is noteworthy. In a technical sense it is termed democratic as the shutdown of a non-functioning module is not initiated by others but by itself. Defective nodes independently decide that they do not want to cause disruptive behavior on the bus.

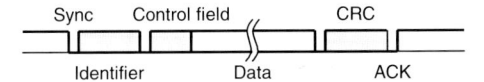

Figure 1.27: CAN Data Frame (Simplified)

A further step to increase error tolerance is that simple errors at least can be blocked by line breaks. CAN uses a differential transmission method that is based on the RS485 standard. If a signal line fails, it is possible to switch from differential to earth-based transmission. Back-up operation is possible but with poorer transmission properties. Not all controllers support this possibility.

In itself, CAN only defines layers 1 and 2 in the OSI model. Although this is sufficient for the exchange of short messages within a closed network, for automation technology it is necessary to define protocols in higher levels. For this reason, the CAN-in-Automation user group (CiA) defined the CAN Application Layer (CAL) and then the CANOpen protocol. Other protocols for automation technology, also based on CAN, are DeviceNet and Smart Distributed System (SDS). The CAN Kingdom protocol has been specially developed for machine controls and safety-critical applications. These higher protocols offer the possibility of exchanging larger volumes of data and of synchronizing end devices. Network management functions solve the problems of node configuration and identifier specification.

1.3.5 EIB

The EIB fieldbus (European Installation Bus) has gained widespread acceptance within Europe. EIBA (EIB Association) was founded in 1990 by leading European companies working in the field of installation technology. This organization then took over the patronage of EIB, guaranteeing standardization and certification independent of individual companies. EIB developed into an open standard which meant that even small companies, by joining EIBA, could take an active part in the development of the EIB fieldbus system.

The main application area of EIB lies in building automation and control and building systems control. As a result of that, the topology of EIB follows this field and has a structure similar to that of a building (Figure 1.28). The fieldbus splits floors and rooms into zones and lines that are separated by couplers. Each of the maximum 15 possible zones is then split into a maximum of 12 or 15 lines depending on the implementation. Each line can contain 256 devices, with an overall maximum count of 57,600 devices or nodes (see chapter 5.1). These individual devices are usually connected via twisted pair lines, but other transmission media such as the power supply line, radio and infrared are currently in the planning stage or have already been implemented.

The main components of the EIB bus are the backbone coupler or line coupler as well as the repeater.

1 Introduction

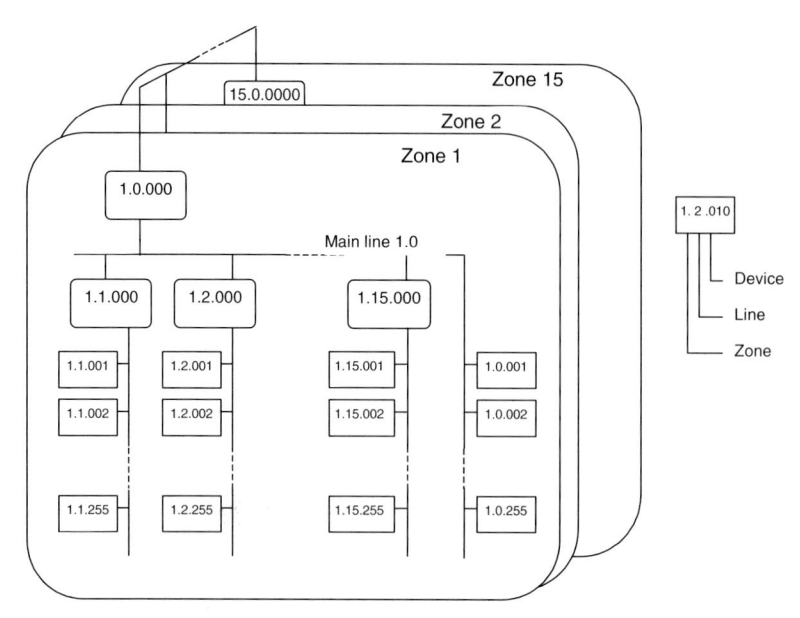

Figure 1.28: Logical Topology with EIB

The backbone and line couplers have the additional task of acting as routers. This means they filter the received data frames and allow only those through that are required in the coupled zone or line. This action saves resources and bandwidth.

The repeater on the other hand has the task of regenerating the electrical signals of the network and separating the bus arbitration. This allows the connection of two electrical segments without any reverse coupling effects of one on the other. The regeneration facilitates the connection of more than 64 end devices per line and an extension of up to 1,000 meters.

An EIB device usually consists of a BCU (bus coupling unit) and an application module. The BCU, which forms the "electrical" contact between the applications and the EIB, represents the universal link between the two "worlds" of the application and the communication system. It comprises a transmission and reception part, a communication controller as well as a PEI (physical external interface) that sets up contact with the external world.

In addition to these external applications, each individual BCU can also contain internal applications. As each BCU has to process the actual EIB protocol stack, the resources for these internal applications are limited.

Overall, the EIB fieldbus system is a pure peer-to-peer network, which uses the CSMA/CA (Carrier Sense Multiple Access with Collision Avoidance) bus access method. This method allows loss-free bus arbitration, as each end device sends priority bits at the beginning of transmission and immediately checks the bus level after each transmitted bit. If it establishes a difference in the applied bit level (0 or 1), the transmission process is stopped. Transmission occurs at a rate of 9,600 kbit/s.

Physically, the EIB devices are contacted via a uniquely assigned address which consists of zone, line and device number and is represented in two bytes. In addition to this "physical addressing", which is mainly used for the initialization, programming and diagnosis of individual devices, there is also "group addressing". The advantage of this method of addressing is the ability to contact several devices at once by setting up a logical connection between individual nodes. The prerequisite is that a sensor may only transmit to one group, whilst an actuator can receive from several groups.

Information exchange between the individual end devices is based on data frames. In addition to the actual data, each frame also contains address, control and check information. With these data frames it is possible to transmit up to 16 data bytes. They can represent simple data types such as Boolean (1 bit), Short (16 bits), Long (32 bits) etc. up to physical values such as temperature, length and energy.

The EIB Interworking Standard, EIS for short, defines a variety of different objects which are specially tailored to the requirements of building automation and control and establish the way in which the individual data types are transmitted. This means that EIB components from different manufacturers are able to communicate with one another.

1.3.6 LonWorks

LonWorks is a fieldbus system developed by the American company of Echelon. It is used within industrial automation as well as in home and building automation. Compared with many other fieldbus systems, particularly in industrial automation, LonWorks can be regarded as an advanced system and should be seen as a network with many of the properties of a LAN. Routers can be used to create any desired topology, guaranteeing a high level of flexibility.

The transmission technology is serial and extremely versatile. Possible media include twisted pair, the power supply line (powerline), radio, infrared and fiber-optic cable [LonM 96]. The choice of medium is made by the so-called transceiver (a word made up from transmitter and receiver). The transceiver connects the medium-independent nodes to the selected medium. In this, it makes the necessary voltage level adjustment and modulates the data. The node application remains unchanged. With LonWorks the data rates range from a few kbit/s to several Mbit/s, depending mainly on the medium and line length. The most common data rate is 78 kbit/s. This is used by the so-called free-topology-twisted-pair-transceiver, which is preferred in all application areas due to its robustness and freely selected topology (Figure 1.29).

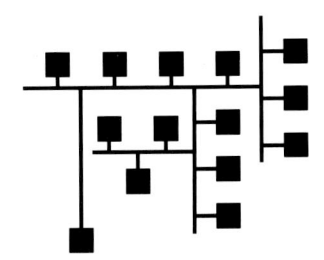

Figure 1.29: Freely Selected Topology

A network can consist of different media, which are connected via routers. Common LonWorks routers can be fitted with any two transceivers and with that connect two network segments of different media. If, for example, the process to be controlled can be split into time-critical and time non-critical parts, a faster medium is used for the time-critical part than

1 Introduction

for the rest of the network. Furthermore, it is possible to connect various segments via a fast segment which then functions as the "backbone". The advantage is that the network can always be managed by a uniform network management even though different media are used. The protocol used in LonWorks (LonTalk) has been standardized in America and Europe since 1998 (EIA 709.1 and ENV 13154-2).

Based on the LonTalk protocol there are a number of various communication objects in use with LonWorks. The most common is the network variable. This represents an object of a distributed application that is automatically transmitted to every distributed application unit in the network. In addition to the network variables, there are also 'explicit message' and 'file transfer' objects.

The majority of LonWorks nodes are based on the so-called neuron-IC. It includes the LonTalk protocol as firmware and can execute applications on-chip. The neuron-IC consists of three CPUs, of which in general terms two are required for the protocol and the third for the application. In addition to the CPUs there is also a ROM, a RAM and an EEPROM for firmware and application on the neuron-IC. The neuron-IC can also be used with external memory (see Figure 1.30).

The neuron-IC is programmed in Neuron-C, an ANSI-C derivative. The command set has been extended in order to use the hardware relevant functions of the neuron-IC, such as for example, the timer and network communication. Neuron-C offers a scheduler, which with so-called "when clauses" allows the execution of various concurrent tasks in a processor.

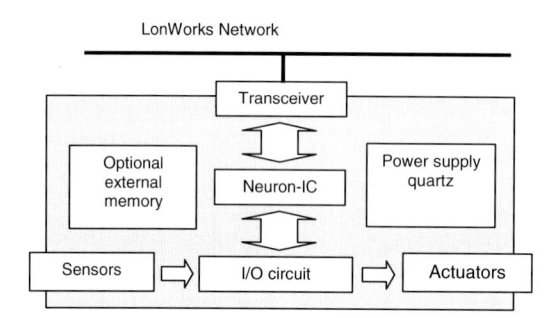

Figure 1.30: A Typical LonWorks Node

In addition to the pins for the connection of the transceiver, power supply and external memory, the neuron-IC has eleven I/O pins that allow various input and output possibilities. Without making any claims on completeness this includes a variety of different serial interfaces, parallel interfaces, I²C bus, an interface for magnetic cards, neurowire, Wiegand input, counter inputs, frequency and pulse width outputs as well as triac outputs and quadrature inputs. The function of the I/O pins is utilized with I/O objects.

The guidelines on interoperability for LonWorks were recorded by the LonMark Interoperability Association. They manage standard network variable types (SNVTs) and functional profiles, published under www.lonmark.org. There are SNVTs for the electrical supply, speed, temperature and all other common parameters [LonM 97). The use of these SNVTs occurs in accordance with profiles such as "8060-Thermostat" or "1030-Pressure Sensor". These profiles have been worked out within "task groups" which include, among others,

"Intrinsic Safety", "HVAC", "Automated Food" or "Home/Utility". Furthermore, [LonM 98] provides instructions on self-describing nodes.

For the PC-side of the applications and for the network management a database system has been created under the name of LNS (LonWorks Network Services). It represents an object-based database that is operated on a PC. It manages all nodes, segments and functions of the network. It is possible to access the database elements with the LNS API [LNS 97, LCA 97]. This API allows Windows applications to be quickly developed, which can then take part in a LonWorks network.

There are two main development tools for creating LonWorks applications: the LonBuilder and the NodeBuilder. NodeBuilder is used to program and debug individual nodes, whilst LonBuilder can manage several nodes simultaneously and with that is more complex and subsequently more expensive.

2 Networks, Disciplines, Applications

Due to the steady increase in the number of interlinked computer systems within office buildings, it has become necessary to structure the layout of these networks in order to be able to handle them more efficiently. The result has been the development of several standards, which are now also used in other areas. In multi-purpose buildings in particular, there are often various network topologies that are coupled via different network components, thereby creating a very heterogeneous communication network.

2.1 Cabling in Functional Buildings

For the cabling within and between buildings a differentiation is made between primary, secondary and tertiary cabling (see Figure 2.1) [CLC 95, IEC 95].

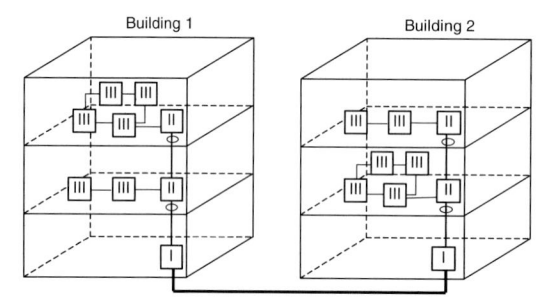

Figure 2.1: Network Structure

The primary cabling is the connection between two buildings, in this case between buildings 1 and 2. It is generally implemented as fiber-optic cable in order to guarantee the required data throughput as well as to provide the necessary protection against electrical interference (lightning protection). Interconnection point I in each building can also include connections to other buildings. The secondary cabling to connect the floors within a building is generally implemented as a backbone, which depending on the requirements, uses either coaxial line or fiber-optic cable. The topology includes line networks and ring networks. The interconnection points on the individual floors are denoted in Figure 2.1 with II. Within each floor, communication occurs via the tertiary cabling in the form of twisted pairs, coaxial lines or fiber-optic cables. Interconnection points denoted with III represent the connections between the floor distributors and the relevant local networks. Line, ring and star network topologies are used.

The EIA/TIA 568 Commercial Building Wiring Standard [EIA 568] recommends four basic types of cable that can be used between the end devices:

- Coaxial cable,
- Shielded twisted pair, STP,
- Unshielded twisted pair, UTP,
- 62.5/125 micron multi-mode gradient fiber.

The standard specifies that there must be two copper-based information paths at every workplace, or at every point where end devices are provided: one for the telephone and one for data transmission. If a fiber-optic cable is to be installed, then in accordance with EIA/TIA 568, this must be done in addition to the two copper-based lines and not as a replacement. This guarantees that in every installation compliant with this standard, there are always at least two copper-based lines. For low frequency cables, the standard has defined the following categories:

- Category 1: economy cable for analog voice transmission and data transmission at rates significantly lower than 1 Mbit/s. Cables of this category are no longer recommended for new installations. The capacity of this category corresponds to that expected from a regular telephone line.
- Category 2: cable to replace category 1 cable. Transmission at rates of up to 4 Mbit/s across mid-range distances, e.g. for small token ring networks and ISDN.
- Category 3: UTP/STP cable for transmissions at rates of up to 10 Mbit/s including category 1/2 applications; capacity of Ethernet 10BaseT up to 100 m.
- Category 4: UTP/STP cable for transmissions at rates of up to 20 Mbit/s across larger distances than category 3 (10BaseT and token ring)
- Category 5: for transmissions at rates exceeding 20 Mbit/s or frequencies up to approx. 100 MHz (for the transmission of FDDI) across distances of up to 100 m.

2.2 Cabling in Domestic Buildings

The progress in cabling within the domestic sector lies way behind that seen in the commercial sector. This is firstly due to the various functions of the different rooms, but it is also a question of cost to the private user. Within CENELEC, European Committee for Electrotechnical Standardization, in the technical committee TC 205, they have been working on a standard for this area [CLC 99] under the title of "Home and Building Electronic Systems (HBES)". This draft standard outlines and classifies the various cables within domestic buildings. These classifications are based on the application and the resulting, necessary bandwidth or transfer rate. Table 2.1 illustrates a selection of possible networks in the domestic sector. The table entries are ordered according to bandwidth or transfer rate.

The current status of cabling is somewhat removed from that listed in Table 2.1. The following networks are currently installed in an average home:

- Main power supply,
- Telephone (analog),
- Cable TV (analog).

2 Networks, Disciplines, Applications

Table 2.1: Networks in Domestic Buildings

Network	Applications	Bandwidth / transfer rate
Power supply	Main power supply	50 Hz
HBES class1	Extra-low safety voltage (SELV)	DC
	Home automation functions	9.6 kbit/s
	Security functions	9.6 kbit/s
Internal communication	Internal telephone, entrance intercom	144 kbit/s
	Transfer of voice information, music, sounds	144 kbit/s
	Entrance monitoring, video system	6—8 MHz
External communication	Telephone (analog)	3.1 kHz
	ISDN	≥144 kbit/s
Data transmission	Mid-range	200 kbit/s
	Long-range	>200 kbit/s
CATV	Radio, TV, video (analog)	3000 MHz
	Radio, TV, video (digital)	>100 Mbit/s

Almost all the applications listed in the above table could be implemented via these networks, but network access is currently only designed for the specific applications and not for the incorporation of "third-party" functions. For example, it is not possible to operate the entrance intercom via the existing analog telephone or to implement room monitoring without making changes. The combining of various applications into a central, broadband network or various, interconnected networks will only be possible in the future, when the network connection costs have been lowered and suitable interfaces for communicating with different applications have become available (see chapter 2.4).

A further aspect is the development in the field of data networks. Firstly, the data rates are becoming higher and higher (in the summer of 1999 the Gigabit-Ethernet Standard was adopted by IEEE and a 10-Gigabit-Ethernet task force founded) and secondly, bus systems are being established in the multimedia branch, such as for example, the IEEE-1394-Standard, also referred to as "FireWire" [Ande 99, Stam 99]. Some properties of the IEEE-1394 standard are as follows:

There is no established topology for setting up a FireWire network. The total length must not exceed 72 m, whereby the maximum distance between the individual nodes is 4.5 m. Devices connected to the FireWire are assigned physical addresses. It is possible to insert and remove any device on the bus during active operation; the reallocation of addresses only occurs in response to a bus reset. The cable is much thinner than the familiar cable (e.g. RJ45, BNC) and therefore allows much greater flexibility. The connectors and easily twisted and looped cables from Nintendo have set a precedent.

In the IEEE 1394 cable, there are two twisted pair lines that are individually shielded. There are also two lines for the power supply which carry 8 to 40 V at a maximum 1.5 A. The entire cable is then shielded again (see Figure 2.2). The transmission rates are 100, 200 and

400 Mbit/s; the optimum speed is selected automatically depending on the device and application.

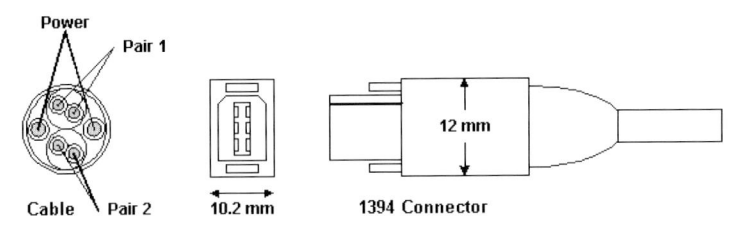

Figure 2.2: FireWire Cable and Connector

2.3 Building Categories

The buildings themselves can also be categorized, according to size and usage. One possible method of categorization is the S/M/L model [Beck 98] (see Table 2.2).

For houses and buildings in the S category, comfort and security are the main reasons for using automation systems, whereby security has become more important than comfort. The security field covers break-in and fire alarms as well as sensors for water and oil leaks, operating and monitoring units, programming units, server stations, data interface units to other levels, systems for special tasks (e.g. warning systems) and emergency calling systems. Comfort involves creating comfortable room temperatures, the simple operation of automation functions and various lighting scenarios for inside and outside the home. The saving of energy and the economical use of our natural resources is a third-place consideration for the use of automation systems within category S buildings. This area is generally assigned to the field of house systems automation.

Table 2.2: S/M/L Model

Category	Examples
S (houses and small buildings)	Private homes (e.g. houses and apartments)
M (mid-sized buildings)	Hospitals, schools, nurseries, old peoples' homes, multi-purpose buildings, small and mid-sized office blocks
L (large buildings and real estates)	Large office complexes, high schools, airports

For mid-sized buildings, the most important argument for the use of automation systems is generally a question of economics. The major considerations are operating costs, energy costs as well as the running costs of the devices and maintenance. The aspects of security (break-in and fire alarms, entrance control, etc.) and comfort (room temperature regulation, lighting control, lighting scenarios etc.) take second place. In such buildings, often run as multi-functional buildings, the use of automation systems from the field of building systems engineering are most common. These are sometimes based on communication standards such as EIB and LonTalk [LonW 93] but also on manufacturer-specific systems.

For larger building complexes, the operating costs or capital returns are the deciding factors for the use of automation systems. These systems must cover the entire spectrum from single room regulation, security systems and energy cost optimization right through to facility management. In these situations, automation systems are used that offer the possibility of linking up to other applications (e.g. accounting systems) and building disciplines (e.g. lift control).

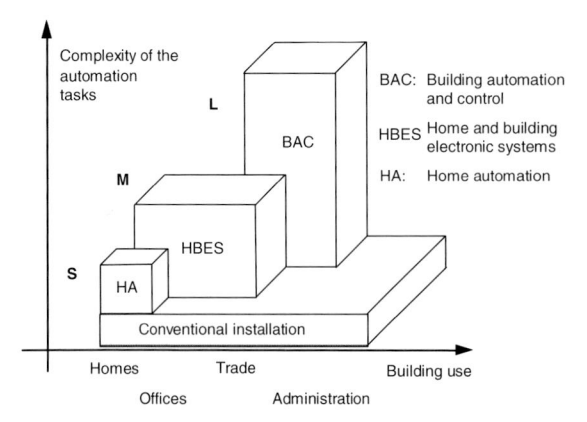

Figure 2.3: Allocation of Building Category to Automation System

The boundaries between the building categories are of course flexible, as is the use of systems from the various areas – home automation, home and building electronic systems and building automation and control. However, this rough division does offer an initial introduction when selecting automation systems for a particular building. Figure 2.3 represents a simplified illustration of the connection between the complexity of the automation tasks and the building categories. Automation systems themselves are divided into three levels – the management level, automation level and field level. Depending on the scope of the tasks, these levels contain operating and monitoring units, programming units, server stations, data interface units to other levels and systems for specific tasks such as for example, warning systems.

A sharp increase in the number of automation functions is expected in the field of home automation, particularly when the components become cheaper and easier to install.

Table 2.3: Levels and Functions in BSA

Level	Functionality, device
Management level	Operating and monitoring units, programming units, server stations, data interface units to other levels and units, systems for specific tasks (e.g. warning systems)
Automation level	Operating and monitoring units, programming units, automation stations with devices that are either connected directly or via communication units, application-specific controllers, data interface units to other levels and units, systems for specific tasks (e.g. warning systems)
Field level	Room automation appliances, direct operating elements (emergency operation level), sensors, actuators

2.4 Building Disciplines

Building disciplines denote the individual areas of a building (mainly in the context of construction) such as, for example, the electrical installation. Buildings of category L now contain several different disciplines, almost all of which are provided with communication interfaces and with that can be incorporated into an automation system; Figure 2.4 shows some typical examples.

Based on the infrastructure within buildings of category S or M, the appliances are not so complex and to a certain extent the requirement profiles are completely different. Take for example an industrial kitchen where the potential for energy savings are vast in comparison with a single electric oven in the domestic kitchen, which can call up recipes and show important information on the TV screen via the communication interface. In this category, everything should be smaller, simpler and more reasonably priced; the fact that this is not yet the case is partly due to the lack of components capable of communication that are able to integrate the respective areas represented in Figure 2.5.

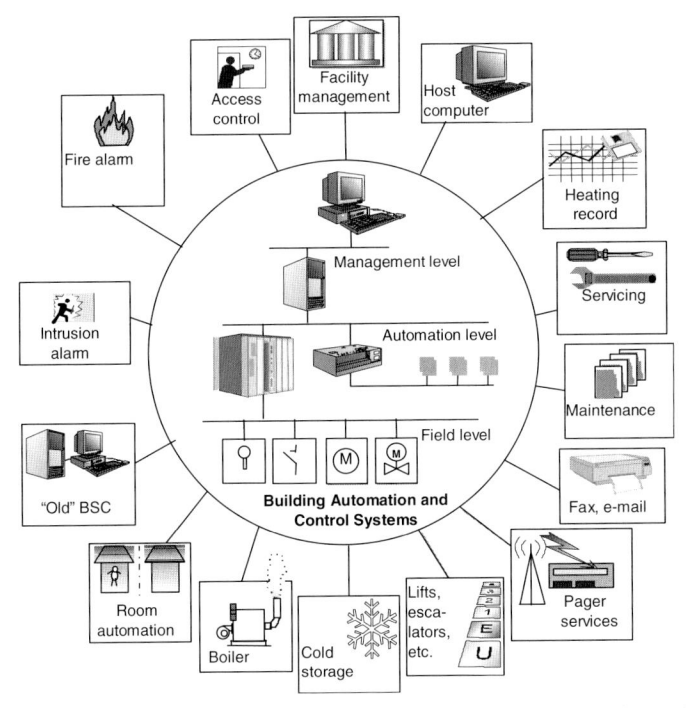

Figure 2.4: Building Disciplines and Building Automation and Control Systems[26]

---

[26] BSC: Building systems control

2 Networks, Disciplines, Applications

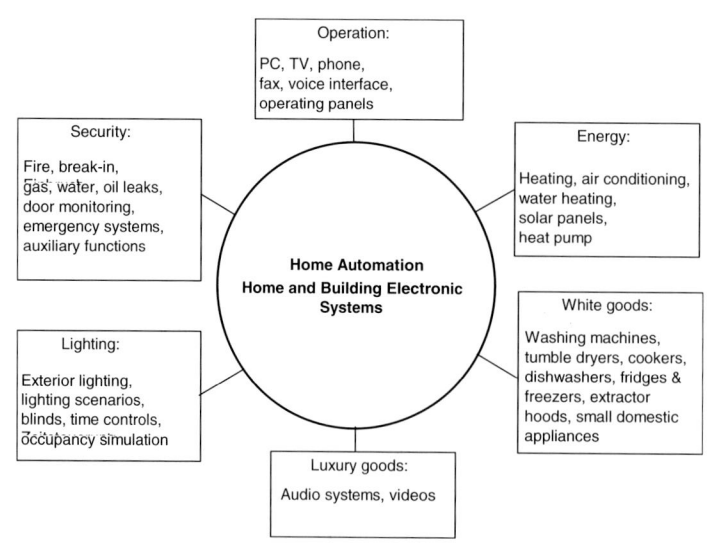

Figure 2.5: Linking Systems in Home Automation and Home and Building Electronic Systems

Figure 2.6 illustrates the future development for the use of EIB in the field of house systems automation. In the home, EIB can cover the entire spectrum from the field level to the management level. The application possibilities of EIB become restricted once the buildings and with that the number and complexity of functions become greater, so that in larger buildings it only covers the field level.

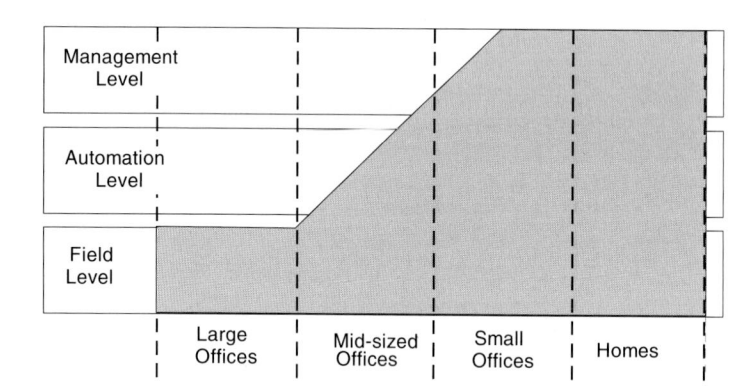

Figure 2.6: Future Application Areas of EIB

3 EIB Protocol

3.1 Introduction

The basis for the following description of the functional methods of EIB is the most commonly used reference model for data communication – the Open Systems Interconnection Model (OSI model for short), conceived in the late seventies by the International Organization for Standardization (ISO). Very briefly, we will first highlight the aspects of this model that are necessary for our further discussions. We will then take a fundamental look at the types of service implemented in EIB and end this introduction with a brief look at the basic principles of data transmission within EIB.

3.1.1 OSI Model

The attempts at creating a reference model for data communication arose from the fact that at the time, there were a series of computer networks all of which were incompatible with one another. The expansion of these networks was therefore limited to a specific circle of users. Data transmission from one network to another was only possible with great investment in specialized hardware and software solutions.

It was the aim of the OSI model to counteract this development. ISO introduced the concept of an open system. Such systems consist of hardware and software components that comply with a given set of standards. These standards guarantee that systems from different manufacturers are compatible with one another and can easily communicate.

As there were many factors to consider, it was decided that the best approach would be to split the complex task of data communication into seven clearly defined sub-areas, hereafter referred to as *layers*, and at the same time to clarify their mutual relationships. The end product was the now familiar OSI model, architecture for connecting open systems. This model however only establishes the effects of each of the layers. Suitable standards were additionally worked out for the layers. These standards are not part of the reference model and were later published in their own right.

The great significance of the OSI model came about due to the consistent implementation of three important concepts:

1. *Service*. This represents any service made available by one layer to the layer directly above it. We call the user of a service the *service user* and the layer that provides the service the *service provider*. Note: For the services of a layer, the OSI model only defines their functionality and not how they are actually implemented.

2. *Interface.* There is an interface between every two layers. This clearly defined interface specifies which services are offered by the lower layer to the upper layer, i.e. how the service user can access the services of the service provider, what parameters need to be transferred and what the expected results are.

3. *Protocol.* The term protocol denotes a set of rules that dictate the communication of layers on the same level. If layer N of open system 1 wishes to contact layer N of open system 2, both systems must adhere to specific rules and conventions. Together, these rules make up the protocol of layer N. We will hereafter refer to layers that lie on the same level as *peer layers*.

The tasks of the individual layers of the OSI model have already been outlined in great detail in the introduction to this book. We will waste no more time on this subject and instead turn our attention to the data that is exchanged between service user and service provider or between two peer layers (Figure 3.1).

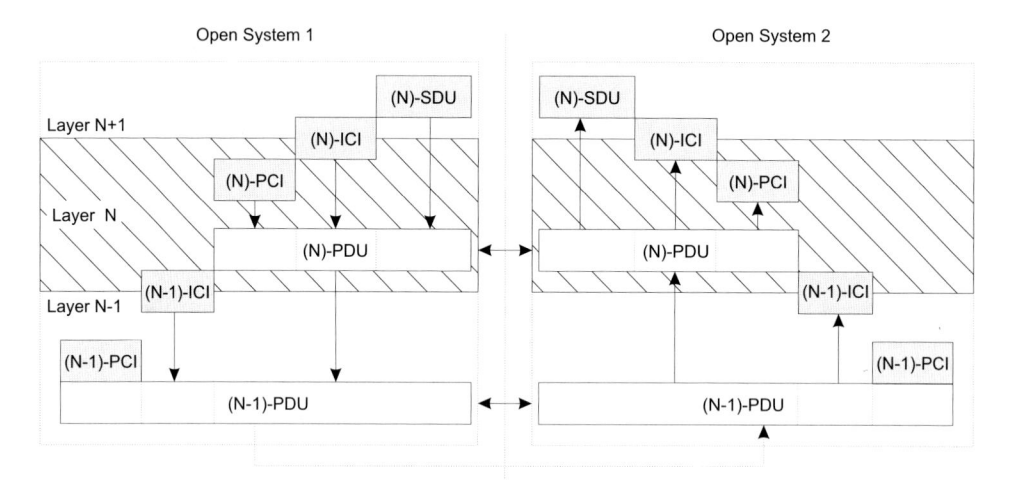

Figure 3.1: Communication in the OSI Model

1. *Service data unit*, SDU. Communication of layer N+1 with the underlying layer N occurs via its services, or more accurately via the interfaces of these services. For layer N, the transferred data represents pure user data that is passed on to the next lowest layer for further processing.

2. *Protocol data unit*, PDU. Communication between two peer layers is implemented via so-called protocol data units. These represent the core element of the rule set that can only be understood and correctly interpreted by the peer layers. A protocol data unit consists of the transmitted user data supplemented with parameters of the interface (*interface control information*, ICI) and unique control information (*protocol control information*, PCI).

With these connections, by no means trivial in themselves, we must now conclude our introduction to the OSI model. We have merely highlighted the aspects that are important to the subsequent chapters. A more comprehensive introduction and explanation would fill a separate book [Tane 96].

3.1.2 EIB in the OSI Model

Not all of the layers defined in the OSI model are always necessary for simple communication systems that are tailored for specific tasks. EIB for example, requires only five of the seven layers (Figure 3.2).

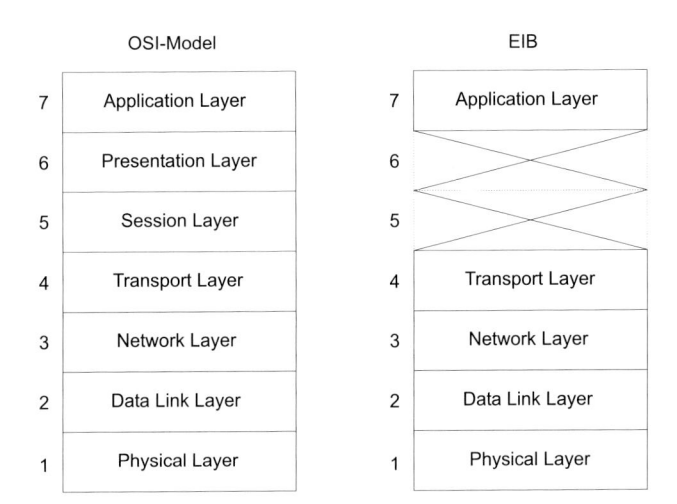

	OSI-Model		EIB
7	Application Layer	7	Application Layer
6	Presentation Layer	6	
5	Session Layer	5	
4	Transport Layer	4	Transport Layer
3	Network Layer	3	Network Layer
2	Data Link Layer	2	Data Link Layer
1	Physical Layer	1	Physical Layer

Figure 3.2: OSI Model and EIB

We will now take a general look at the interplay between the individual layers. This is achieved on the basis of EIB services that have been adapted from the services in the OSI model. Strictly speaking, every EIB service is composed of a series of operations. Hereafter, these operations will be referred to as *service primitives*.

There are basically four different service primitives: *Request*, req, *indication*, ind, *confirmation*, con and *response*, res. However, services need not always comprise each of these four service primitives. More often than not, services are classified according to the number of service primitives and their mutual effect.

1. *Locally confirmed service.* A locally confirmed service comprises a request, an indication and a confirmation. Each service primitive that is called up by the service user is termed a request. With the request, layer N receives the order to execute a specific task. The respective task and corresponding data is converted into a corresponding PDU. In accordance with the OSI model, the service provider uses the services of the underlying layer N-1, in order to carry out its task. This interaction continues until the lowest layer entrusts the data to the physical medium.

   On the receiver side, the peer layer is activated via an indication (or sometimes even a whole series of indications). After the remote layer N has decoded the PDU and extracted the control information, the user data is passed on to the above lying layer (i.e. the layer directly above or in the case of the application layer, the actual user) also by means of an indication.

   Layer N of the sender then receives a confirmation from local layer N-1. Based on the parameters transmitted, the service provider can tell whether its underlying layer was able to accept and process the request accordingly. After a brief pre-processing, this local con-

3 EIB Protocol

firmation is also made known to the service user by means of a confirmation service primitive.

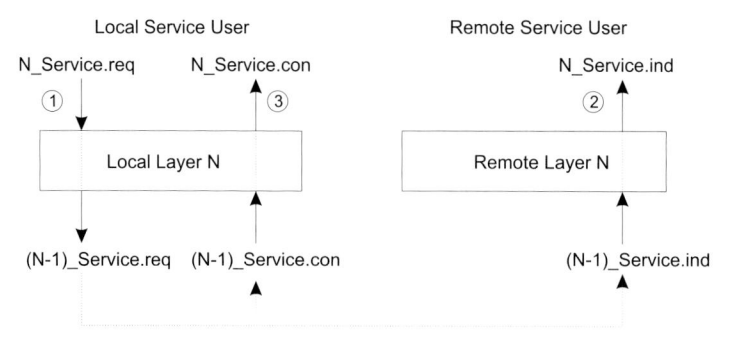

Figure 3.3: Locally Confirmed Service

2. *Confirmed service.* This type of service also consists of a request, an indication and a confirmation. With a confirmed service however, the peer layer generates an acknowledgement immediately after receiving the indication. The acknowledgement is returned to the sender via a service primitive of the remote layer N-1 and signaled to the service provider via an indication of local layer N-1. Note: Contrary to the procedure outlined in point 1 above, the transmitted parameters and data come from the partner side!

From the indication, layer N can conclude whether or not the service requested by layer N-1 was executed without error. If necessary (i.e. if errors have occurred) this service is used again. This depends on the protocol used in layer N. In all cases, the service user is informed of the output with a suitable confirmation (positive or negative). From this confirmation it is possible to derive whether or not the originally requested service has been satisfactorily fulfilled by layer N.

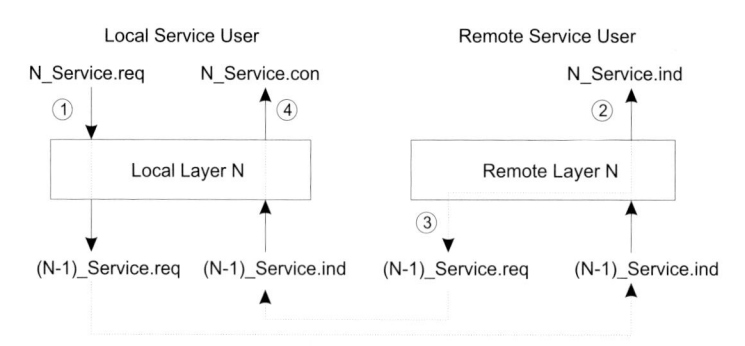

Figure 3.4: Confirmed Service

3. *Answered service.* This represents a special case in which after an indication on the partner side a response is generated by the remote service user. The response is transmitted via a service primitive of layer N-1 and passed on to the service user from which the original service request came as a confirmation. This mechanism permits data traffic in both directions. Contrary to the other two services, an answered service always consists of request, indication, response and confirmation.

3.1 Introduction

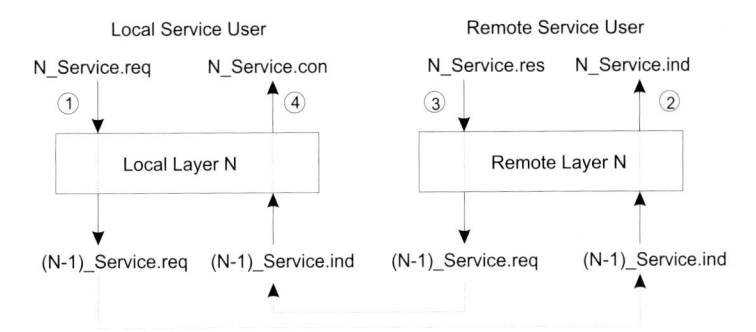

Figure 3.5: Answered Service

In short: Request and response are always called up by the service user, the resulting confirmation and indication originate from the corresponding layer.

The interaction between the individual layers is best understood if you imagine that the layers lying on top of one another are primarily inactive. The service provider is only "awoken" after a request from the service user. In order to perform this service, it then activates the underlying layer by calling up the appropriate service primitive. This interaction continues down to the lowest layer, which in turn accesses the transmission medium. On the remote side, the indication of a low lying layer informs the layer lying directly above that a service is to be executed. This indication is then processed in accordance with the service type – it causes either a confirmation or a response.

Having now given an overview of the various service types and the sequence of the service primitives, the time has come to examine the individual EIB layers in more detail, including their services and protocols.

First a note on nomenclature. In each of the subsequent chapters we have assigned unique names to the protocol data units of the respective layers. The PDUs of layer 2 are frames, the PDUs of layer 3 packets. In layer 4 they are referred to as telegrams and the word "message" is reserved for protocol data units that originate from the application layer.

And now a brief word on the layout of the subsequent chapters. Each chapter begins with a brief look at the functionality of the corresponding layer. This is followed by a detailed outline of the available services. In order to substantiate the actual working methods, we then specify a corresponding routine (a so-called handler) for each service primitive. These routines have been separated for the local layer (sender) and remote layer (receiver). In reality, such a handler would always be executed when a service primitive is called up by the service user or the underlying layer. We would like to point out that the program fragments given in programming language C have more of a didactic purpose than a practical one. Many technical details have been omitted for the sake of simplicity, and the purpose of listing the selected fragments of program code is to achieve a basic understanding rather than an exact efficiency (this is also the reason for the separation into corresponding handlers for sender and receiver). However, we believe this is the best approach to give the reader a good understanding of the way in which the services work and of the corresponding protocols. The interested reader will gain an invaluable insight into the cooperation of the EIB layers. The discussion of each layer usually ends with the concrete layout of its protocol data unit.

Readers who are just interested in the basic principles of data transmission in EIB are referred to the next chapter and for the rest – let us begin!

3.1.3 A Brief Overview of EIB

EIB is characterized by a clear, hierarchic structure that has been specially designed for home and building electronic systems [Fran 97, Rose 00]. EIB end devices are topologically arranged in lines and zones. Each line is separated from the other lines by line couplers. Up to 12 or 15 of such lines (depending on the implementation) can be combined into a zone. Backbone couplers can combine a further 15 zones into one compound system. As each line can accommodate a maximum of 256 devices, a complete extended EIB system can comprise a total of 57,600 end devices (see chapters 3.2.1 and 5.1).

In the "classic" EIB system the network nodes are connected via twisted pair lines. Other media however are in planning or to a certain extent already in use (coaxial cable, supply mains and even infrared or radio). The twisted pair remains the most common (typical for fieldbus systems). Access to this transmission medium in EIB occurs via loss-free bus arbitration, better know by the name *Carrier Sense Multiple Access with Collision Avoidance* (CSMA/CA). Each transmitting device sends its data to the bus bit by bit and after every bit checks whether the bus level matches the applied bit. Any difference indicates that a collision has occurred and the end device must stop the transmission procedure immediately.

Information is exchanged via data frames. In addition to the control and address fields, each data frame contains 1-16 bytes of user data and a check field of 1 byte.

The telegram priority is recorded in the control field. The address field of approximately 4 bytes contains the addresses of the source and the message destination. The subsequent data includes configuration data or messages that originate from the application. To detect transmission errors, the sender forms a check character and adds it to the data frame. This check character is recalculated on the receiver side. If differences are established, retransmission of the data frame is activated, otherwise a positive confirmation is returned to the sender.

In the EIB network, it is possible to address single devices, device groups or all devices. A differentiation is made between the physical address and the group address. The physical address uniquely identifies the EIB end device. It has the format zone/line/device. The group address defines the allocation between communicating application processes and is split into 15 main groups, each with up to 2,048 sub-groups. It has the format main group/sub-group.

In order to integrate a new device into the EIB system, the EIB user must bring the device into the programming status (this generally involves pressing a specific programming key on the device itself). Afterwards, the physical address can be programmed via the EIB. This is achieved using a telegram that is fundamentally addressed to all devices but only processed by the device that is in programming mode. Once the address has been set, the corresponding device can be identified by its physical address and is able to receive further configuration data (applications, parameters and group addresses) via the EIB.

The actual applications communicate via group addresses, which involves an exchange of so-called communication objects. The information contained in these objects activates analog and digital functions in the addressed devices. For example, after reading a switch setting, lights can be controlled or temperature regulated. The layers, services and protocols outlined in detail in the following chapters are responsible for the transportation of this data.

So much for the overview – we would now like to examine the details of the EIB layer model [EIBA 95, EIBA 98].

3.2 Physical Layer

The *physical layer* establishes all electrical, mechanical, functional and procedural parameters of the physical connection between end devices in a distributed system. In this context, it is clearly of the utmost importance to agree on the type of transmission path used (should the data be transferred via a twisted pair, infrared or radio?), the layout/topology of the transmission path as well as the signal coding.

Roughly speaking, the main task of the physical layer is the transmission of the individual bits. There are certain rules governing how the bits are to be applied to the physical medium. These conventions however must always be viewed in the context of the medium and to a large extent determine the protocol of layer 1. The overlying *data link layer* notices nothing of these processes. It simply uses services that allow an entire stream of bits to be transported via the respective physical medium.

In its classic version, EIB uses a twisted, shielded two-wire line (*twisted pair*) as the transmission medium. Other media, which are becoming increasingly popular, are radio and infrared. Even the conventional mains supply (*powerline*) can be used as an EIB system, equipped with the corresponding end devices. However, we will restrict our examinations in this and subsequent chapters to the "classic" EIB system.

As the EIB is designed as a decentralized system, all end devices can communicate directly (i.e. without a central control center) via the twisted pair (hereafter referred to as the bus). With regard to bus access, every end device has the same authorization and can access the bus whenever it wants as long as the bus is not busy.

With this type of access, i.e. without a controlling instance (known as a *master*), it is not possible to exclude the possibility of collisions. Such collisions occur when several end devices are simultaneously of the opinion that the bus is free. These collisions must be detected (*collision detection*) and removed by means of a special solution procedure (*collision avoidance*). If there is simultaneous multiple access, only one end device should be successful. This end device is then free to use the bus for a certain period of time, before there is any decision regarding new access (*bus arbitration*).

In order to transport this bit stream without interruption, another important task of the physical layer is therefore to control access to the bus. In addition, the bit stream must be converted into suitable impulses that can be transmitted via the network (analog). On the receiver side, digital bits are reformed from the analog signal and passed on to the data link layer. To provide a better means of solving these two part functions it is advisable to separate the primary "digital" functions from the pure "analog" ones. For this reason, layer 1 comprises two part layers. The upper layer, termed the *logical unit* (LU), works closely with the data link layer, the lower layer, termed the *medium attachment unit* (MAU), is responsible for the connection to the medium (Figure 3.6).

We will continue with a look at the topology of the EIB system. This is followed by a presentation of the LU part layer and associated protocol. To clarify some of the details of the service primitives of layer 1, we have used short sections of program code. Whilst the LU part layer is dominated by pure digital functions, we need to consider the analog reality of the transmission medium when it comes to the lower layer, MAU. As the analog functions are difficult to represent in pseudo code, we must deviate from the procedure used for all other layers and avoid the use of program fragments.

3 EIB Protocol

```
P_Data.req        P_Data.con                              P_Data.ind
    │                 ▲                                       ▲
    ▼                 │                                       │
┌─────────────────────────────┐              ┌─────────────────────────────┐
│   Local Logical Unit (LU)   │              │   Remote Logical Unit (LU)  │
└─────────────────────────────┘              └─────────────────────────────┘
                  ▲                                          ▲
MAU_Trans   │   MAU_Read                              MAU_Rec │
    ▼       │                                                │
┌─────────────────────────────┐              ┌─────────────────────────────┐
│ Local Medium Attachment Unit (MAU) │       │ Remote Medium Attachment Unit (MAU) │
└─────────────────────────────┘              └─────────────────────────────┘
            │                    Medium                      ▲
            ▼                                                │
──────────────────────────────────────────────────────────────────
```

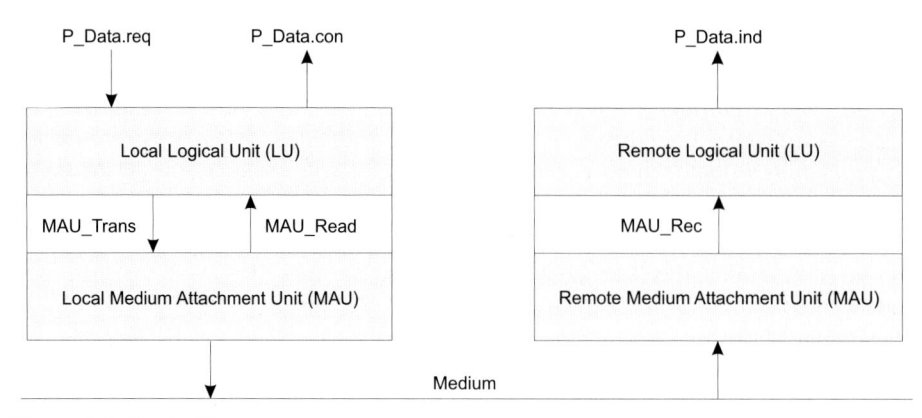

Figure 3.6: Physical Layer

3.2.1 Topology

In the twisted pair version, EIB consists of bus-compliant sensors, actuators and other system components. The bus line is used to transmit the supply for the electronics and for the exchange of information (to carry out the necessary tasks). The EIB is distinguished by a clear, hierarchic structure. This makes it attractive for use in homes as well as larger functional buildings.

The maximum permitted line length depends on the data transmission rate. With a baud rate of 9.6 kbit/s the maximum line length is 1,000 m. The maximum separation between two end devices is 700 m. Two different bus lines are generally used. The requirements are listed below (see Table 3.1).

Table 3.1: EIB Line Material

Type	Structure			Laying
PYCYM 2x2x0.8	Basis:	EIBA guidelines according to DIN VDE 0207 and DIN VDE 0815		Internal: Fixed installation in dry, humid and wet areas. External: Protected from direct sunlight. Surface mounted, flush mounted and in pipes.
	Wires:	red black yellow white	+bus -bus free free	
	Wiring:	Shielding with supplementary earth wire		
J-Y(ST) 2x2x0.8	Basis:	DIN VDE 0815		Internal: Fixed installation in dry, humid and wet areas. External: Protected from direct sunlight. Surface mounted, flush mounted and in pipes.
	Wires:	red black yellow white	+bus -bus free free	
	Wiring:	Shielding with supplementary earth wire		

3.2 Physical Layer

The smallest functional EIB unit is the *line* (or *sub-line*). Depending on the implementation, a line comprises a maximum of 64 (TP64) or 256 (TP256) end devices. It must be provided with a separate power supply (*bus power supply unit,* BPSU).

Using *line couplers,* LC, individual lines can be combined along a *main line.* This creates larger sections called *zones.* In turn, these zones – consisting of up to 12 (TP64) or 15 (TP255) lines – can be combined via a common *backbone.* This requires special devices known as *backbone couplers,* BC. The backbone also requires a separate power supply.

In its fullest configuration, it is theoretically possible to connect 11,520 (TP64) or 57,600 (TP256) end devices to the EIB. There are also links to other fieldbus systems as well as the Internet.

Figure 3.7 shows an overview of the hierarchic structure of the EIB system.

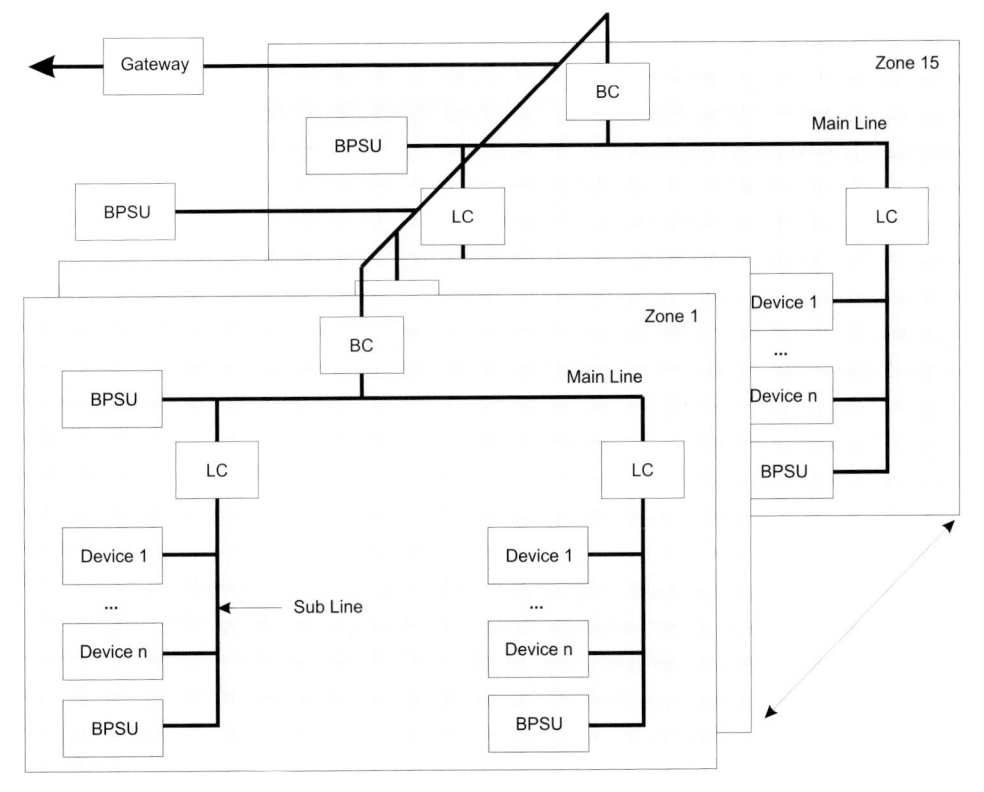

Figure 3.7: Topology

3.2.2 Logical Unit

The logical unit (LU) part layer more or less represents the medium-independent part of the physical layer. On the sender side, it accepts a data byte (with fieldbus systems a byte is often referred to as an octet) from the above-lying data link layer, adds extra information to it and finally converts it into a sequence of individual bits. These signals, which are still logical, are then converted by the lower lying medium attachment unit into analog signals. The actual transmission involves standardized UART characters.

One reason for the use of UART characters is that the end devices connected to EIB do not have a common timing signal. The bit clock pulse must be derived from the data signal (*self-clocked*); otherwise it would be necessary to use an extra line for transmitting the timing signal. A prerequisite for the self-clocked variant (put simply) is that there is an occasional change in the logic signal. Otherwise there is the danger that the timing at the sender and receiver drifts too far apart. Such a UART character counteracts this possibility and consists of a total of 11 bits:

- 1 start bit (ST). The start bit is always logic 0 and because there is no common cycle for the end devices, is used for synchronization.
- 8 information bits (D0-D7). The data bits correspond to the octet to be transmitted. The *least significant bit*, LSB, is coded first.
- 1 parity bit (P). The physical layer checks whether the data bits D0 to D7 consist of an even or odd number of "ones". In the first case, the parity bit is set to zero and in the latter, i.e. with an odd number of "ones", it is set to one (even parity). With this strategy it is possible to very quickly detect a *single* transmission error within a UART character, but not to remove it – double errors within a UART character remain undetected. It is necessary to use other mechanisms to exclude these. These are to a certain extent found in the protocols of the higher layers. These mechanisms will be examined in chapter 3.3.4.
- 1 stop bit (SP). The stop bit concludes the actual character and is always logic 1. With that, it is also used for the purpose of synchronization, like the start bit.

The start and stop bits can now be brought into play for the purpose of synchronization. At the very least, they guarantee a level switch after every transmitted octet. After the transmission of a UART character, the physical layer waits for a further two bit times before the next character is applied to the medium. This means that contrary to other common fieldbus systems, transmission of the individual characters is not free from gaps (if so, the start bit of the (n+1)th character would follow immediately after the stop bit of the nth character).

The medium attachment unit is equally capable of carrying out the conversion of the octet into UART characters as well as the parity check (and this is the case in many implementations). For didactic reasons, we have decided to view the conversion as a task of the logical unit, which is why it is represented here.

On the receiver side, the first step is to reconstruct the logic signals from the received analog signals. The remote LU layer retrieves the user data from the UART characters which it then passes on to its data link layer.

Up till now we have made no mention of access control. It is now time to examine this aspect (Figure 3.8).

As mentioned earlier, this asynchronous accessing method means that several end devices may want to access the bus at the same time. This leads to access conflicts. To detect such conflicts, every end device during transmission rereads the signal that has been applied to the bus. As long as this signal corresponds to the one it released for transmission, it can continue to use the bus. If, however, an end device detects that the signal on the bus is no longer the one it sent (i.e. a collision has occurred), it must end transmission immediately (in other

3.2 Physical Layer

words withdraw from the bus). This end device can try again as soon as its "opponent" has completed its transmission and the bus has been idle for a specific period of time.

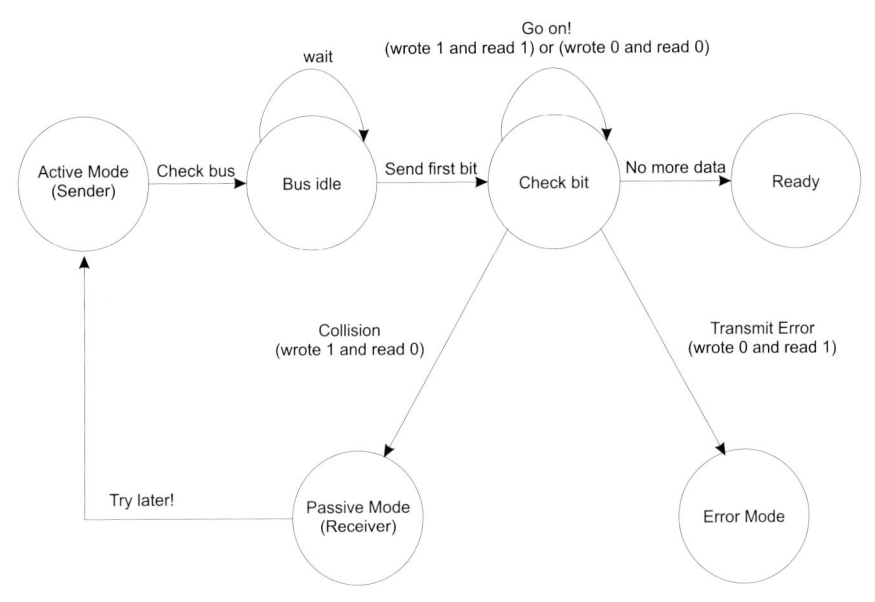

Figure 3.8: CSMA/CA Protocol

The basis of this access method, called bit-wise arbitration, is the differentiation between two physical bus levels, i.e. a dominant (logic 0) level and a recessive level (logic 1). As long as the bus is not occupied by an end device it is on the recessive level. Every end device that wants to transmit begins to apply the first bit after the bus has been on the recessive level for at least a period equal to 50 bit times. Please remember that the first bit is the start bit of the first character! This start character is always logic 0 and with that dominant. This causes the bus level to switch from the original recessive status into the dominant status. All other partners, with nothing to transmit, know that the bus is busy.

Transmission of the start bit opens the so-called arbitration phase. During this phase each transmitting end device compares the signal it has applied with the actual bus level. Every end device that has sent a recessive bit but detects a dominant bit on the bus immediately cancels the arbitration and becomes the receiver of the character currently being transmitted by another end device.

This type of bit-wise, loss-free arbitration is named accordingly – *Carrier Sense Multiple Access with Collision Avoidance*, CSMA/CA. If we consider that the messages to be transmitted are also assigned priorities which in turn influence the bus arbitration, then the underlying protocol could also be referred to as *Carrier Sense Multiple Access with Collision Detection and Arbitration on Message Priority*. This is discussed in more detail in subsequent chapters.

The CSMA/CA protocol can be implemented in the data link layer, the physical layer or in both. In our case the LU part layer is primarily responsible for putting this method into practice. It provides the data link layer with a single service, called P_Data. This service builds on the services of the underlying medium attachment unit.

The P_Data service comprises the service primitives of P_Data.req, (request), P_Data.ind (indication) and P_Data.con (local confirmation). The interaction of these services is as follows:

If the data link layer wants to transmit data, it activates the physical layer with the P_Data.req operation. With this service primitive, layer 2 hands over the data to be transmitted piece by piece, byte-wise and specified with the `class` parameter regardless of whether it is the first character of a (new) message, a middle part or a special message that we have named confirmation. At the beginning of "proper" messages, the system decides whether it involves high priority information or data that needs to be transmitted again due to transmission errors (access class 1) or whether it is the first character of a "regular" message. Characters of access class 1 can be sent after a bus idle period of 50 bit times, characters of access class 2 must wait a period of 53 bit times. This means that end devices with important or "old" data are given priority.

As the transmission of the characters is not gap-free, a wait of 2 further bit times is imposed for every other character of a message for which the bus must remain recessive. The special message type of "confirmation" can only be sent after a wait of 15 bit times. The exact connections of the `class` parameter, which is brought into play for the purpose of temporal access control, can only be properly understood after studying the data link layer. For now though, we must make do with this brief overview.

```
typedef enum {class1, class2,           // access classes and
              innerframe, ack}          // other character types
             class_type;
```

In order to comprehend the following program fragment, we must define another type for the return value of the P_Data.con service primitive.

```
typedef enum {ok, medium_busy, collision}
             status_type_layer1;        // status information layer 1
```

The interface to the MAU part layer has intentionally been kept simple. After calling up the `MAU_check_bus()` operation, the sender waits for the specified idle time and is then informed whether the medium is free. The individual bits are passed on to the MAU part layer via the `MAU_Trans()` routine, which in turn uses the `MAU_Rec()` routine (details later) to inform the LU part layer of the arrival of a new bit. The sender is also able to read the current bus level using the `MAU_Read()` operation.

To start with, the octet coming from the data link layer is converted into a valid UART character. After waiting for the period of the bus idle time and having established that the bus is free, the bits are passed on to the medium attachment unit and at the same time the bus status is checked. Then the occurring "recessive" bits must be handed over for transmission and must not be overwritten by dominant bits, otherwise the end device has lost and with the P_Data.con service primitive returns the `collision` value to layer 2. The case in which a dominant bit has just been sent but the bus level remains recessive is interesting. This could be due to an error in the medium attachment unit. Just to be on the safe side, the end device is withdrawn from the bus. If the transmission was successful, the value `ok` is returned to the data link layer (via the P_Data.con local confirmation).

3.2 Physical Layer

```
P_Data.req_handler(class, data)
  class_type class;                          // priority of the character
  char data;                                 // character to be transmitted
{
  char i;                                    // auxiliary variable
  status_type_layer1 cont;                   // status information MAU
  uart_type uart;                            // UART character

  uart=char2uart(data);                      // conversion octet into byte
  switch (class) {
  case class1:                               // access class 1
      cont=MAU_check_bus(50);                // 50 bit times bus idle status
      break;
  case class2:                               // access class 2
      cont=MAU_check_bus(53);                // 53 bit times bus idle status
      break;
  case innerframe2:                          // access class 2
      cont=MAU_check_bus(2);                 // 2 bit times bus idle status
      break;
  case ack:
      cont=MAU_check_bus(15);                // 15 bit times bus idle status
      break;
  }
  if (cont == ok) {                          // if the medium was free
      for (i=0; i < 11; i++) {       // hand over UART bits one by one
          if ((uart >> i) & 0x01) {          // find bit to be sent
              MAU_Trans(1);                  // transmit logic 1
              if (MAU_Read() != 1) {
                  cont=collision;            // collision
                  break;                     // eliminated!
              }
          }
          else {
              MAU_Trans(0);                  // transmit logic 0
              if (MAU_Read() != 0)
                  STOP;                      // MAU error!
          }
      }
      if (cont == ok)
          P_Data.con(ok);                    // transmission successful
      else
          P_Data.con(collision);             // collision occurred
  }
  else
      P_Data.con(medium_busy);               // medium was busy
}
```

The aim of the remote LU part layer is to create a complete UART character by assembling the individual bits transmitted by the MAU level. The incoming bits can be the start of a new message or parts thereof or even confirmations. The MAU can tell which it is by the idle time of the medium. In order to further simplify the interface, we assume that the medium attachment unit has a separate routine for each of the possibilities described above that must be started by a handler in the LU layer (MAU_Rec_class1(), MAU_Rec_class2(), etc.)

3 EIB Protocol

The operations that need to be executed in each of these routines are almost the same. Firstly, the bit stream must be combined into a UART character. Then using the parity bit it is possible to check whether the data has been transmitted without error. If so, the restored octet is handed over to the data link layer. Depending on the previously activated handler, the `class` parameter of the P_Data.ind service primitive must be set (possible values are `class1`, `class2`, `ack` and `inner_frame`).

For the situation that involves the start of a message, the operations that need to be carried out in the handler could look as follows (the other handlers behave in a similar way and have been omitted to save space):

```
MAU_Rec_class1()
{
    uart_type uart;                            // UART character
    char data;                                 //octet

    init_uart(uart);                           // initialize UART character
    for (i = 0; i < 11; i++)                   // ... and restore it
        if (MAU_Rec() == 1)                    // assemble UART character
            uart= uart | (0x01 << i);
    if (check_uart(uart) == OK) {              // check parity
        data=uart2data(uart);                  // restore octet
        P_Data.ind(class1, data);  // generate indication for layer 2
    }
}
```

3.2.3 Medium Attachment Unit

The signal coding represents the basis for transmitting signals on the EIB. The medium attachment unit (MAU) forms the interface between the digital world of the protocols and the transmission medium. As the sender, its task is to convert the bits into suitable impulses that can be transmitted via the network. As the receiver, this part layer must restore the digital model from the analog signal in order to make it available to the above lying protocol layers.

As the MAU acts as the physical interface to the transmission medium, it has a number of other important tasks in EIB. It must make the necessary power supply available to the remotely supplied bus nodes. These tasks can also include the monitoring of the power supply and the generation of a corresponding warning signal (for saving sensitive data). And finally, it must be tolerant of any faulty device connections to the bus.

Obviously, the MAU needs to be matched to the transmission medium in use. We will therefore restrict our discussions to the most common transmission medium of the twisted pair. The translation of this theory into practice is discussed in more detail in chapter 5 along with the functionality of the medium attachment unit for other media.

To use the CSMA/CA method on the twisted pair we need two different signal forms to represent a zero bit and a one bit. With EIB the zero bit is dominant, i.e. it can overwrite the recessive one bit. The previously described collision resolution is based on this property.

A logic zero is denoted by a short negative impulse that is applied to the power supply. Technically, it is implemented in the MAU by the temporary generation of a current pulse, which in connection with the input impedance of the device corresponds to the required voltage pulse. An equalizer pulse connects to this active phase, which is defined by the

3.2 Physical Layer

impedance of the bus line and primarily by the obligatory choke (see chapter 5.5). Overall, this yields a DC-free pulse (Figure 3.9).

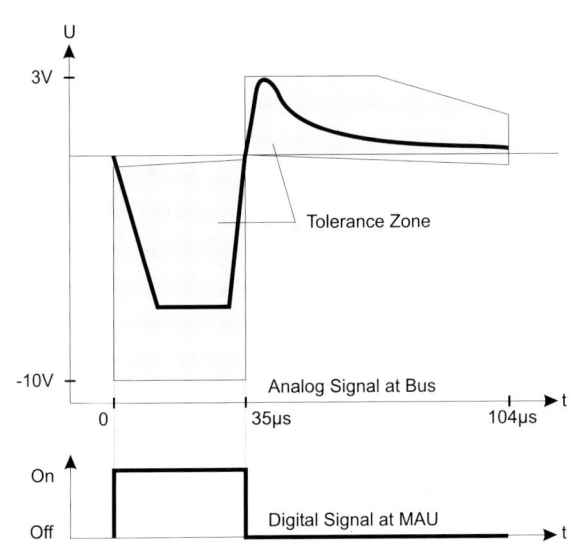

Figure 3.9: Zero Coding

The important signal parameters are given in Table 3.2. The voltages refer to the start of the negative voltage pulse.

Table 3.2: Signal Parameters for the Zero Bit

Parameter	Min	Max
Bit time	104 µs (typically)	
Pulse duration (sender)	34 µs	36 µs
Pulse duration (receiver)	25 µs	70 µs
DC part	21 V	32 V
Pulse height (start)	-0.6 V	-10 V
Pulse height (end)	-0.3 V	-10 V
Height of the equalizer pulse (start)	0 V	3 V
Height of the equalizer pulse (end)	-0.5 V	1.8 V

The long permitted pulse duration at the receiver requires further explanation. Zero bits can overlap. This occurs during the arbitration phase for example or when sending common confirmations. If the transmitting bus devices are far apart, there can be significant delays as a result of the signal propagation times. The receiver therefore sees an overlapping of two pulses (Figure 3.10). That is why the digitized pulse created by the MAU becomes longer. Depending on the size of the delay there could also be a short spike. Such effects must be detected and eliminated in the MAU by corresponding signal processing.

3 EIB Protocol

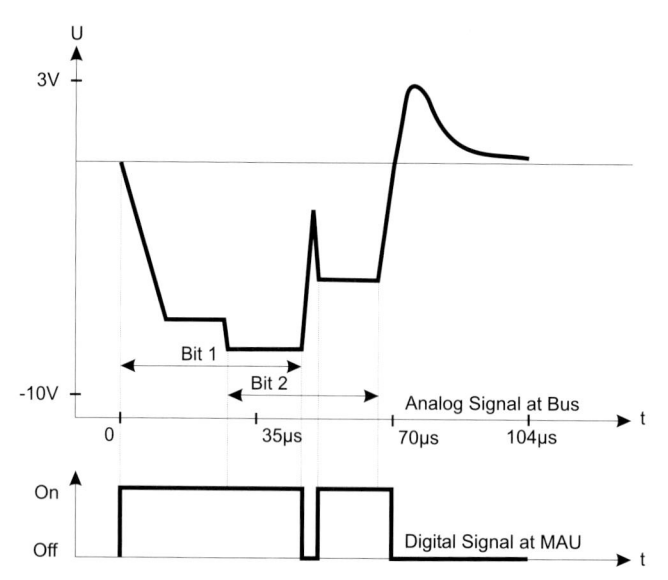

Figure 3.10: Zero Coding with Signal Overlapping

The one bit is distinguished by the lack of activity on the bus. In principle therefore, it is impossible to detect whether a one bit is currently being transmitted or whether the bus is currently idle. Figure 3.11 illustrates the definitions of the tolerance limits with regard to the start of the bit. The signal level can drop during the period of the bit time if immediately prior to this a zero bit has been sent and the equalizer pulse is not complete.

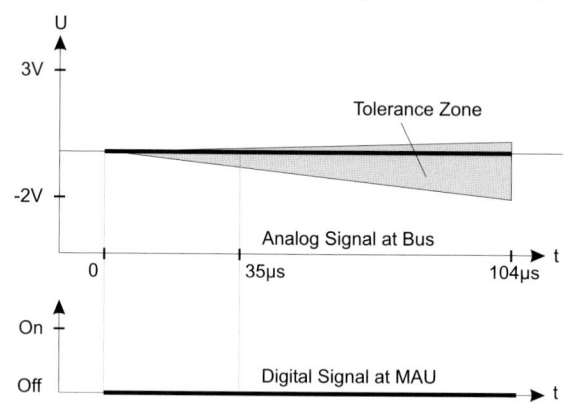

Figure 3.11: One Coding

The MAU of the receiver now digitizes the analog signal on the bus line and generates impulses from the zero bits. The bit detection unit of the MAU decodes these impulses into a serial data stream, as shown in Figure 3.12.

As we already know, the data is transmitted on the bus as UART characters with start, parity and stop bits. In order to be accepted by the receiver as valid, the edges of the individual bit pulses must lie within a defined time window with respect to the start bit. The pulse duration is also monitored. The characters thus decoded are then made available to the logical unit.

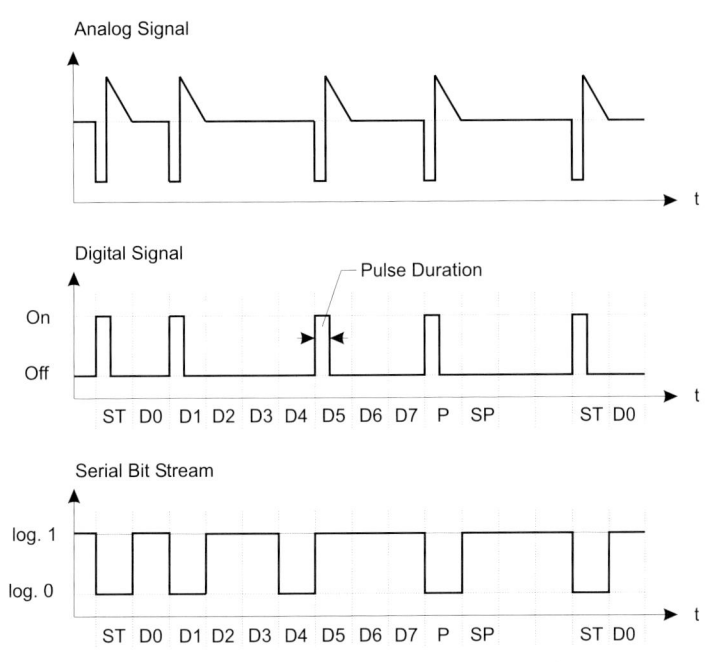

Figure 3.12: Signal Pattern on the Bus and Restored Logic Representation

This concludes our discussion of the physical layer. We would like to point out that your understanding of this layer will improve once you have read the next chapter. Unfortunately, it is not always possible to view layers 1 and 2 separately from one another!

3.3 Data Link Layer

The main task of the *data link layer* is to transfer user data, which has originated from the *network layer*, from one sender station to one or more receiver stations totally free from error. The solution to this may seem trivial, but in reality the data link layer has many problems to solve. Some of these problems can only be resolved in close cooperation with the underlying layer (and the transmission medium). Other problems on the other hand can be managed totally separately. For this reason, the data link layer of the EIB consists of two part layers, which we will take a quick look at before going into the details (Figure 3.13).

The upper half of the data link layer is called the *logical link control layer* (LLC). This part layer contains the services that are offered to the network layer.

On the transmission side, the input data is compiled into a *data frame*. The second half of the data link layer passes this data frame on to the *physical layer* character by character or bit by bit, by calling up the corresponding services.

At the target station, the individual characters are then recompiled into a complete frame and passed on to the local LLC layer, which returns an *acknowledgement frame* to the sender. Of course, the transmission provided by the medium is not one hundred percent reliable. Parts of the data frames can be distorted or destroyed. In the worst case, entire frames can be lost! The protocol used in the LLC layer must be able to detect a damaged data frame and return a negative acknowledgement, which initiates retransmission of the user data. Such measures come under the heading of *error control*.

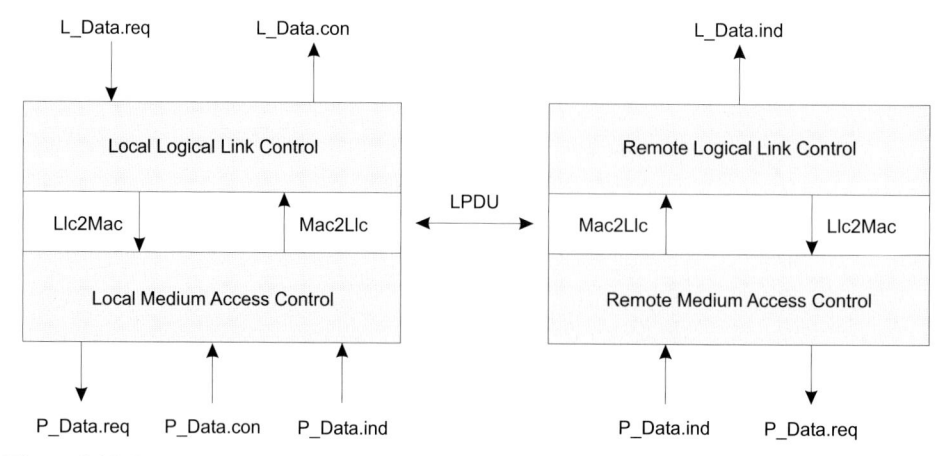

Figure 3.13: Data Link Layer

Another important aspect of the LLC layer is the provision of mechanisms for so-called *flow control*. Without these mechanisms, a faster sender would, for example, be able to "swamp" a slower (possibly otherwise occupied) receiver with data frames. There is a whole range of commonly used mechanisms that manage flow control (and at the same time error control). The LLC layer in the EIB is only ready to transmit more user data when a positive acknowledgement frame has been received for the previously transmitted data frame (*stop-and-wait*). More details on this are given in the following chapter.

The lower half of the data link layer is called the *medium access control layer* (MAC). Its services are used by the above lying LLC layer. On the sender side it splits up the data frames received from the LLC layer and passes them on piece by piece to the physical layer. On the receiver side it compiles the individual characters to form a complete data unit.

In EIB, access to the transmission medium is basically asynchronous. This means that the end devices can send their data whenever they need to and are not restricted to specific times for using the transmission medium. If two or more end devices want to send data at the same time, the access method implemented in layer 1, i.e. CSMA/CA decides which end device is actually authorized to send its data.

This access method is primarily based on the priority of the data frame. In order to further limit collisions, high priority data frames are sent immediately and lower priority data frames are put last. The data frames are acknowledged by the target stations but not immediately, only after the transmission medium has remained unused for a short, specified period of time. The control of this temporal sequence is a further task of the medium access control layer. More details on this are given in chapter 3.3.2.

3.3.1 Logical Link Control

The LLC layer provides a reliable datagram service to the network layer. A *datagram* is a data unit that can be transmitted from a sender to one or more receivers without requiring the

3.3 Data Link Layer

prior setup of a connection. To send this data unit it must first be converted into a *data frame*.

As there is no connection setup, every data frame must include the full address of the receiver. It would therefore be sensible to outline the various types of addressing used in EIB before we examine the individual service primitives of the LLC layer in more detail.

3.3.1.1 Types of Addressing

There are basically two modes of addressing in EIB:

1. Device addressing (physical addressing). Every EIB device is provided with a unique address during the project-engineering phase. Unique means that this address must only occur once within the entire EIB network. A data frame, which has been physically addressed, is only acknowledged by a single LLC layer (assuming it was addressed properly) and passed on to the local network layer. In the "classic" EIB the physical address comprises

   a) Device number (0-255), which denotes the device within its line,

   b) Line number (0-15), which specifies the line in which the device is located,

   c) Zone number (0-15), which defines the zone in which the device has been installed.

2. Group addressing (logical addressing). Several devices can be combined into a group. Note: This type of addressing is independent of the actual locations of the individual devices. This means that the members of a group can be located in different areas of the EIB network, but functionally they depend on one another. Data frames that have been logically addressed are acknowledged by the LLC layers of all devices of the group and passed on to the respective network layers. The user is free to define the group addresses. In order however to maintain a certain consistency within the EIB system, a logical address consists of:

   a) Sub-group number (0-2047), which denotes the function of the group,

   b) Main group number (1-14), which specifies the task area of the group.

In EIB, the same format has been selected for both modes of addressing. The address type therefore has the following layout:

```
typedef union {                          // address in format:
        struct {
                char addr;               // device number (0-255)
                char line :4;                      // line (0-15)
                char zone: 4;                      // zone (0-15)
                } physical;              // physical addressing
        struct {
                char subnumber_low;               // sub-group
                char subnumber_high :3;           // sub-group
                char mainnumber: 4;               // main group
                char reserved :1;                  // reserved
                } logical;               // logical addressing
        } address;
```

As the address formats are the same, a further parameter must be incorporated to determine which mode is actually being used. We will refer to this parameter from now on as the destination address flag (DAF), whose function is as follows:

- With physical addressing the destination address flag is set to zero. The LLC layer must then compare the address of the data frame with the actual device address.
- With logical addressing the destination address flag is set to one. The LLC layer must then establish whether the end device is a member of the addressed group.

At this point we should briefly mention a special addressing form of layer 2, the *polling group*. In this special case, one end device (the so-called *master*) is able to request data from a series of up to 15 devices (*slaves*). The data that is returned by the slaves can only consist of a single octet and must be returned within a specific time slot shortly after reception of the initial prompt. This therefore is a type of group addressing. Contrary to the logical addressing however (which uses destination address flags), this is based on special address tables (*polling group tables*) of layer 2, requires special services (on the master side the L_Poll_Data service and on the slave side the L_Poll_Data_Update service) and with that uses separate protocol data units.

3.3.1.2 Datagram Service (L_Data)

The service of layer 2 that is responsible for the reliable transmission of data frames is established in the upper half of the data link layer and is called L_Data. This is an acknowledged service. It consists of the three service primitives L_Data.req, L_Data.ind and L_Data.con.

With the L_Data.req service primitive the network layer requests the transmission of a packet. The LLC layer forms the data frame from the user data (*link layer service data unit, LSDU*) and the given parameters. This data frame (extended with a check character) yields a protocol data unit of layer 2 (*link layer protocol data unit*, LPDU).

The exact layout of the data frame is described in chapter 3.3.3.1. At this point it is enough to know that a data frame consists of the following fields:

- *Source address*: The source address is the physical address of the transmitting end device, which allows easy identification of the data origin (particularly useful during servicing).
- *Destination address*: As we already know from the previous chapter, the receiver can be addressed physically or logically. Differentiation is achieved via the destination address flag. Every LLC layer must recognize its own physical address and know which groups the device belongs to. The group addresses can be saved in tables or lists. These however are implementation details that are of secondary interest.
  ```
  address device;                        // device address
  table group;                           // group addresses
  ```
- *Priority*: Data frames in EIB can be sent with different priorities. Or more accurately speaking, one of the upper layers establishes the priority with which the user data is to be sent. User data that is transmitted with priority 1 is reserved for EIB system functions. The other priorities can be used as required. This includes alarm functions (priority 2) and data frames that can either be sent with preference (priority 3) or normally (priority 4).
  ```
  typedef enum {alarm, system, high, low}
              class_type;                // priority of the data frame
  ```

- *User data*: The user data is composed of data that originates from the application and layer-specific control information. For the LLC layer however it is pure user data that needs to be transmitted. So that the receiver knows how much data to expect, a specification of length precedes the user data area.
- *Check character*: To allow the detection of transmission errors, the parameters and user data are extended with a check character. On the reception side, this allows the detection of potential changes in the data frame.

For the sake of simplicity we will assume that the LLC layer uses the `assemble_frame()` (or `assemble_rep_frame()`) routine to construct the complete data frame, which also simultaneously calculates the necessary check character which is added to the frame.

This data frame is passed on to the MAC layer with the Llc2Mac.sender routine. The MAC layer then passes on the frame to the physical layer piece by piece via the P_Data.req service primitive.

As the LLC layer of the sender cannot know in advance whether the data frame will reach the receiver error-free, measures must be initiated that allow the renewed transmission of the data. This includes, among others, setting a *timer*, initializing two repetition counters and saving the last transmitted data. For this, the LLC layer requires additional auxiliary variables:

```
char counter_nack;                        // auxiliary counter
char counter_busy;                        // auxiliary counter
timer llc;                   // timer for checking acknowledgement
char[MAXLEN] repeated_frame;              // last data frame
```

The concrete implementation of the handler for the L_Data.req service primitive could therefore have the following layout:

```
L_Data.req.sender_handler(source, destination, daf, class, l_sdu)
  address source;                    // physical address of sender
  address destination;               // (group) address of receiver
  char daf;                          // destination address flag
  class_type class;                  // priority of the message
  char[MAXLEN] l_sdu;                // user data layer 2
{
  char[MAXLEN] frame;                           // user data MAC

  frame=assemble_frame(class, source, destination, daf, l_sdu);
  Llc2Mac.sender(frame);      // hand over data frame to MAC layer
  start_timer(llc);                         // set LLC layer timer
  counter_nack=0;             // initialize repetition counter
  counter_busy=0;             // initialize repetition counter
  repeated_frame=assemble_rep_frame(frame);    // save data frame
}
```

In this way, the individual parts of a data frame reach the target station in succession. There they are passed on to the MAC layer via the P_Data.ind service primitive. It is the task of the remote MAC layer to recombine the received characters to form a complete data frame. Once this is done, the upper half of the data link layer is activated.

The remote LLC layer decides whether the data frame is intended for the device. If the data frame was not addressed to that particular end device it is rejected immediately. Otherwise, the LLC layer must now determine whether the data has arrived without error. The procedure is as follows:

1. A faulty data frame is acknowledged with a *negative acknowledgement frame (nack)*.
2. If the data frame is free from error, the flow control mechanisms are brought into play.
    a) If the receiver is currently busy, a *busy acknowledgement frame (busy)* is returned,
    b) Otherwise the LLC layer generates a *positive acknowledgement frame (ack)* and passes on the user data to the network layer with the L_Data.ind service primitive.

In the subsequent program fragment, the check to determine whether the data frame has been correctly transmitted is carried out by the `parity()` routine. It is obvious from the name that this involves a parity check on the basis of the transmitted check character. We will only go into the details of the calculation of this check character once we know the concrete frame format – just a little more patience!

To transmit the individual acknowledgement frames (in principle this involves layer 2 PDUs) the LLC layer accesses the MAC layer, which then uses the services of the physical layer.

```
typedef enum {ack, nack, busy, collision, medium_busy}
              status_type_llc;        // status information LLC layer

Mac2Llc.receiver_handler(frame)
  char[MAXLEN] frame;
{
  class_type class;                           // message priority
  address source;                             // physical address of sender
  address destination;                        // (group) address of receiver
  char daf;                                   // destination address flag
  char[MAXLEN] l_sdu;                         // layer 2 user data
  char check;                                 // check character

  class=decode_class(frame);                  // extracted message priority
  source=decode_address(frame);               // extracted sender
  destination=decode_adress(frame);           // extracted receiver
  daf=decode_daf(frame);       // extracted destination address flag
  l_sdu=decode_sdu(frame);                    // extracted user data
  check=decode_parity(frame);                 // extracted check character
  if (((daf == 0) &&                          // physical address
      (destination.physical.zone == device.physical.zone) &&
      (destination.physical.line == device.physical.line) &&
      (destination.physical.addr == device.physical.addr)) ||
     ((daf == 1) && (in(destination, group)))) {  // logical address
    if (check != parity(class,source,destination,daf,l_sdu))
      frame=assemble_frame(nack);                       // case 1
    else
      if (receiver_status == busy)
        frame=assemble_frame(busy)                      // case 2a
      else if (receiver_status == ok) {                 // case 2b
        L_Data.ind(source, destination, class, daf, l_sdu);
        frame=assemble_frame(ack);
    }
```

3.3 Data Link Layer

```
        Llc2Mac.receiver(frame);
    }
}
```

On the sender side, the MAC layer passes the received acknowledgement frame on to the LLC layer. A positive acknowledgement is immediately indicated to the network layer via the L_Data.con service primitive. Upon reception of a negative acknowledgement or a busy frame, the LLC layer tries to re-transmit and passes the frame with its original contents to the MAC layer. In the case of faulty frame transport or if the target station was busy, this procedure is repeated up to max_nack or max_busy times. If despite these repetitions the sender station still fails to receive a positive acknowledgement, the network layer must be informed via the L_Data.con service primitive, this time however with a negative acknowledgement (not_ok). This stops any further transmission attempts.

There is one last situation to be discussed, namely when transmission is not possible due to another end device wishing to transmit at the same time and whose message has a higher priority. The network layer is again informed via the L_Data.con service primitive (with negative acknowledgement).

```
Mac2Llc.sender_handler(status)
    status_type_llc status;            // status information LLC layer
{
    switch (status) {
    case ack:                          // character transmission successful
        L_Data.con(ok);
        stop_timer(llc);                                       // stop timer
        break;
    case busy:
        counter_busy++;                // increment repetition counter
        if (counter_busy < max_busy) {
            Llc2mac.sender(repeated_frame);   // new transmit attempt
            start_timer(llc);                             // reset timer
        }
        else             // number of permitted repetitions exceeded
            L_Data.con(not_ok);
        break;
    case nack:
        counter_nack++;                // increment repetition counter
        if (counter_nack < max_nack) {
            Llc2mac.sender(repeated_frame);   // new transmit attempt
            start_timer(llc);                             // reset timer
        }
        else             // number of permitted repetitions exceeded
            L_Data.con(not_ok);
        break;
    case collision:                                 // collision or ...
    case medium_busy:                               // medium busy
            L_Data.con(not_ok);
    }
}
```

It will not have escaped the observant reader that data and acknowledgement frames can also be lost. To ensure the LLC layer does not wait endlessly for the arrival of an acknowledgement, it uses the timer mentioned earlier.

3 EIB Protocol

The timer is started as soon as the sender transmits a data frame. Under normal conditions, an acknowledgement or busy frame is received before the timer expires (see the Mac2Llc.sender_handler routine). If the timer does expire however, this is (in the truest sense) an alarm signal for the sender. It then sends the data frame again or if this has already been tried several times without success, returns a negative acknowledgement to the network layer.

If a data frame has been sent more than once, there is a danger that the receiver also accepts the data more than once and in so doing repeatedly makes the user data available to the network layer. To avoid this situation, "repeated" data frames are specially marked in the EIB (more details on this in chapter 3.3.3.2). We have taken this situation into account by saving the data frame with the `assemble_rep_frame()` auxiliary routine when implementing the handler of the L_Data.req service primitive.

The subsequent routine of the LLC layer, which is always executed when the timer expires, could take over the management of these cases.

```
Llc_timeout()
{
   counter_nack++;                      // increase repetition counter
   if (counter_nack < max_nack) {
      Llc2mac.sender(repeated_frame);   // further transmit attempts
      start_timer(llc);                 // reset timer
   }
   else                                 // number of permitted repetitions exceeded
      L_Data.con(not_ok);
}
```

We will now summarize the processes and with that the protocol of the LLC layer (Figure 3.14).

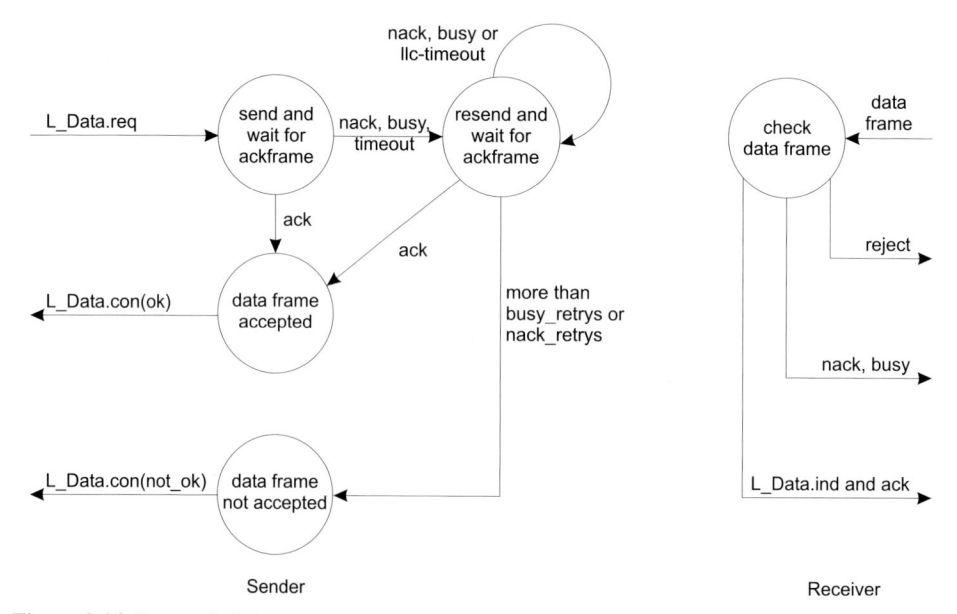

Figure 3.14: Protocol of the LLC Layer

86

On the sender side, the user data and parameters are combined into a data frame along with a check character. The LLC part layer then waits for an acknowledgement frame. If this does not arrive, the data frame is retransmitted after the timer has expired. The sender reacts in a similar way when a negative acknowledgement is received or a busy frame. If the number of permitted repetitions has been exceeded the transmission is cancelled and the network layer informed accordingly.

Upon reception of a positive acknowledgement, it is possible to assume that the user data has been successfully transmitted. This situation is also made known to the network layer by means of the corresponding service primitive.

On the reception side, the LLC layer waits for the reception of a data frame. It then determines whether or not this can be passed on, i.e. whether it has reached its destination without error. Depending on the result of this check, there is either no generation of acknowledgement, or a positive or negative acknowledgement respectively is returned to the original sender (or in some cases even a busy frame).

3.3.2 Medium Access Control

The MAC layer is firstly responsible for the breaking up and putting together of the data and acknowledgement frames, and on the other hand it is also responsible for time-controlled access to the transmission medium. Therefore, its implementation must be tailored to the respective layer 1. Due to this close cooperation, it is often impossible to define a strict modularization of the MAC layer and physical layer. The coding of the individual characters and the (time) controlled access to the medium in particular could just as easily be implemented in either layer. In this chapter, we will try to concentrate on the important aspects of the medium access control layer.

The "classic" EIB uses the *twisted pair*, TP, medium. The bus access method for this medium is loss-free, bit-wise arbitration, better known as *Carrier Sense Multiple Access with Collision Avoidance* (CSMA/CA) (see chapter 3.2.2). Although CSMA/CA is designed for the asynchronous transmission of data, with EIB there are additional temporal procedures to be taken into account. The control of these procedures is one of the tasks of the MAC layer.

At the beginning on the sender side, the LPDU is received by the MAC layer and transferred byte wise (i.e. as octets) to the physical layer via the P_Data.req service primitive. Layer 1 then codes this octet into a UART character and converts it into electrical signals.

Before the first character of a data frame is actually sent however, the bus must have been free for at least 50 bit times. The remaining waiting time is firstly dependent on the priority of the data frame (obvious from the first field of the LPDU), and secondly on whether the data frame is undergoing repeated transmission (this is the case when there has been no acknowledgement for a sent data frame):

1. "Repeated" transmissions and data frames with alarm or system priority can be sent immediately after the previously mentioned 50 bit times. These data frames are datagrams of access class 1.

2. Data frames of priority 3 or 4 are subject to an additional wait of 3 bit times before being sent. These data frames are datagrams of access class 2.

3 EIB Protocol

On the sender side, a data frame is transferred from the LLC layer with the Llc2Mac.sender routine. The MAC layer establishes the access class and passes on the first character to the physical layer.

```
typedef enum {class1, class2,                // access classes and
              innerframe, ack}               // other character types
              class_type_layer1;
char s_next;                 // index to next character of the data frame
char[MAXLEN] actual_frame;                   // current data frame

Llc2Mac.sender_handler(frame)
  char[MAXLEN] frame;
{
  switch (class(frame)) {
  case class1:
      P_Data.req(class1, frame[0]);
      break;
  case class2:
      P_Data.req(class2, frame[0]);
      break;
  }
  s_next=1;                                  // initialize index
  actual_frame=copy(frame);                  // save data frame
}
```

If it was possible to apply the individual bits of the frame to the underlying medium, the local layer 1 returns a positive acknowledgement to the MAC layer via the P_Data.con service primitive. The next octet can then be assigned to the physical layer. If transmission was not possible, because the medium was busy or a collision occurred (and the sender lost bus access), the MAC layer receives a negative acknowledgement. This stops the current transmission of the data frame and a corresponding acknowledgement is passed on to the LLC layer.

```
typedef enum {ok, medium_busy, collision}
              status_type_layer1;        // status information layer 1

P_Data.con.sender_handler(status)
  status_type_layer1 status;            // status information from layer 1
{
  switch (status) {
  case ok:                                      // everything OK
      while (s_next < sizeof(actual_frame))     // then next byte
          P_Data.req(innerframe, actual_frame[s_next++]);
      break;
  case medium_busy:                             // medium busy
      Mac2Llc.sender(medium_busy);
      break;
  case collision:                               // collision occurred
      Mac2Llc.sender(collision);
  }
}
```

3.3 Data Link Layer

On the receiver side, the octets arrive via the P_Data.ind service primitive. A remote MAC layer recombines the individual bytes to form a complete frame. Once all characters have been received, the complete data frame is transferred to the LLC layer for further processing.

We must now risk a jump forward, as otherwise it would be difficult to understand the following program fragment. The sixth character of a data frame contains, among other things, information about the quantity of expected user data. This information is the only way a receiver can determine whether the data frame is finished. But note – this length specification does not cover the check character that completes the data frame (this check character does not represent user data of layer 2 but control information for the LLC layer!).

```
char r_next;                                // index to set up a data frame
char[MAXLEN] received_frame;                // received data frame
char len;                                   // quantity of expected user data
P_Data.ind.receiver_handler(class, data)
  class_type_layer1 class;                  // information on character type
  char data;                                // user data
{
  switch (class) {
  case class1:                              // first character of access class 1
  case class2:                              // ... or 2
      received_frame[0]=data;               // start frame set up
      r_next=0;                             // initialize index
      len=5;                                // read to at least the 6^th character
      break;
  case innerframe:                          // other data frame characters
      if (r_next == len) {                  // information on amount of user data
         len=((data & 0x0F) + 6);
      else if (r_next <= len)               // continue frame set up
         received_frame[r_next++]=data;
      if (r_next > len)                     // transfer data frame to LLC layer
         Mac2Llc.receiver(received_frame);
      break;
  }
}
```

If necessary (i.e. if the data frame was directed at the station), an acknowledgement is generated by the LLC layer and returned to the MAC layer. This acknowledgement is transferred to the physical layer via the P_Data.req service primitive, which then permits transmission after a further 15 bit times.

```
Llc2Mac.receiver_handler(ackframe)
  char ackframe;                            // acknowledgement frame
{
  P_Data.req(ack, ackframe);                // return to MAC layer of the sender
}
```

On the sender side, the acknowledgement is indicated to the MAC layer by the P_Data.ind service primitive. The MAC layer then transfers this unchanged to the LLC layer. All subsequent actions are triggered by the LLC layer, as described in the previous chapter.

```
P_Data.ind.sender_handler(class, data)
  class_type_layer1 class;                  // information on character type
  char data;                                // user data (acknowledgement frame)
{
```

3 EIB Protocol

```
    Mac2Llc.sender(data);           // acknowledgement to LLC layer
}
```

Figure 3.15 is a greatly simplified representation of the temporal behavior.

	data frame		ack	
class waiting period		line idle	line idle
(0-3 bit times)		(15 bit times)	(50 bit times)

Figure 3.15: Protocol of the MAC Layer

3.3.3 Frame Formats (LPDU)

Having now gained an understanding of the functionality and individual services of the physical layer, we can now look at the corresponding frame formats. It should again be pointed out that data and acknowledgement frames represent protocol data units of the data link layer (*link layer protocol data unit*, LPDU).

3.3.3.1 Data Frame

A data frame consists of 6 fields of different lengths (Figure 3.16).

C-Field	Source-Field	Destination-Field	RL-Field	Data-Field	P-Field
(8 Bit)	(16 Bit)	(17 Bit)	(7 Bit)	(1-16 Byte)	(8 Bit)

Figure 3.16: Data Frame

The first position of the data frame is occupied by a 1-byte control field, which is used to establish the priority of the frame and its type (new or – due to a negative acknowledgement – repeated frame).

The next 2 bytes specify the address of the sender station, i.e. the station that wishes to send the frame. After this, there are two more bytes that provide more details on the receiver(s). If the data frame is directed at a single receiver, the contents of the field are interpreted as the physical address (or device number). If the data frame is addressed to a group of receivers, the content of the two bytes is interpreted as the logical address. This differentiation is made on the basis of the destination address flag, which is accommodated in the subsequent octet.

In addition to the destination address flag, the sixth octet of the data frame contains two counters. One counter is used by the network layer to establish whether a data frame can be passed on to another network section. The other specifies the number of transported user data bytes. Up to 16 bytes can be transmitted in one data frame. After the user data, a 1-byte check character completes the data frame.

We would now like to take a closer look at the structure of the individual fields, in so far as they are relevant to layer 2.

3.3 Data Link Layer

Control field

In the control field, bits P0 and P1 decide the priority of the data frame. Data frames with priority 1 are reserved for EIB system functions (and with that the transport layer). The user is free to use the remaining priorities as desired. This includes alarm functions (priority 2) and data frames, which are either sent with preference (priority 3) or as normal (priority 4).

In order to understand the 5th bit of the control field, the so-called repeat flag and its effect, we will imagine the following scenario: A data frame was addressed to a group of receivers and has been negatively acknowledged by at least one of these receivers. The sender must now repeat this data frame. With the renewed transmission it is necessary to ensure that target stations having correctly received the original data frame do not pass on the user data (e.g. an adjustment command for an actuator) to the upper layer again (and with that unintentionally reactivate the corresponding actuators). It is therefore the task of the sender station to mark such "repeated" data frames accordingly. On the basis of the repeat flag, the receiver can determine whether the data frame is new or repeated and then decide whether to pass on the user data or reject the frame.

C-Field (8 Bit)	Source-Field (16 Bit)	Destination-Field (17 Bit)	RL-Field (7 Bit)	Data-Field (1-16 Byte)	P-Field (8 Bit)

Control-Field

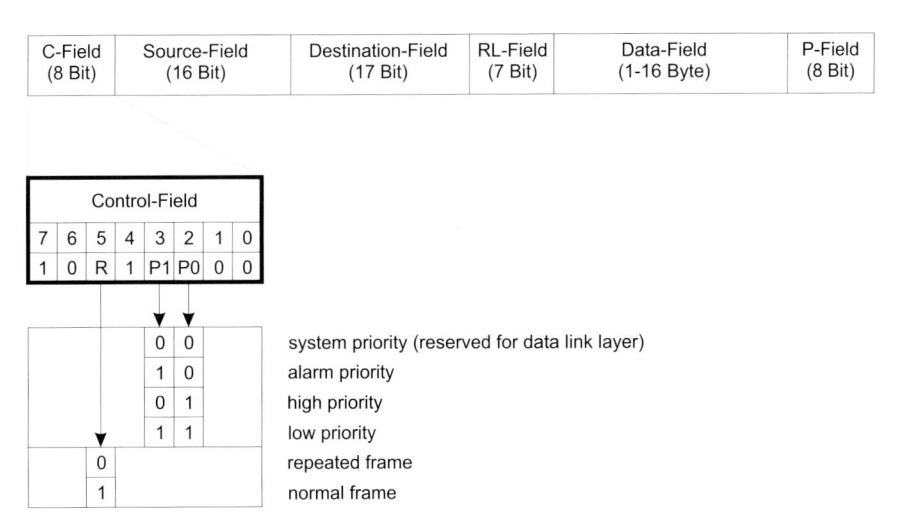

7	6	5	4	3	2	1	0
1	0	R	1	P1	P0	0	0

P1	P0	
0	0	system priority (reserved for data link layer)
1	0	alarm priority
0	1	high priority
1	1	low priority

R	
0	repeated frame
1	normal frame

Figure 3.17: Control Field

For the second last time we will now recall the CSMA/CA bus access method and the time controlled access to the transmission medium. In the bit-wise arbitration phase, an end device, which has just applied a dominant level to the bus, overwrites a recessive level, which has possibly been applied to the bus by another end device. As the application of a data frame begins with bit 0, i.e. the *least significant bit*, LSB, of the control field, the coding of the individual priorities, as shown in Figure 3.17, must have been selected. The end device that wins through is that which is the first to apply a dominant level to the bus – its data frame has the highest priority!

If two or more end devices wish to send data frames with the same priority, the end device that wins through is that with the lowest device address (this is transmitted in the subsequent octet). Because EIB device addresses must be unique, this does not result in any irresolvable access conflicts.

3 EIB Protocol

If we also take into account the time-controlled access to the transmission medium, the ranking for the data frames is as shown in Table 3.3.

Table 3.3: Priority of the Data Frame

Rank	Class	Priority	Repeated data frame	User	Control field (LSB – MSB) 0123 4567
1	1	System	Yes	Layer 4	0000 1001
2	1	System	No	Layer 4	0000 1101
3	1	Alarm	Yes	Application	0001 1001
4	1	Alarm	No	Application	0001 1101
5	1	High	Yes	Application	0010 1001
6	1	Low	Yes	Application	0011 1001
7	2	High	No	Application	0010 1101
8	2	Low	No	Application	0011 1101

Source address

The source address is always the physical address of the end device that wishes to send the data frame. As every EIB device must be assigned a unique address during installation, later identification of the message source is guaranteed.

The physical address also gives the location of the respective end device. The four high bits of the first byte (Z0-Z3) specify the zone and the lower four bits (L0-L3) the line in which the end device is located. For devices that are located in the main zone or in the main line, the corresponding bits must be set to zero. The actual device number (D0-D7) can be freely selected, but as already mentioned, must be unique. In general, backbone couplers and line couplers should be assigned device number zero.

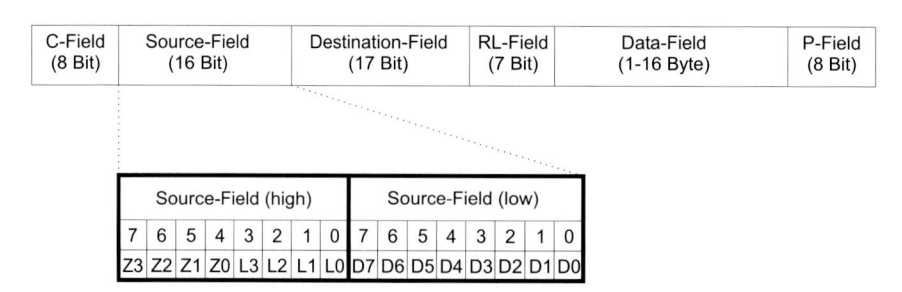

Figure 3.18: Source Address

Destination address

If a data frame is only addressed to one receiver, this field must contain the physical address of the corresponding end device. If the user wishes to send a data frame to several devices, which may be located anywhere within the network but which have been combined into a logical group during the installation phase, this field is interpreted as the logical address. In principle, the user is able to freely assign the group addresses. Division into main group number (M0-M3) and sub-group number (S0-S10) is however advisable. Every data frame that corresponds to the selected logical address is received, evaluated and acknowledged by all end devices of the group.

The destination address flag, which is accommodated in the highest bit of the next field, determines whether the address is physical or logical. The destination address therefore comprises 17 bits.

C-Field (8 Bit)	Source-Field (16 Bit)	Destination-Field (17 Bit)	RL-Field (7 Bit)	Data-Field (1-16 Byte)	P-Field (8 Bit)

Destination-Field (high)	Destination-Field (low)	Destination Address Flag
7 6 5 4 3 2 1 0	7 6 5 4 3 2 1 0	7
Z3 Z2 Z1 Z0 L3 L2 L1 L0	D7 D6 D5 D4 D3 D2 D1 D0	0 physical address
R M3 M2 M1 M0 S10 S9 S8	S7 S6 S5 S4 S3 S2 S1 S0	1 logical address

Figure 3.19: Destination Address

Destination address flag, routing counter and length specification

The next field is also 1 byte large and in bit 7 contains the previously mentioned destination address flag. This is followed by a counter (R0-R2), which is used by the network layer to establish whether the user data transported in the frame can be passed on via couplers. Don't panic! An exact description of the functionality follows in a later chapter. For now, it is enough to know that only data frames with counter values greater than zero are passed on. Bits L0-L3 of this field specify how much user data follows in the data frame. If all bits are set to zero, a single user data octet is transmitted.

C-Field (8 Bit)	Source-Field (16 Bit)	Destination-Field (17 Bit)	RL-Field (7 Bit)	Data-Field (1-16 Byte)	P-Field (8 Bit)

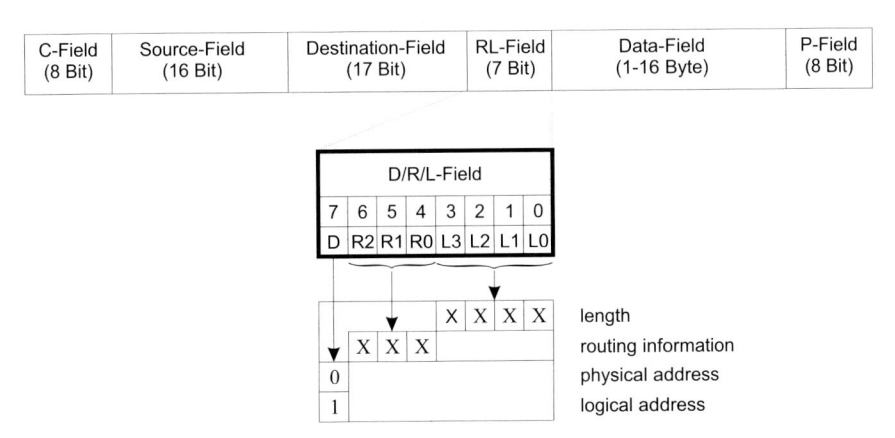

D/R/L-Field

7	6	5	4	3	2	1	0
D	R2	R1	R0	L3	L2	L1	L0

X X X X length
X X X   routing information
0   physical address
1   logical address

Figure 3.20: Destination Address Flag, Routing Counter and Length Specification

Check character

In order to detect data that has been transmitted incorrectly, the data frame is completed with a check character. Using the same process, the receiver generates the check character, compares it with the original and in this way can detect any changes. If a defective data frame is established on the receiver side, it is the task of the LLC layer to automatically repeat the transmission. Calculation of the check character is illustrated in the next chapter.

C-Field (8 Bit)	Source-Field (16 Bit)	Destination-Field (17 Bit)	RL-Field (7 Bit)	Data-Field (1-16 Byte)	P-Field (8 Bit)

Parity-Field

7	6	5	4	3	2	1	0
C7	C6	C5	C4	C3	C2	C1	C0

Figure 3.21: Check Character

3.3.3.2 Acknowledgement Frame

Acknowledgements and busy frames consist of one byte.

Acknowledge-Field

7	6	5	4	3	2	1	0
N1	N0	0	0	B1	B0	0	0

N1 N0	B1 B0	
1 1	1 1	positive acknowledgement frame
1 1	0 0	busy frame
0 0	1 1	negative acknowledgment frame

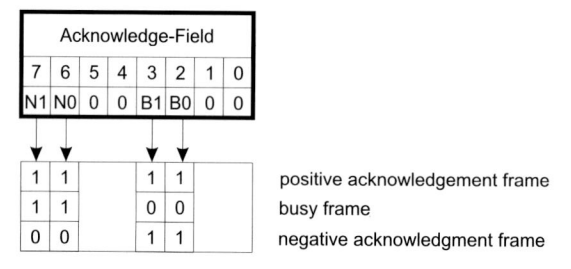

Figure 3.22: Acknowledgement Frame

If we again consider the CSMA/CA bus access method (for the last time!), the selected coding should become immediately obvious:

– If a logically addressed data frame is wrongly received by some end devices, the dominant bits (N0 and N1) of the negative acknowledgement win through against the recessive bits of the positive acknowledgement.

– If a logically addressed data frame cannot be processed by an end device, this device generates a busy frame. On the basis of the dominant bits (B0 and B1) of the busy frame, the sender can tell that at least one addressee was otherwise occupied at the time of reception (bits B0 and B1 of a positive acknowledgement would be logic 1).

– If the data frame was wrongly addressed (or destroyed so that it was unrecognizable) no acknowledgement is returned. The line remains "open", i.e. the acknowledgement "mutates" to seven successive recessive bits.

3.3 Data Link Layer

In all cases, the data frame is automatically transmitted again (caused by the protocol of the LLC layer).

3.3.4 Error Detection

The EIB protocol places great value on the detection of transmission errors. The necessary mechanisms are implemented in layers 1 and 2. Some have already been detailed in previous chapters and others, for didactic reasons, only mentioned in passing. We would now like to summarize all these mechanisms.

- Bit monitoring (layer 1). Every transmitting end device monitors whether the bus level it requires actually exists on the bus. If the two values do not match, there is a so-called bit error. The overwriting of a recessive applied bus level by a dominant level is only tolerated during the arbitration phase and when returning acknowledgement frames. Otherwise, a bit error has occurred. The end device must automatically withdraw from the bus and afterwards can only be reactivated manually.

- Character coding (layer 1). Every octet is converted into a UART character by the physical layer. This character must start with a dominant (start) bit, contain an (even) parity bit and end with three recessive bits (1 stop bit + 2 pause bits).

- Parity check (layers 1 and 2). Actual data protection occurs in layers 1 and 2. The physical layer creates the checksum of the individual bits of a character and establishes whether it is even or odd. With an even checksum the parity bit of the UART character is set to zero, with an odd checksum it is set to one. This strategy only allows errors to be detected; it does not provide a means of rectifying them. Any duplicate errors within a character also remain undetected. For this reason, a vertical parity (this time odd) is formed in addition to the block check, which is transmitted in the concluding check character of a data frame.

  To clarify this situation we will represent a complete datagram in the form of a matrix, in which every line of the matrix corresponds to a byte of the data frame (Table 3.4). The columns of this matrix contain the individual bits (D0-D7) of the fields. From this it is clear to see that the (even) block check is accommodated in the parity bit of each individual character, whilst the (odd) vertical parity is saved in the check character.

  This allows errors within characters to be localized and even removed. However, there is no error removal process as such in EIB. If an error is detected, a negative acknowledgement is sent which causes the data frame to be re-transmitted.

- Time monitoring (layers 1 and 2). A sender can only transmit a data frame after the bus has remained unused for a minimum of 50 bit times. For frames of access class 2 there is a further delay of 3 bit times.

  Target stations must send the acknowledgement frame after 15 bit times (after receiving the last character of the data frame). If the sender fails to receive an acknowledgement (recognizable by the fact that the acknowledgement automatically mutates to seven recessive bits as the line remains "open"), it must be assumed that the data frame was lost or wrongly addressed.

- Frame format (layer 2). EIB datagrams must maintain the previously described frame formats. This means that in data frames or acknowledgement frames, bits in certain positions must be set to zero or one and also that the data frame must be complete. Data

3 EIB Protocol

frames must comprise at least 8 bytes and have a maximum size of 23 bytes. An acknowledgement frame can only comprise 1 byte.

Table 3.4: Parity Check

Field	D0	D1	D2	D3	D4	D5	D6	D7	P
Control field	1	0	1	1	0	0	0	0	1
Source (1st byte)	0	0	0	0	0	0	0	1	1
Source (2nd byte)	0	0	0	0	1	0	0	1	0
Target (1st byte)	0	0	0	0	0	0	0	1	1
Target (2nd byte)	0	0	0	0	0	1	1	1	1
Routing/length	0	0	0	1	0	0	0	1	0
User data	0	0	0	0	0	0	0	0	0
User data	1	0	0	0	0	0	0	1	0
Checksum	1	1	0	1	0	0	0	1	0

3.3.5 Bridges and Routers

To finish with we would like to present the physical layer in devices that are used to link together (different) network parts.

A *bridge* is responsible for the amplification and regeneration of the bus signal. This allows transmission distances to be increased and in older EIB systems greater numbers of end devices to be connected. The bridge does not have its own device address. Its data link layer accepts data or acknowledgement packets from one side of the EIB network and passes them on to the other. Note that although – according to the ISO/OSI definition – this is exactly the function of a bridge, this coupling element is more commonly known as a *repeater* (please refer to section 7.1).

Using *line and backbone couplers* it is possible to create branching and with that EIB-typical topologies. The main task of the coupler (line and backbone couplers are hereafter referred to by the generic term *router*) is to establish whether or not a data frame can be fed into another line or zone. Its data link layer therefore passes on a received data frame to the network layer without processing it further. Here, there is a check to see whether the addressee is stationed on the other side of the network. With physical addressing this check is made using the actual device address of the coupler, with logical addressing the coupler uses a routing table. Additional strategies are used to avoid a wrongly addressed data frame being widely distributed. With this little taster of the following chapter we will end our discussions of EIB layer 2.

3.4 Network Layer

The task of the *network layer* is to set up end-to-end connections between end devices and to transport packets, with either logical or physical addresses, from the source to the target. For this it is necessary to connect network paths together and to channel the data through the

3.4 Network Layer

network along these paths. Devices, which link together parts of the network (including those with different structures) and which have implemented layer 3 of the OSI model, are called routers.

The network layer provides the above lying *transport layer* with a connectionless packet exchange. Depending on the destination address, a packet is either sent to one receiver (*unicast*), to a group of receivers (*multicast*) or to all nodes (*broadcast*) within the EIB network. The services that manage these tasks are N_Data, N_Group and N_Broadcast. In order to carry out the required task, the network layer accesses the (single) L_Data service of the data link layer (Figure 3.23).

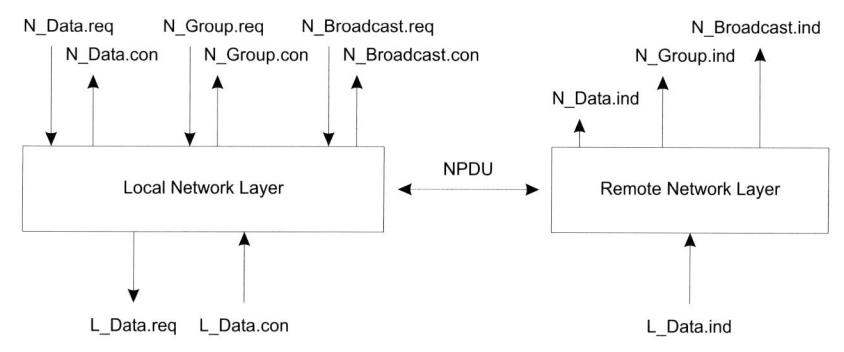

Figure 3.23: Network Layer

Before we take a proper look at the individual services, we will first list the types and variables that have already been established in the data link layer (to reiterate and also to provide a better understanding of the subsequent program fragments).

```
/* Types and variables from layer 2 */
typedef union {                              // address in format:
          struct {
                  char addr;                 // device number (0-255)
                  char line:4;                     // line (0-15)
                  char zone: 4;                    // zone (0-15)
                  } physical;                // physical addressing
          struct {
                  char subnumber_low;                // sub-group
                  char subnumber_high:3;             // sub-group
                  char mainnumber: 4;                // main group
                  char reserved :1;                  // reserved
                  } logical;                 // logical addressing
          } address;
typedef enum {alarm, system, high, low}
          class_type;                        // priority of the message
typedef enum {not_ok, ok}
          status_type_layer2;    // acknowledgement from layer 2
address device;                   // address of an EIB device or router
```

The network layer additionally requires the following types and variables:

```
/* Types and variables of layer 3 */
```

3 EIB Protocol

```
typedef enum {unicast, multicast, broadcast}
            service_type;                   // services of layer 3
service_type last_service;         // last requested layer 3 service
address receiver;                      // note destination address
char counter_routing;              // counter for passing on packets
```

3.4.1 Unicast (N_Data)

If the transport layer wishes to transmit a packet to a single receiver (*unicast*), it uses the local confirmed service, N_Data. This service consists of the N_Data.req request, the N_Data.ind indication and the N_Data.con local confirmation.

On the sender side, the N_Data.req service primitive is used to transmit the priority of the message, the physical address of the target station and the user data (*network service data unit*, NSDU) to the network layer. The network layer adds the routing counter to the user data, which is important for the router for passing on packets, and from this forms a layer 3 *network protocol data unit*, NPDU. Together with the rest of the parameters, the NPDU is passed on to layer 2 by means of the L_Data.req service primitive. One of these parameters is the previously discussed destination address flag, which is used by the data link layer to distinguish between physical and logical addressing.

As the N_Data service is used for packet exchange between a sender and receiver, the destination address flag must be set to zero. To finish with, the type of call and the destination address are saved. These two parameters are required to implement the handler that is activated when the L_Data.con local confirmation is called up.

```
N_Data.req.sender_handler(destination, class, n_sdu)
  address destination;                 // address of the receiver
  class_type class;                    // priority of the message
  char[MAXLEN] n_sdu;                  // layer 3 user data
{
  char[MAXLEN] l_sdu;                  // layer 2 user data
  char daf;                            // destination address flag

  l_sdu=insert_routing(n_sdu);  // add routing counter to user data
  daf=0;                                              // unicast
  L_Data.req(source, destination, daf, class, l_sdu);
  last_service=unicast;                // note last call and
  receiver=destination;                // ... last destination address
}
```

On the receiver side, a received data packet is transferred to the network layer via the L_Data.ind operation. The routing counter must now be extracted from the received NPDU – as we have already explained the SDU of layer N-1 is identical to the PDU of layer N. Afterwards, the user data can be passed on to the transport layer along with the other parameters via the N_Data.ind operation.

```
L_Data.ind.handler_receiver(source, destination, daf, class, l_sdu)
  address source;                      // address of sender
  address destination;                 // address of receiver
  char daf;                            // destination address flag (0)
  class_type class;                    // priority of the message
  char[MAXLEN] l_sdu;                  // user data layer 2
```

3.4 Network Layer

```
{
  char[MAXLEN] n_sdu;                          // user data layer 3

  n_sdu=delete_routing(l_sdu);                 // extracted routing counter
  N_Data.ind(source, destination, class, n_sdu);
}
```

On the sender side, the confirmation of layer 2 activates the network layer. For the network layer, the confirmation only states whether or not the packet has reached the receiver without error. The latter also includes the possibility that the receiver was busy for a long period of time and therefore unable to receive the packet or (due to a collision) there was no transmission. The acknowledgement is transferred to the transport layer unchanged via the N_Data.con service primitive.

```
L_Data.con.sender_handler(status)
  status_type_layer2 status;                   // status information layer 2
{
  switch (last_service) {
  case unicast:
      N_Data.con(receiver, status);
      break;
  case multicast: ...                          // to be completed later
  case broadcast: ...
  }
}
```

3.4.2 Multicast (N_Group) and Broadcast (N_Broadcast)

The locally confirmed services for transmitting a message to several receivers (*multicast*) or all nodes (*broadcast*) have a very similar structure and for this reason are dealt with together. As with the previously mentioned N_Data service, there are operations for requests (N_Group.req, N_Broadcast.req), for indications (N_Group.ind, N_Broadcast.ind) and for local confirmations (N_Group.con, N_Broadcast.con).

On the sender side, the priority of the message, the logical address of the target station (this parameter is omitted for broadcast telegrams) and the user data are transferred to the network layer with the N_Group.req (or N_Broadcast.req) operation. The network layer then adds the routing counter to the user data and from this forms a layer 3 network protocol data unit. Together with the remaining parameters, the NPDU is passed on to layer 2 via the L_Data.req service primitive.

On the basis of the destination address flag, the data link layer decides whether the user data needs to be passed on to one or more receivers. As the N_Group and N_Broadcast services are used for telegram traffic between one sender and several receivers or all nodes, the network layer must now set the corresponding parameters to one.

```
N_Group.req.sender_handler(destination, class, n_sdu)
  address destination;                         // address of receiver
  class_type class;                            // priority of the message
  char[MAXLEN] n_sdu;                          // layer 3 user data
{
  char[MAXLEN] l_sdu;                          // layer 2 user data
  char daf;                                    // destination address flag
```

3 EIB Protocol

```
  l_sdu=insert_routing(n_sdu);    // add routing counter to user data
  daf=1;                                          // multicast or broadcast
  L_Data.req(source, destination, daf, class, l_sdu);
  receiver=destination;                       // note group address
}
```

N_Broadcast.req.sender_handler(class, n_sdu)
```
  class_type class;                     // priority of the message
  char[MAXLEN] n_sdu;                         // layer 3 user data
{
  char[MAXLEN] l_sdu;                         // layer 2 user data
  char daf;                              // destination address flag

  l_sdu=insert_routing(n_sdu);    // add routing counter to user data
  daf=1;                                          // multicast or broadcast
  receiver.phyiscal.zone=0;       // zone address for broadcast is 0!
  receiver.physical.line=0;       // line address for broadcast is 0!
  receiver.physical.addr;         // device address for broadcast is 0!
  L_Data.req(source, receiver, daf, class, l_sdu);
}
```

On the receiver side, an arriving data packet is passed on to the network layer via the L_Data.ind operation. The routing information must be extracted from the received NPDU. The user data together with the other parameters can then be passed on to the transport layer.

The service primitive that needs to be called up is determined in accordance with the destination address flag and the destination address. If a single end device only was addressed, this must be indicated to the transport layer via the N_Data.ind operation; with group addressing the N_Group.ind operation must be selected; if the packet was directed at all end devices N_Broadcast.ind is used. We can therefore extend the original functionality of the network layer as follows:

```
L_Data.ind_handler_receiver(source, destination, daf, class, l_sdu)
  address source;                             // address of sender
  address destination;                        // address of receiver
  char daf;                              // destination address flag
  class_type class;                     // priority of the message
  char[MAXLEN] l_sdu;                         // layer 3 user data
{
  n_sdu=delete_routing(l_sdu);           // delete routing counter
  if (daf == 0)                                              // unicast
     N_Data.ind(source, destination, class, n_sdu);
  else
     if ((destination.physical.zone == 0) &&         // broadcast
         (destination.physical.line == 0) &&
         (destination.physical.addr == 0))
        N_Broadcast.ind(source, class, n_sdu);
     else                                                    // multicast
        N_Group.ind(source, destination, class, n_sdu);
}
```

On the sender side, the confirmation of layer 2 now reaches the network layer. Depending on the original call, the N_Group.con or N_Broadcast.con operation informs the transport layer of the success of packet transmission.

3.4 Network Layer

With that, we must now complete the handler for the L_Data.con service primitive that was started earlier as follows:

```
L_Data.con.sender_handler(status)
  status_type_layer2 status;           // status information layer 2
{
  switch (last_service) {
  case unicast:                        // unicast
      N_Data.con(receiver, status);
      break;
  case multicast:                      // multicast
      N_Group.con(receiver, status);
      break;
  case broadcast:                      // broadcast
      N_Broadcast.con(status);
  }
}
```

3.4.3 Routers

A router is a device that connects separate network parts and acts in (or on) layer 3 of the OSI model. Contrary to a bridge, i.e. a device that is concerned with the regeneration and amplification of signals, a router only passes packets onto the parts of the network in which they are actually required. In order to permit this filter function, the network layer of a router must know the structure of the neighboring network parts in order to be able to decide whether or not the packet can be passed on.

In an EIB network, routers can connect lines or zones. Depending on where they are used, the corresponding devices are thus referred to as *line couplers, LC* or *backbone couplers, BC*. Line and backbone couplers have the same structure, the different functionalities in the passing on of packets arises from their assigned location and associated physical address.

Table 3.5: Physical Address of the Router

Coupler type	Physical address		
	Zone	Line	Device number
Line coupler	1-15	1-15	0
Backbone coupler	1-15	0	0

The router checks the destination address of the incoming packets. On the basis of the destination address flag, it decides whether the packet is addressed to one end device or a group of end devices. These two cases are detailed below.

3.4.3.1 Routers and End Device Addressing

If the destination address flag is set to zero, the packet is addressed to one end device. If the destination address matches the address of the router, the user data was intended for the router itself. This allows configuration and parameter data to be transferred to the routers after the installation phase.

If the packet is addressed to another end device, the router decides on the basis of its place of installation whether or not the packet can be passed on. The possible situations are as follows:

1. Sub-line → line coupler → main line: The packet is passed on when the receiver is stationed outside the sub-line.
2. Main line → line coupler → sub-line: The packet is allowed through if the receiver is stationed in the line of the line coupler.
3. Main line → backbone coupler → backbone: The packet is passed on if the receiver is stationed in another zone.
4. Backbone → backbone coupler → main line: The packet is allowed through if the receiver is stationed in the zone of the backbone coupler.

The only packets therefore that should arrive at the main line of a zone are those addressed to end devices within this main line or end devices in the various associated sub-lines. The only packets that should arrive at the backbone are those addressed to end devices in this backbone or end devices in different zones.

To summarize: A packet that originates within a line (or zone) and has been addressed to a device in the same line (or zone) is never passed on by a line coupler (or backbone coupler) to an adjoining sub-network. By virtue of this mechanism, information exchange between end devices only ever occurs within limited networks – this is an important factor for increasing the data throughput!

We would now like to clarify this theory using an example (Figure 3.24).

In the diagram, the physical addresses of the EIB end devices and couplers are recorded in the format zone/line/device number. The network consists of two zones that are connected via the backbone couplers 1/0/0 and 2/0/0. There are also sub-lines within the individual zones. Packet exchange between the various end devices is represented with arrowed lines.

1. Packets within a sub-line, e.g. between end devices 1/1/2 and 1/1/3, only run along the sub-line as they are not allowed through by the line coupler, in this case 1/1/0.
2. Packets between end device 1/1/1 (stationed in sub-line 1) and end device 1/2/2 (stationed in sub-line 2) are passed on by line couplers 1/1/0 and 1/2/0 but refused by backbone coupler 1/0/0.
3. Packets from end device 1/0/1 in main line 1 addressed to end device 2/1/1 are allowed through by backbone couplers 1/0/0 and 2/0/0 before reaching their target via line coupler 2/1/0.

The actions to be executed depend on whether the router is a line or backbone coupler, this is given by the address, and on which side of the router the packet is received. The corresponding actions are listed in the following function. The function returns the value

- `false`, if the packet cannot be passed on,
- `true`, if the packet is to be passed on and
- `local`, if the packet was actually addressed to the router.

3.4 Network Layer

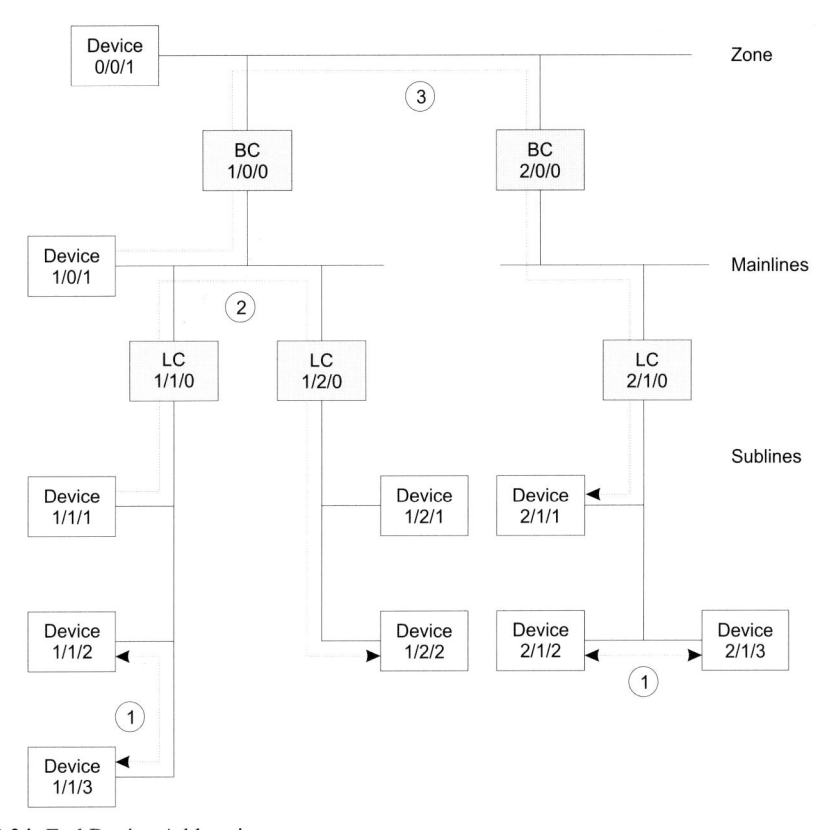

Figure 3.24: End Device Addressing

Sections 1 and 2 of the function represent the actions of a line coupler and sections 3 and 4 the actions of a backbone coupler.

```
/* Types for the definition of the let_pass function */
typedef enum {true, false, local}
            r_status;          // return value of the route function

typedef enum {mainline2subline, subline2mainline,
              zone2mainline, mainline2zone}
            direction;                              // packet direction

/* As a reminder, the most important variables of layers 2+3 */
address device;                 // address of an EIB device or router

char counter_routing;                        // for passing on packets

/* additional variables of layer 3 of a router */
direction dir;                        // current direction of packet

r_status let_pass(destination)
   address destination;                           // destination address
{
   switch (dir) {
```

103

```
    case mainline2subline:                  // section 1: line coupler
        if ((destination.physical.zone == device.physical.zone) &&
            (destination.physical.line == device.physical.line))
            if (destination.physical.addr == 0)
                return local;          // packet refers to line coupler
            else
                return true;                // end device in sub-line
        else return false;             // end device not in sub-line
    case subline2mainline:                  // section 2: line coupler
        if ((destination.physical.zone != device.physical.zone) &&
            (destination.physical.line != device.physical.line))
            return true;               // end device outside sub-line
        if ((destination.physical.zone == device.physical.zone) &&
            (destination.physical.line == device.physical.line) &&
            (destination.physical.addr == 0)
            return local;          // packet refers to line coupler
        else return false;                  // end device in sub-line
    case backbone2mainline:            // section 3: backbone coupler
        if ((destination.physical.zone == device.physical.zone))
            if ((destination.physical.line==device.physical.line) &&
                (destination.physical.addr == 0)
                return local;    // packet refers to backbone coupler
            else return true;               // end device in zone
        else return false;             // end device in another zone
    case mainline2backbone:            // section 4: backbone coupler
        if ((destination.physical.zone != device.physical.zone)
            return true;           // end device in another zone
        else
            if ((destination.physical.line == 0) &&
                (destination.physical.addr == 0))
                return local;    // packet refers to backbone coupler
            else return false;              // end device in same zone
    }
}
```

3.4.3.2 Routers and Group Addressing

If the destination address flag is set to one, the packet is addressed to a group of end devices. At commissioning, every line or backbone coupler receives a list of the group addresses that refer to its line or zone respectively. The couplers only allow packets with these group addresses to pass through.

We wish to combine the actions that are necessary for scrolling through the list of group addresses into one function (this function is not explained further). The `group(destination)` function returns the value:

- `false`, if the packet cannot be passed on,
- `true`, if the packet is to be passed on,
- `local`, if the packet was addressed to the router and
- `local_and_true`, if the packet was addressed to the router but should also be passed on.

3.4 Network Layer

Figure 3.25 illustrates the previous case study. In the EIB end devices and couplers the individual components of the physical address are recorded in the format zone/line/device number.

If we now assume that end device 1/1/3 sends a packet to a group (comprising end devices 1/1/2, 1/0/1, 2/1/2 and 2/1/3), then this packet is passed on by line couplers 1/1/0 and 2/1/0, as well as backbone couplers 1/0/0 and 2/0/0. Line coupler 1/2/0 on the other hand filters out the packet as no members of the group are stationed in its line.

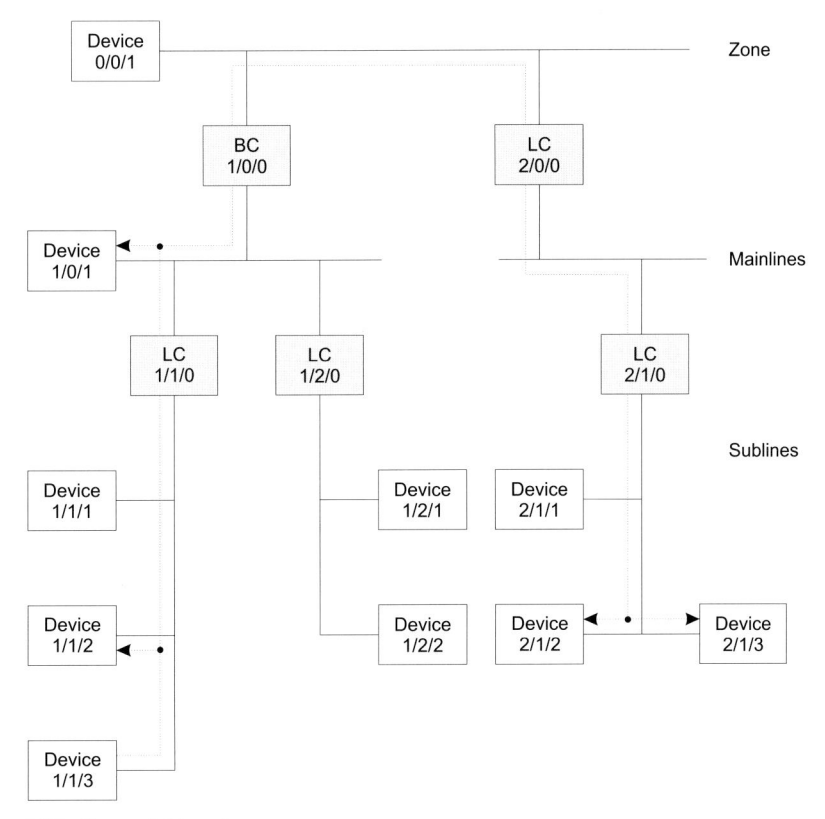

Figure 3.25: Group Addressing

3.4.3.3 Services

Until now we have avoided the fact that when deciding whether or not to let packets pass, routers use not only the address but also the current value of the routing counter. Accurately speaking, the only packets that are passed on are those in which the counter is greater than zero and whose addressees are stationed in the subsequent communication system.

Before a packet is passed on to the next sub-network, the router reduces the current value of the counter by one. Exchange across more than 6 stages is not generally provided for in EIB. If however, it is necessary to cover greater distances (e.g. in a heavily branched EIB network), the routing counter must be initialized with the value 7. Packets with this special value are passed on without limit and the value of the routing counter is never changed.

105

3 EIB Protocol

It should be remembered that bridges also reduce the routing counter by one when passing on a packet – a functionality that should actually be reserved for layer 3 of the OSI model (and with that routers explicitly)! We don't however wish to examine the subject in any more detail. Suffice to say that this mechanism also allows bridges to avoid passing on wrongly addressed packets. This slight deviation from the strict OSI model can therefore be accepted with an easy conscience.

The groundwork laid out in the previous chapters now provides an easy understanding of the services of the network layer of routers. The N_Data.req, N_Group.req and N_Broadcast.req requests are only directed at its layer 3 very rarely (e.g. for a new router in the configuration and test phase). In the same way as the subsequent local acknowledgements N_Data.con, N_Group.con and N_Broadcast.con they have the same parameter type profile and a similar type of implementation as that previously discussed. We can therefore avoid going over this again.

The reaction of a router to the arrival of the L_Data.ind indication however is quite different (Figure 3.26).

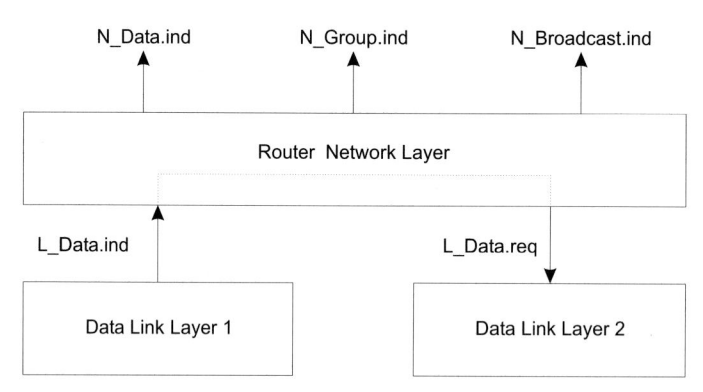

Figure 3.26: Router Network Layer

1. Physical addressing:

   a) If the user data is meant for the router alone, it passes it on to the above lying layer with the N_Data.ind operation,

   b) Otherwise, the router establishes whether or not the packet can be passed on. If necessary the counter is reduced accordingly and the packet passed on to the corresponding sub-section of the network with the L_Data.req service primitive.

2. Logical addressing:

   a) If the user data is directed at all stations in the network, the router passes it on to its above lying layer with the N_Broadcast.ind service primitive. After adjusting the routing counter accordingly, it also passes it on to the next sub-section in the network with the L_Data.req operation.

   b) In the case of a group address, the router determines whether it is actually a member of the group and if necessary signals this status to its above lying layer with the N_Group.ind service primitive. There is then a check to see whether members of the group are stationed in other network sections. If in addition, the router counter is equal

3.4 Network Layer

to zero, the packet is filtered out, otherwise it is allowed through with the L_Data.req service primitive.

The reaction of the network layer to the L_Data.ind event could therefore be implemented as follows (the auxiliary functions of `group()`, `get_routing()`, `delete_routing()` and `insert_routing()` have not been described in detail):

```
L_Data.ind_router_handler(source, destination, daf, class, l_sdu)
  adress source;                              // address of sender
  adress destination;                         // address of receiver
  char daf;                                   // destination address flag
  class_type class;                           // priority of the message
  char[MAXLEN] l_sdu;                         // layer 2 user data
{
  char[MAXLEN] n_sdu;                         // layer 3 user data

  counter_routing=get_routing(l_sdu);   // extract routing counter
  n_dsu=delete_routing(l_sdu);          // delete routing counter
  if (daf == 0)                         // physical addressing
      switch (let_pass(destination))
         case local:                    // packet target is the router
             N_Data.ind(source, destination, class, n_sdu);
             break;
         case true:
             if (counter_routing == 7) {   // for very big networks!
                 l_sdu=insert_routing(n_sdu);   // user data counter
                 L_Data.req(source, destination, daf, class, l_sdu);
             }
             else if (0 < counter_routing < 7) {
                 counter_routing--;         // decrement routing counter
                 l_sdu=insert_routing(n_sdu);   // user data counter
                 L_Data.req(source, destination, daf, class, l_sdu);
             }
         case false: break;
      }
  else
      if ((destination.physical.zone == 0) &&        // broadcast
          (destination.physical.line == 0) &&
          (destination.physical.addr == 0) {
         N_Data.ind(source, destination, class, n_sdu);
         if (counter_routing == 7) {       // for very big networks!
             l_sdu=insert_routing(n_sdu);      // user data counter
             L_Data.req(source, destination, daf, class, l_sdu);
         }
         else if (0 < counter_routing < 7) {
             counter_routing--;            // decrement routing counter
             l_sdu=insert_routing(n_sdu);      // user data counter
             L_Data.req(source, destination, daf, class, l_sdu);
         }
      }
      else                                          // multicast
         switch (group(destination)) {
             case local:                   // router is in the group
                 N_Data.ind(source, destination, class, n_sdu);
                 break;
```

3 EIB Protocol

```
        case local_and_true:        // router is in the group and ...
            N_Data.ind(source, destination, class, n_sdu);
        case true:                  // packet goes on to other devices
            if (counter_routing == 7) {
                l_sdu=insert_routing(n_sdu);   // user data counter
                L_Data.req(source, destination, daf, class, l_sdu);
            }
            else if (0 < counter_routing < 7) {
                counter_routing--;             // decrement routing counter
                l_sdu=insert_routing(n_sdu);   // user data counter
                L_Data.req(source, destination, daf, class, l_sdu);
            }
        case false: break;                     // packet rejected
        }
}
```

3.4.4 Network Protocol Data Unit (NPDU)

To finish with, we will highlight the concrete implementation of the network layer protocol data unit. This is composed of control information from layer 3 and the user data. In order to save space in the data frame, the control information (or more accurately the routing counter) is coded in the 6th byte of the frame in bits R0-R2.

We have now earned a brief pause. Figure 3.27 reviews the tasks and services of the network layer. The subsequent discussion of the transport layer will require full concentration – don't lose heart!

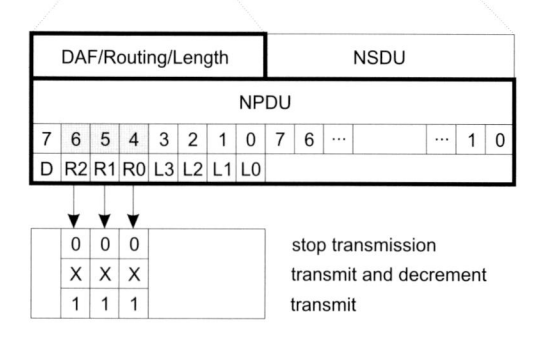

C-Field (8 Bit)	Source-Field (16 Bit)	Destination-Field (17 Bit)	RL-Field (7 Bit)	Data-Field (1-16 Byte)	P-Field (8 Bit)

```
              DAF/Routing/Length          NSDU
                              NPDU
              7 6 5 4 3 2 1 0 7 6 ...    ... 1 0
              D R2 R1 R0 L3 L2 L1 L0

                  0 0 0        stop transmission
                  X X X        transmit and decrement
                  1 1 1        transmit
```

Figure 3.27: NPDU

3.5 Transport Layer

The EIB *transport layer* offers the user – generally the *application layer* – two different methods of transferring data: connection-oriented and connectionless.

Connection-oriented communication is related to our telephone system. In order to speak with someone, it is necessary to pick up the receiver and dial a number. If the desired conversational partner is not available or otherwise engaged, it is not possible to talk and the connection is terminated by hanging up the receiver. Otherwise there is an exchange of information. Of course, there is always the possibility of occasional noise on the line (think of mobile phone conversations) or the partner is temporarily distracted and the message misunderstood. Misunderstandings of this kind are very easily avoided if every message is confirmed by the partner on the other side. A conversation lasts as long as it takes to exchange all necessary information and is quit by one of the partners hanging up. This generally occurs when the partners have politely said their goodbyes. An abrupt end to the conversation could also occur if, for example, one of the partners were to inadvertently drop the receiver back onto its cradle. These situations of everyday life, described here in a light-hearted manner, are implemented on a one-to-one basis in the transport layer using the services of connection-oriented communication.

Connectionless communication is similar to our postal system. If one needs to send a message to a friend or possibly a whole group of friends (e.g. an invitation to a birthday party), the message must be put into an envelope, correctly addressed and then deposited into a post box. As we are all too well aware, the postal system is not always 100% reliable. Letters can be lost or be so delayed that their content is no longer accurate or relevant, i.e. in our case the birthday party has long since passed. Usually however, the postal system works really well. For the birthday child it is enough to know that all the invitations have been posted and that they have every possibility (but no guarantee) of arriving. In addition, the postal path remains the cheapest. Just imagine doing the same by telephone – phoning each guest individually and explaining to each the details of the party. The phone bill would be much higher. Such connectionless communication is implemented in the transport layer by a group of locally confirmed services that are responsible for delivering messages to one partner (*unicast*), a group of partners (*multicast*) or all partners in the EIB system (*broadcast*).

3.5.1 Connection-Oriented Communication

Connection-oriented communication has three phases. To start with, a connection is set up between two partner instances of layer 4. Data is securely transmitted on this connection and finally the link is terminated again. There are separate services for each of these phases. These services are described in detail below. We will first present an overview of the protocol and define the necessary types and variables.

Correct protocol procedure is guaranteed by control information, which is coded into the user data of the transport layer (*transport service data unit*, TDSU). The resulting information packet is called a *transport protocol data unit*, TPDU.

Specific protocol data units can be assigned to each phase of connection-oriented communication:

1. Connect-PDUs are used to set up the connection,

2. Data-PDUs are used to transmit user data,

3. Ack-PDUs confirm the correct reception of user data,

4. Nack-PDUs are sent when an error has occurred during the transmission of user data

5. Disconnect-PDUs lead to the termination of the connection.

As already mentioned, phase 1 of the protocol involves the setup of a connection between sender and receiver. To manage this connection, the transport layer on the sender side must note the receiver address and on the receiver side the sender address.

```
address s_partner;           // variable for layer 4 of the sender
address r_partner;           // variable for layer 4 of the receiver
```

In phase 2 the sender begins to transmit data. It is important that the receiver is able to distinguish between telegrams that it is seeing for the first time and those it has already received. The simplest way of achieving this is if the sender transmits a sequence number along with the user data. The receiver can then decide whether the data is new or duplicated, in which case it needs to be rejected.

On the sender side, this number is called the send counter, and on the receiver side the receive counter.

```
char sent;                                         // send counter
char received;                                     // receive counter
```

After the sequence number has been selected, the sender activates the so-called acknowledgement timer, transmits the entire telegram and waits to receive an acknowledgement. If a positive acknowledgement is returned, the sender increases the send counter and begins with the transmission of another telegram. If, on the other hand, the timer expires without an acknowledgement having been received, it retransmits the data with the same sequence number. This procedure can be repeated up to a maximum of three times. If there has still been no acknowledgement, the sender quits the connection.

We define three further variables for the sender side of the transport layer in order to be able to carry out these tasks.

```
timer ack;                                   // acknowledgement timer
char counter_repeat;                         // repetition counter
char act_sdu[MAXLEN];                        // last transmitted data
class_type act_class;         // priority of last transmitted data
```

As soon as the receiver has reason to believe that an error has occurred, it sends a negative acknowledgement. The receiver is mistrustful when it receives a protocol data unit with an unexpected sequence number. If a negative acknowledgement is returned the sender quits the connection. Data can only be transmitted again after a new connection is set up.

If no data has been transmitted for a specific period of time, the protocol automatically switches to phase 3 and the connection is terminated. This requires extra timers on the sender and receiver sides. These timers are referred to as connection timers.

```
timer s_connection;                       // sender connection timer
timer r_connection;                       // receiver connection timer
```

3.5 Transport Layer

Protocols that operate in such a way are often referred to as PAR (*Positive Acknowledgement with Retransmission*) or ARQ (*Automatic Repeat Request*) protocols. Note: Although the data flows in one direction from sender to receiver, the protocol data units move in both directions!

In order to accurately describe the subsequent services and the protocol, it is necessary to be able to establish the actual phase of sender and receiver:

```
typedef enum {closed, connect, idle, wait,
              repeat, disconnect, error}
              sender_state;      // possible statuses of the sender

sender_state s_state;             // current status of the sender

typedef enum {closed, wait, ack, error}
              receiver_state;    // possible statuses of the receiver

receiver_state r_state;           // current status of the receiver
```

This brief overview completes the groundwork necessary for a detailed description of the services associated with connection-oriented communication in EIB. As can be seen from Figure 3.28, to fulfill these tasks the transport layer only accesses the (unicast) N_Data service of the network layer.

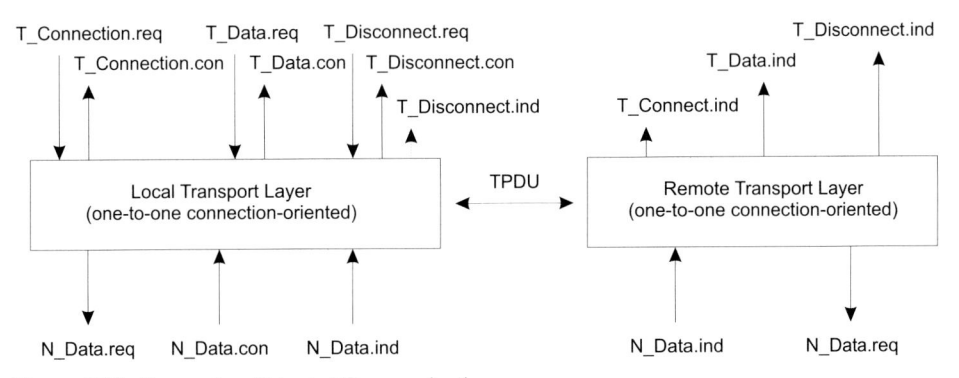

Figure 3.28: Connection-Oriented Communication

3.5.1.1 Connection Setup (T_Connect)

The locally confirmed T_Connect service is responsible for setting up the connection between sender and receiver. As usual, this service consists of the T_Connect.req request, the T_Connect.ind indication and the T_Connect.con local confirmation.

To begin with, both sender and receiver are in the closed status. From the application layer, the T_Connect.req operation informs the sender that a new connection with a receiver is required. It should be noted that a sender can only be connected to a single receiver at any one time. The transport layer therefore rejects successive T_Connect.req requests.

The sender assembles a connect-PDU (`Connect_Pdu`), sets the priority of the telegram to system message and hands over the assembled parameters to the network layer using the

3 EIB Protocol

N_Data.req operation. The connection timer is then set and the sender switches to the connect status. It is now ready to send data to the receiver and to accept acknowledgements.

```
T_Connect.req.sender_handler(destination)
   address destination;                  // physical address of receiver
{
   char[MAXLEN] n_sdu;                   // layer 3 user data

   s_partner=destination;                // save physical address
   sent=0;                               // still no data telegrams sent
   n_sdu=encode_pdu(Connect_Pdu);        // code connect-PDU
   N_Data.req(s_partner, system, n_sdu);
   start_timer(s_connection);            // start connection timer
   s_state=connect;                      // next status: connect
}
```

With the N_Data.ind service primitive, the receiver is shown that the sender has opened a connection. This indication is passed on to the application layer with the T_Connect.ind operation. The receiver then starts its connection timer and switches to the wait status. It is now ready to receive data and return acknowledgements.

```
N_Data.ind.receiver_handler(source, destination, class, n_sdu)
   address source;                       // physical address of sender
   address destination;                  // physical address of receiver
   class_type class;                     // priority of the message
   char[MAXLEN] n_sdu;                   // layer 3 user data
{
   switch (decode_pdu(n_sdu)) {
   case Connect_Pdu:                     // announcement of connection setup
         r_partner=source;               // save physical address
         received=0;                     // still no data packets received
         T_Connect.ind();
         start_timer(r_connection);      // start connection timer
         r_state=wait;                   // next status: wait
         break;
   /* to be completed later */
   }
}
```

Back to the sender. In the connect status there are two possibilities:

1. Arrival of local confirmation N_Data.con – a reaction of the L_Data.con confirmation of layer 2. The sender signals this condition to its application layer and switches to the idle status. The connection is opened!

```
   N_Data.con.sender_handler(destination, status)
      address destination;               // physical address of sender
      status_type_layer2 status;         // confirmation of layer 3 and 2
   {
      switch (s_state) {
      case connect:
            T_Connect.con();
            s_state=idle;                // next sender status: idle
            break;
      }
   }
```

3.5 Transport Layer

2. The connection timer expires. The sender reverts to the closed status. The application layer is informed of this status via the T_Disconnect.ind indication. Connection setup has failed!

We will demonstrate the treatment of connection timeouts using the example of the `sender_connection_timeout()` routine. Of course, this routine will need to be completed in subsequent chapters but for now, here is the first installment:

```
sender_connection_timeout()
{
    switch (s_state) {
    case connect:           // protocol error during connection setup
        T_Disconnect.ind();                         // no connection!
        s_state=closed;                 // next sender status: closed
        break;
    /* to be completed later */
    }
}
```

First, a brief summary of the processes (Figure 3.29).

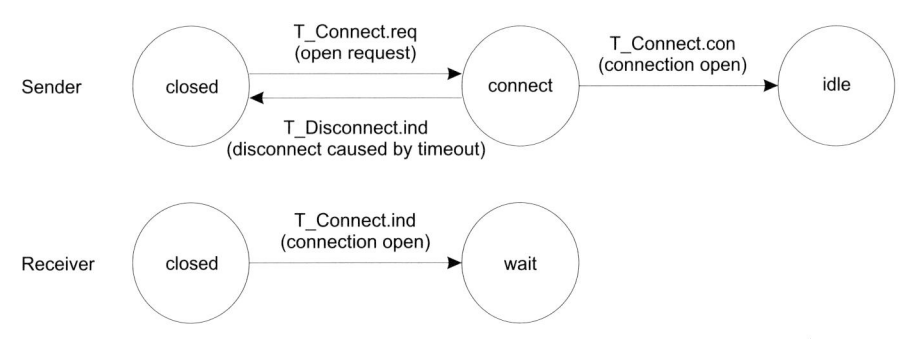

Figure 3.29: Connection Setup

With the T_Connect.req operation, the application layer of the sender requests the opening of a connection with a partner (open request). After receiving the T_Connect.con local confirmation, the sender views the connection as open (connection open). If, however, the confirmation is not received, there is no possibility of regular data traffic. After expiration of the connection timer (disconnect caused by timeout), the transport layer returns the T_Disconnect.ind indication to the application layer, which then tries again.

The receiver informs its own application layer of the request for opening a connection with the T_Connect.ind operation. It then waits for the arrival of the first data (connection open).

3.5.1.2 Data Transfer (T_Data)

Data can be transmitted once the connection has been opened. For the transmission of telegrams the transport layer provides the confirmed service T_Data. This service consists of the T_Data.req request, the T_Data.ind indication and the T_Data.con confirmation.

The sequence of service elements is as follows: On the sender side, the T_Data.req request is passed on to the transport layer for the transmission of the user data. A data PDU

113

3 EIB Protocol

(`Data_Pdu`) is then put together, the current "sequence number" is coded and the acknowledgement timer is set.

If this timer expires without the sender having received a confirmation from the peer layer of the receiver, the same data PDU is sent again. This requires the prior saving of the message priority and original content. Finally, the telegram is passed on to the local layer 3 with the N_Data.req request and the connection timer is reinitialized. The sender switches to the wait status and from now on waits for a confirmation.

The observant reader will see immediately that the destination address is missing from the next program fragment. It should not be difficult to solve this riddle. Would a clue help? After successfully setting up a connection, its management is a task of the transport layer. As there can only be one connection to another partner at any one time, there is no magic when it comes to the allocation of address.

```
T_Data.req.sender_handler(class, t_sdu)
    class_type class;                       // priority of the message
    char[MAXLEN] t_sdu;                     // layer 4 user data
{
    char[MAXLEN] n_sdu;                     // layer 3 user data

    n_sdu=encode(sent, Data_Pdu, t_sdu);   // code sequence number+PDU
    act_class=class;                                        // save priority
    act_sdu=copy(n_sdu);                                    // save user data
    counter_repeat=0;                       // initialize repetition counter
    start_timer(ack);                       // start acknowledgement timer
    N_Data.req(s_partner, class, n_sdu);
    start_timer(s_connect);                 // start connection timer
    s_state=wait;                           // next sender status: wait
}
```

On the receiver side, the possibilities are as follows:

1. The N_Data.ind indication activates the transport layer

   a) If it receives a data PDU with expected sequence number a positive confirmation is returned to the sender and the receive counter is increased. Afterwards, the user data is passed on to the application layer with the T_Data.ind service primitive and the connection timer is reinitialized.

   b) If it gets the last received data PDU, then the receiver assumes that its confirmation was lost and it has received a copy. It sends another positive acknowledgement, but the user data must not be retransmitted to the application layer.

   c) If a data PDU is received with an unexpected sequence number, the receiver returns a negative acknowledgement.

   The outgoing confirmations are always sent as system messages. We will now expand the previously started handler for the N_Data.ind service primitive as follows:

```
N_Data.ind.receiver_handler(source, destination, class, n_sdu)
    address source;                 // physical address of sender
    address destination;            // physical address of receiver
    class_type class;               // priority of the message
    char[MAXLEN] n_sdu;             // layer 3 user data
{
```

```
        char seqnumber;                            // sequence number
        char[MAXLEN] t_sdu;                        // layer 4 user data

        switch (decode_pdu(n_sdu)) {
        case Data_Pdu:                             // create data PDU
            seqnumber=decode_seq(n_sdu);    // decode sequence number
            t_sdu=decode_sdu(n_sdu);              // retrieve user data
            if (seqnumber == received) {                  // case (a)
              n_sdu=encode(received, Ack_Pdu);    // code seq.no.+PDU
              N_Data.req(source, system, n_sdu);
              received++;                  // increase receive counter
              T_Data.ind(class, t_sdu);
            }
            else if (seqnumber == received-1) {           // case (b)
              n_sdu=encode(seqnumber, Ack_Pdu);
              N_Data.req(source, system, n_sdu);
            }
            else {                                        // case (c)
              n_sdu=encode(seqnumber, Nack_Pdu);
              N_Data.req(source, system, n_sdu);
            }
            start_timer(r_connection);     // start connection timer
            break;
        case Connect_Pdu:
        /* see connection setup */
}
```

2. The connection timer expires. With that, the transport layer of the receiver views the connection as terminated.

   In the same way as with the sender, we will begin with the definition of the valid routine `receiver_connection_timeout()`.

```
receiver_connection_timeout()
{
    char[MAXLEN] n_sdu;                        // layer 3 user data

    switch (r_state) {
    case wait:
        n_sdu=encode(Disconnect_Pdu);            // code disconnect
        N_Data.req(r_partner, system, n_sdu);
        T_Disconnect.ind();
        r_state=closed;            // next receiver status: closed
        break;
    }
}
```

If we now return to the sender and consider the following scenarios – depending on the reaction of the receiver:

1. A confirmation is received with the N_Data.ind operation.

   a) If it is a positive acknowledgement (`Ack_Pdu`) of the last transmitted telegram, the sender can transmit new data. It informs the application layer of this status with the T_Data.con service primitive. Note: T_Data.con occurs on the basis of the activity of

3 EIB Protocol

the peer layer, quite the opposite to the local confirmation T_Connect.con when setting up connections!

b) If the acknowledgement was negative (Nack_Pdu), the transport layer of the sender quits the connection. It first tries to make the situation clear to the receiver by means of a disconnect-PDU (Disconnect_Pdu).

```
N_Data.ind.sender_handler(source, destination, class, n_sdu)
    address source;                     // physical address of sender
    address destination;                // physical address of receiver
    class_type class;                   // priority of the message
    char[MAXLEN] n_sdu;                 // layer 3 user data
{
    char seqnumber;                     // sequence number of the PDU
    char[MAXLEN] t_sdu;                 // layer 4 user data

    switch (decode_pdu(n_sdu)) {
    case Ack_Pdu:                                       // case (a)
        seqnumber=decode_seq(n_sdu);
        stop_timer(ack);                // stop acknowledgement timer
        if (seqnumber == sent) {
            sent++;                     // increase sequence number
            T_Data.con(class, t_sdu)
            start_timer(s_connection);  // start connection timer
        }
        s_state=idle;                   // next sender status: idle
        break;
    case Nack_Pdu:                                      // case (b)
        n_sdu=encode_pdu(Disconnect_Pdu);   // disconnect-PDU
        N_Data.req(r_partner, system, n_sdu);
        T_Disconnect.ind();
        s_state=closed;                 // next sender status: closed
        break;
    }
}
```

2. The acknowledgement timer expires (i.e. no confirmation is received from the receiver within a specific time interval). The sender now has the possibility of repeating the previously sent data-PDU to the receiver a further three times.

If a valid acknowledgement is received, the sender switches to the idle status and is ready for new telegrams. If the receiver fails to return an acknowledgement, the connection is interrupted and the receiver has either failed or is faulty. The sender transmits a disconnect-PDU and stops data traffic.

The following ack_timeout() routine can manage the execution of these tasks. This routine is always called into play when the acknowledgement timer expires.

```
ack_timeout()
{
    char[MAXLEN] n_sdu;                             // layer 3 user data

    switch (s_state) {
    case wait:
        if (counter_repeat < 3) {
```

3.5 Transport Layer

```
            N_Data.req(s_partner, act_class, act_sdu);
            counter_repeat++;                  // increase repetition counter
            start_timer(ack);                  // start acknowledgement timer
            start_timer(s_connection);         // start connection timer
        }
        else {
            n_sdu=encode_pdu(Disconnect_Pdu);  // code disconnect
            N_Data.req(s_partner, system, n_sdu);
            T_Disconnect.ind();
            s_state=closed;                    // next sender status: closed
        }
    }
}
```

We will now summarize the most important steps (Figure 3.30). With the T_Data.req operation, the application layer of the sender activates the transport layer, which then arranges for the transmission of the user data (send Data_Pdu). Afterwards, it waits for the arrival of positive acknowledgements (get Ack_Pdu), which are indicated by the network layer with the N_Data.ind event. If no acknowledgement is received (acknowledgement timeout), the transport layer tries again up to three times to retransmit the original data (resend Data_Pdu).

On the receiver side, the transport layer data PDUs are indicated with the N_Data.ind operation (get Data_Pdu). If they also have the expected sequence numbers, they are positively acknowledged by means of the N_Data.req service primitive (send Ack_Pdu).

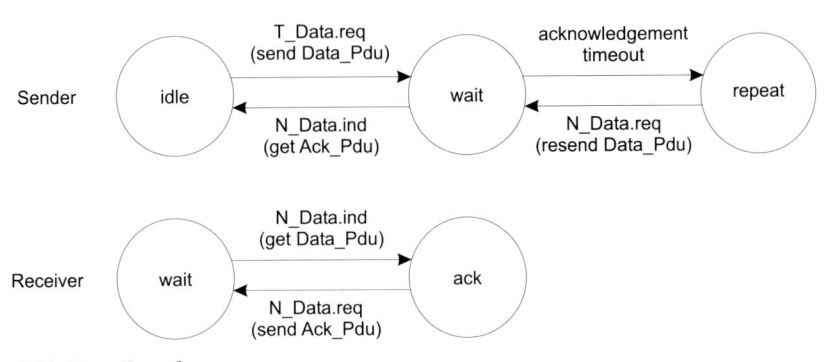

Figure 3.30: Data Transfer

3.5.1.3 Termination (T_Disconnect)

It now remains to take a closer look at the procedures involved in the termination of connections. This is either achieved by the T_Disconnect.req request of the locally confirmed T_Disconnect service or by the expiration of a connection timer.

Under normal circumstances, the application layer quits the connection after all user data has been transmitted. For this it uses the T_Disconnect.req service primitive. The transport layer of the sender assembles a disconnect-PDU and for transmission of this system message uses the N_Data.req service primitive of the network layer. Afterwards, the sender switches to the disconnect status.

`T_Disconnect.req.sender_handler()`

3 EIB Protocol

```
{
  char[MAXLEN] n_sdu;                         // layer 3 user data

  n_sdu=encode_pdu(Disconnect_Pdu);           // code disconnect
  N_Data.req(s_partner, system, n_sdu);
  s_state=disconnect;                         // next sender status: disconnect
}
```

With the N_Data.ind service primitive, the receiver is shown that the sender wants to close the connection. This indication is passed on to the application layer with the T_Disconnect.ind operation. With that, the connection is quit for the receiver.

```
N_Data.ind.receiver_handler(source, destination, class, n_sdu)
  address source;                   // physical address of sender
  address destination;              // physical address of receiver
  class_type class;                 // priority of message
  char[MAXLEN] n_sdu;               // layer 3 user data
{
  switch (decode(n_sdu)) {
  case Disconnect_Pdu:              // request to cancel connection
      T_Disconnect.ind();
      r_state=closed;               /next status: closed
  case Connect_Pdu: ...             // see connection setup
  case Data_Pdu: ...                // see data transfer
  }
}
```

We will now take another look at the sender. In the disconnect status there are two possible scenarios:

1. The N_Data.con local confirmation – a reaction of the L_Data.con confirmation of layer 2 – arrives. The sender maps it onto the T_Disconnect.con service primitive and with that "wakes" the application layer. It then switches to the closed status. With that, the connection has been closed!

   ```
   N_Data.con.sender_handler(destination, status)
        address destination;               // physical address of sender
        status_type_layer2 status;         // ack'ment from layer 3 and 2
   {
        switch (s_state) {
        case disconnect:                              // disconnect
            T_Disconnect.con();
            state=closed;                  // next sender status: closed
            break;
        case connect: /* see connection setup */
        }
   }
   ```

2. The connection timer expires. The sender signals this status to its application layer with the T_Disconnect.ind operation and switches to the closed status.

   We can now extend the `sender_connection_timeout()` routine as follows:

   ```
   sender_connection_timeout()
     {
        char[MAXLEN] n_sdu;                        // layer 3 user data
   ```

```
    switch (s_state) {
    case disconnect:              // protocol error during disconnect
    case connect:
        T_Disconnect.ind();
        state=closed;             // next sender status: closed
        break;
    /* to be completed later */
    }
}
```

In order to completely understand the protocol we must now consider every case that can arise due to a faulty sender or receiver. Some of these have already been touched on.

− Expiration of the connection timer at the sender. At this point in time, the sender can only be in the idle status (in the wait status an acknowledgement timeout would have occurred first and as a result the connection timer reset or stopped). With that, the transport layer of the sender transmits a disconnect PDU to the peer layer – as the last message so to speak before the imminent end. It then informs its own application layer that an error has occurred with the T_Disconnect.ind service primitive. We can therefore complete the `sender_connection_timeout()` function as follows:

```
sender_connection_timeout()
{
    char[MAXLEN] n_sdu;                   // layer 3 user data

    switch (s_state) {
    case disconnect:        // protocol error during disconnection
    case connect:           // protocol error during connection setup
        T_Disconnect.ind();
        break;
    case idle:                            // no more data-PDUs to send?
        n_sdu=encode(Disconnect_Pdu);     // code disconnect
        N_Data.req(s_partner, system, n_sdu);
        T_Disconnect.ind();
        break;
    }
    s_state=closed;                       // next sender status: closed
}
```

− Expiration of the connection timer at the receiver. If the connection timer of the receiver expires (early), the remote transport layer takes the initiative and transmits a disconnect-PDU to the sender. This situation is made known to the application layer of both sender and receiver by means of an indication (i.e. by the T_Disconnect.ind service primitive) – a true rarity in the EIB protocol!

The `receiver_connection_timeout()` routine now acts in a similar way to the `sender_connection_timeout()` function.

```
receiver_connection_timeout()
{
    char[MAXLEN] n_sdu;                   // layer 3 user data

    switch (r_state) {
    case wait:                            // where are the data PDUs?
```

3 EIB Protocol

```
            n_sdu=encode(Disconnect_Pdu);      // code disconnect PDU
            N_Data.req(r_partner, system, n_sdu);
            T_Disconnect.ind();
            break;
    }
    r_state=closed;                             // next receiver status: closed
}
```

The actions must therefore be adapted as follows with the arrival of the N_Data.ind indication:

```
N_Data.ind.receiver_handler(source, destination, class, n_sdu)
    address source;                 // physical address of sender
    address destination;            // physical address of receiver
    class_type class;               // priority of the message
    char[MAXLEN] n_sdu;             // layer 3 user data
{
    switch (decode(n_sdu)) {
    case Disconnect_Pdu:            // request for disconnect
        T_Disconnect.ind();
        r_state=closed;             /next status: closed
    case Ack_Pdu: ...               // see positive acknowledgement
    case Nack_Pdu: ...              // see negative acknowledgement
    }
}
```

The following is a summary of the services involved in disconnection (Figure 3.31).

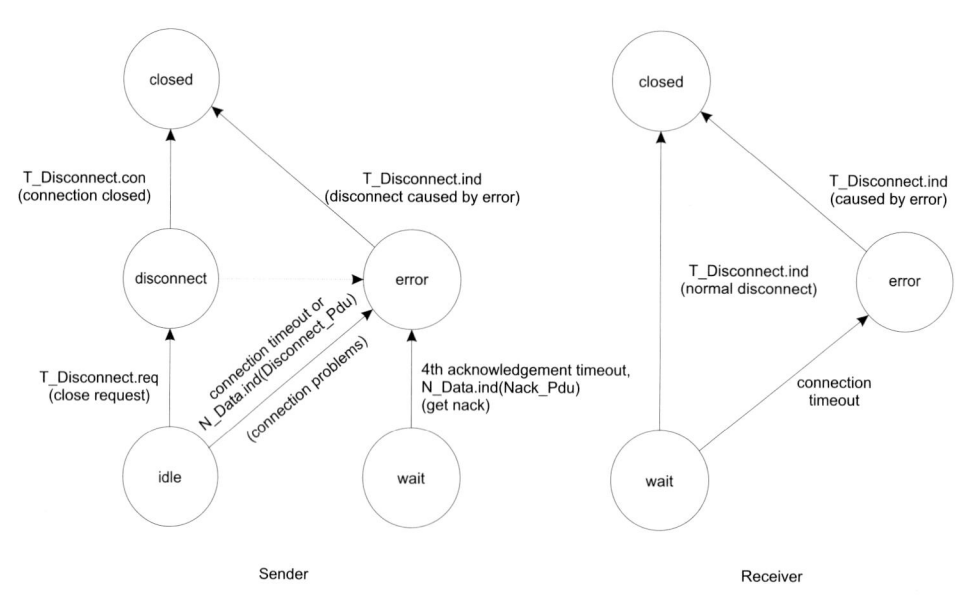

Figure 3.31: Disconnection

With all this information, the reader should now have a good understanding of the protocol of connection-oriented communication on layer 4 (Figure 3.32).

120

3.5 Transport Layer

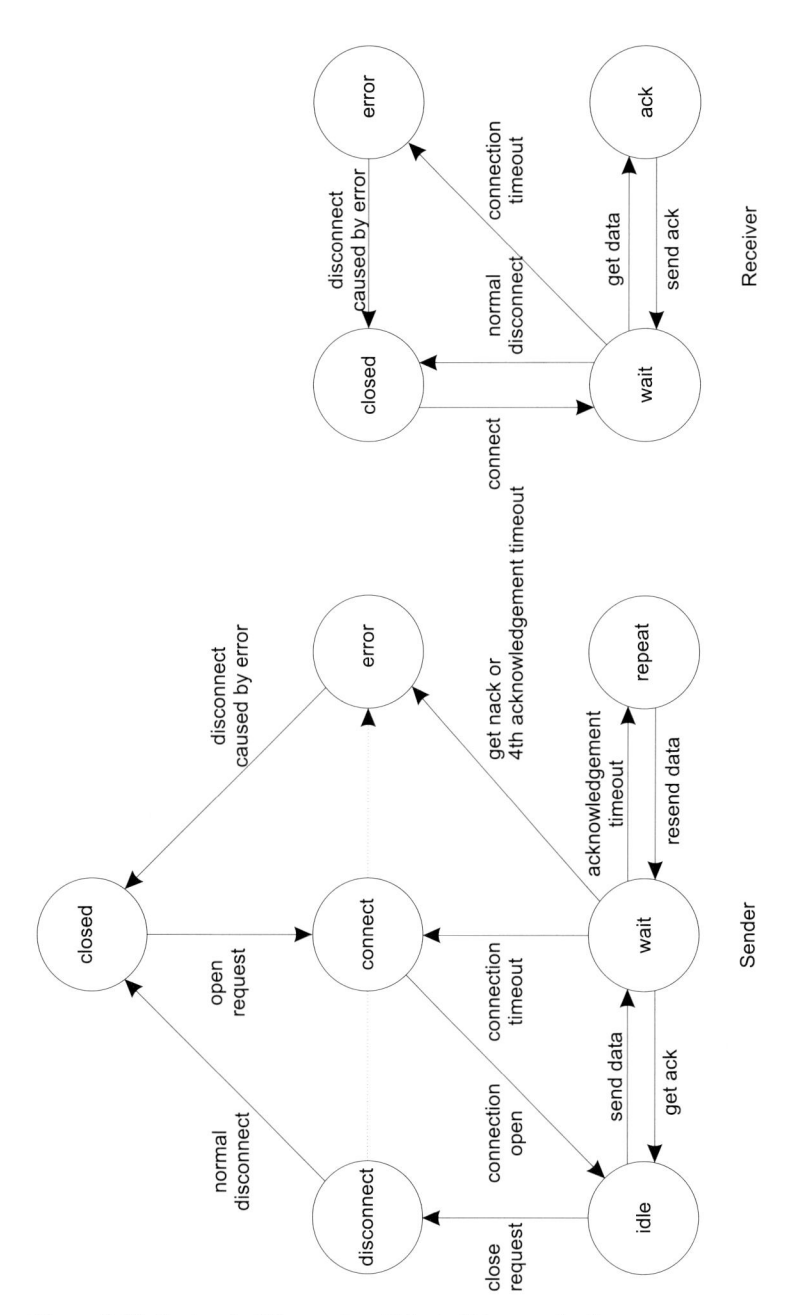

Figure 3.32: Protocol of Connection-Oriented Communication

If a T_Disconnect.ind is signaled to the application layer of the sender then it must assume that a protocol error has occurred (disconnect caused by error). Reasons may include:

– Connection setup failed (disconnect caused by timeout),
– Disconnection failed,

121

3 EIB Protocol

- The sender has received a negative acknowledgement (get nack),
- Despite several attempts to send a data-PDU to the receivers, the sender has received neither ack-PDUs nor nack-PDUs in return (4th acknowledgement timeout),
- The connection timer on the sender or receiver side has expired (connection problems).

On the receiver side, the application layer is always informed of the disconnection (either controlled or due to a protocol error) by means of the T_Disconnect.ind service primitive.

3.5.2 Connectionless Communication

We will now turn our attention to the services that facilitate the exchange of telegrams between two devices (*unicast*), within a group (*multicast*) and between all stations (*broadcast*) in EIB. These functions are covered by locally confirmed services of connectionless communication. In order to satisfy its task in the best possible way, the transport layer uses the relevant services of the underlying network layer (Figure 3.33).

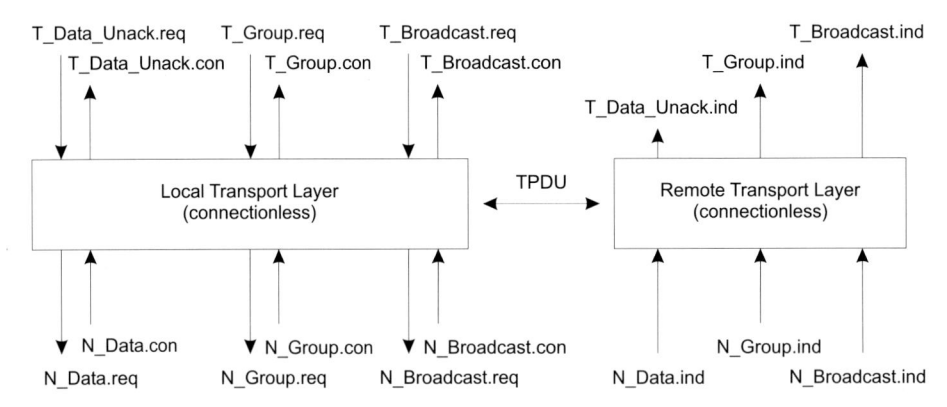

Figure 3.33: Connectionless Communication

3.5.2.1 Unicast (T_Data_Unack)

The T_Data_Unack service is requested by the user of the transport layer when data is to be transmitted to a single device within the EIB network. The T_Data_Unack.req request is not acknowledged by the remote peer layer. The incoming T_Data_Unack.con acknowledgement originates from the L_Data.con service primitive of the local layer 2.

The procedure is as follows: The user transfers the user data and desired destination address to the transport layer. As the N_Data service of the network layer is used for the further telegram transport, it is necessary to create a new protocol data unit. This TPDU is subsequently referred to as Data_Unack-PDU (in code as `Data_Unack_Pdu`).

```
T_Data_Unack.req.sender_handler(destination, class, t_sdu)
  address destination;              // physical address of receiver
  class_type class;                 // priority of the message
  char[MAXLEN] t_sdu;               // layer 4 user data
{
  char[MAXLEN] n_sdu;               // layer 3 user data
```

3.5 Transport Layer

```
    n_sdu=encode_pdu(Data_Unack);                        // code PDU
    N_Data.req(destination, class, n_sdu);
}
```

On the receiver side, the transport layer is activated by the N_Data.ind service primitive. After the Data_Unack-PDU has been decoded, the user data is passed on to the application layer via the T_Data_Unack.ind operation.

```
N_Data.ind.receiver_handler(source, destination, class, n_sdu)
    address source;                      // physical address of sender
    address destination;                 // physical address of receiver
    class_type class;                    // priority of message
    char [MAXLEN] n_sdu;                 // layer 3 user data
{
    char [MAXLEN] t_sdu;                 // layer 4 user data

    switch (decode_pdu(n_sdu)) {         // decode PDU
    case T_Connect_Pdu: ...
    case T_Disconnect_Pdu: ...
    case Data_Pdu: ...       // see connection-oriented communication
    case Data_Unack:         // connectionless telegram exchange
        t_sdu=decode_sdu(n_sdu);         // extract user data
        T_Data_Unack.ind(source, destination, class, t_sdu);
        break;
    }
}
```

Finally, the sender passes the incoming confirmation from layer 3 to the application layer unchanged with the T_Data_Unack.con service primitive. To reiterate – this is a local acknowledgement. In other words, if a positive acknowledgement is passed on to the application layer, the last activation of the T_Data_Unack.req operation was successful, the application layer however does not know whether the data has been transmitted to the receiver free from error.

With the specification of the parameter type profile of this service primitive we can conclude our discussions on the connectionless unicast service.

```
T_Data_Unack.con(status)
    status_type_layer2 status;  // local confirmation from layers 2+3
```

3.5.2.2 Multicast (T_Group)

In order to send a message to a group of devices, the application layer uses the T_Group service of layer 4. This service consists of the T_Group.req request, the T_Group.ind indication and the T_Group.con local confirmation.

The user data is passed on to the transport layer together with the group address, using the T_Group.req service primitive. The transport layer assembles the so-called group-PDU and activates the network layer with the N_Group.req operation.

```
T_Group.req.sender_handler(destination, class, t_sdu)
    address destination;                 // logical address (group)
    class_type class;                    // priority of the message
    char[MAXLEN] t_sdu;                  // layer 4 user data
{
```

3 EIB Protocol

```
  char[MAXLEN] n_sdu;                           // layer 3 user data

  n_sdu=encode_pdu(Group_Pdu);                  // code PDU
  N_Group.req(destination, class, n_sdu);
}
```

On the receiver side, the transport layer reacts to the N_Group.ind indication as follows:

```
N_Group.ind.receiver_handler(source, destination, class, n_sdu)
  address source;                     // physical address of sender
  address destination;                // physical address of receiver
  class_type class;                         // priority of message
  char [MAXLEN] n_sdu;                         // layer 3 user data
{
  char[MAXLEN] t_sdu;                          // layer 4 user data

  t_sdu=decode_sdu(n_sdu);                       // extract user data
  T_Group.ind(destination, class, t_sdu);
}
```

With the T_Group.con service primitive the application layer of the sender then receives a confirmation originating from the local layer 2.

```
N_Group.con.sender_handler(destination, status)
  address destination;                // physical address of receiver
  status_type_layer2 status;     // status information from layer 3
{
  T_Group.con(destination, status);
}
```

3.5.2.3 Broadcast (T_Broadcast)

The T_Broadcast service for transmitting a message to all stations can be explained very quickly. The T_Broadcast.req request is mapped on to the N_Broadcast.req service primitive of the network layer. On the receiver side, the user data reaches the transport layer via the N_Broadcast.ind indication and from there is passed on to the application layer via the T_Broadcast.ind operation. The incoming confirmation on the sender side originates, as with all other services of connectionless communication, locally.

We will now list the parameter type profiles of the individual service primitives. Implementation of the (extremely simple) handler for the request that originates from the application layer or for the indication and confirmation that are activated by the network layer is left as an exercise for the reader.

```
T_Broadcast.req(class, t_sdu)
  class_type class;                         // priority of the message
  char[MAXLEN] t_sdu;                          // layer 4 user data

T_Broadcast.ind(source, class, t_sdu)
  address source;                     // physical address of sender
  class_type class;                         // priority of the message
  char[MAXLEN] t_sdu;                          // layer 4 user data

T_Broadcast.con(status)
  status_type_layer2 status;    // status information from layers 2+3
```

3.5.3 Transport Protocol Data Unit (TPDU)

A protocol data unit of the transport layer is composed of control information, which is accommodated in the 7th byte of the data frame, and of a maximum of 15 subsequent user data bytes (Figure 3.34).

| C-Field (8 Bit) | Source-Field (16 Bit) | Destination-Field (17 Bit) | RL-Field (7 Bit) | Data-Field (1-16 Byte) | P-Field (8 Bit) |

Transport Control Field | TSDU

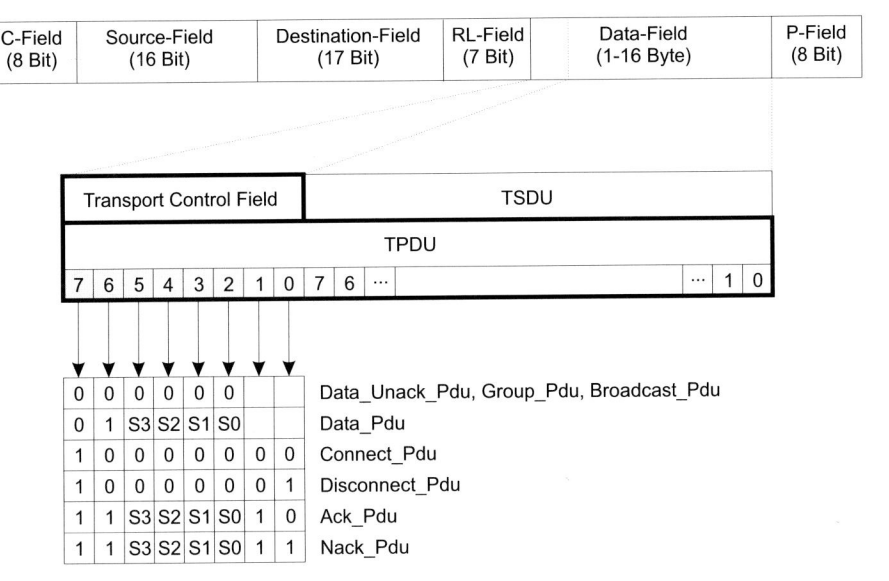

TPDU

| 7 | 6 | 5 | 4 | 3 | 2 | 1 | 0 | 7 | 6 | ... | ... | 1 | 0 |

0	0	0	0	0	0			Data_Unack_Pdu, Group_Pdu, Broadcast_Pdu
0	1	S3	S2	S1	S0			Data_Pdu
1	0	0	0	0	0	0	0	Connect_Pdu
1	0	0	0	0	0	0	1	Disconnect_Pdu
1	1	S3	S2	S1	S0	1	0	Ack_Pdu
1	1	S3	S2	S1	S0	1	1	Nack_Pdu

Figure 3.34: TPDU

A brief word on the protocol data units of connection-oriented communication: With a data-PDU the send counter is located in bits S0-S3. The receive counter is also located here for ack-PDUs and nack-PDUs. The remaining protocol data units are used for setting up and terminating connections.

And now to the protocol data units of connectionless communication. As can be seen from Figure 3.34 they all have the same coding. The observant reader will immediately know that further distinction can be made on the basis of the destination address flag and destination address. Table 3.6 clarifies the situation.

Table 3.6: Destination Address Flag and Destination Address

PDU Type	Destination Address Flag	Destination Address
Data_Unack-PDU	0	0-256
Group-PDU	1	1-256
Broadcast-PDU	1	0

If a protocol data unit is used to transmit "proper" data in addition to the control information (i.e. data, data_unack, group and broadcast PDUs), then it is the responsibility of the application layer to correctly set bits 0 and 1 in the control field and if necessary to complete the user data. The next section examines which messages and commands can actually be transmitted via the EIB, which services are available to achieve this and it also looks at the structure of the protocol data units of layer 7.

3.6 Application Layer

As we already know, the underlying EIB layer model is based on the OSI model, but without the *session layer* or *presentation layer*. The levels discussed so far guarantee the transportation of data but they do not actually carry out any "proper" functions for the user. Data can be sent and received via broadcast and multicast, as well as connectionless and connection-oriented unicast services.

The *application layer* now comprises all higher-order services. This includes services that are required for the device configuration and network management, and others that are responsible for communication between application processes. If the first two mentioned services access broadcast and unicast operations of the transport layer, communication between pure application processes is achieved via multicast service primitives of layer 4. The services of the application layer are presented in more detail in the following sections.

3.6.1 Communication Objects

The basis of common communication between two or more application processes are so-called communication objects. Objects in the sense of information technology have attributes (such as for example data type, access rights and access codes) and methods that define the operations that can be applied to these objects.

Communication objects are adapted from "conventional" information technology objects. They too comprise attributes and data. If one wishes to access the contents of a remote communication object, it is necessary to use certain services of layer 7. Viewed in an abstract way therefore, these objects can be compared with the common memory area of an EIB system. Depending on the above-lying application, the "global" memory area can either be read or written to. It is the task of the application layer to provide a suitable service that transports the content of a communication object via the network. As the consistency of the global memory area (to stay with the abstract view) must be maintained, this service must reach all devices that are interested in the corresponding communication object. How can we solve this tricky problem? The answer is quite easily by using group addresses to identify the communication objects! There is however one significant restriction – although a communication object can receive data via several group addresses, it can only send its data via one group address!

We will first examine the situation in which a communication object wishes to write data (via the bus). One of the tasks of the application layer is to now find the corresponding group address and to transmit the desired data with the help of the underlying transport layer. For this, layer 7 uses the so-called association table, which (simply stated) records the connection between communication object and group address. For the "write" case and as we already know, there can only be one such entry in the table per communication object as data can only be transmitted via a single group address.

We will now look at the arrival of data, which is targeted at the communication objects. Here, the application layer must use the association table to establish which communication

3.6 Application Layer

objects are concerned and with that ensure the data reaches each one in turn. Figure 3.35 illustrates these situations.

Association Table	
Group Address	Communication Object (SAP)
0/0/1	0
0/0/3	1
0/0/2	2
0/0/3	2

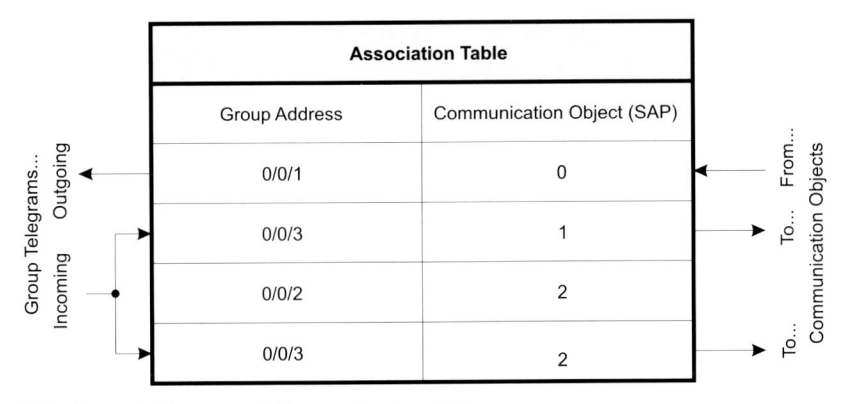

Figure 3.35: Group Addresses and Communication Objects

We will now turn our attention to the relevant services. The application layer provides the user (or more accurately an application process) with access to the communication objects. In order to be able to read a communication object, the application must use the A_Read_Group service, and to write to a specific communication object the A_Write_Group service. This is perhaps a little confusing and it would have been better to call the services A_Read_Commobj and A_Write_Commobj. We must however try to come to terms with the first named variants. We will now take a closer look at these two services.

The A_Read_Group service consists of an A_Read_Group.req request, an A_Read_Group.ind indication, an A_Read_Group.res response and an A_Read_Group.con confirmation. On the sender side, the A_Read_Group.req request activates the application layer. On the basis of the communication object (in the program fragment below referred to as SAP – the data type of an object is irrelevant for further discussions and for this reason has been called SAP_type) and the association table, layer 7 establishes which group address is relevant. A corresponding Read_Group_Req-PDU is then coded and the message transmitted to the local layer 4 with the T_Group service. The sender then waits for the arrival of data.

```
A_Read_Group.req_handler(SAP, class)
  SAP_type SAP;                              // communication object
  class_type class;                          // priority of the message
{
  char[MAXLEN] t_sdu;                        // layer 4 user data
  address destination;                       // group address

  destination=SAP2group(SAP);                // find group address
  t_sdu=encode(Read_Group_Req);              // code PDU
  T_Group.req(destination, class, t_sdu);    // multicast
}
```

On the receiver side, the application layer is activated by the arrival of T_Group.ind. After decoding the user data, the A_Read_Group indication is passed on to the communication object established in the association table.

3 EIB Protocol

```
T_Group.ind.receiver_handler(destination, class, t_sdu)
  address destination;                          // group address
  class_type class;                      // priority of the message
  char[MAXLEN] t_sdu;                         // layer 4 user data
{
  SAP_type SAP;                             // communication object
  switch (decode_pdu(t_sdu)) {
  case Read_Group_Req:
      SAP=group2SAP(destination);             // find group address
      A_Read_Group.ind(SAP, class);
      break;
  }
  /* to be completed later */
}
```

At least one remote user must now respond with A_Read_Group.res (which one has already been established in the configuration phase). With this response, the remote layer 7 receives the communication object data. A maximum of 14 octets can be transferred. Beforehand however, it is necessary to create a PDU, which is called Read_Group-PDU and is newly transmitted via the T_Group service primitive.

```
A_Read_Group.res_handler(SAP, class, data)
  SAP_type SAP;                             // communication object
  class_type class;                      // priority of the message
  char[14] data;                              // layer 7 user data
{
  char[MAXLEN] t_sdu;                         // layer 4 user data
  address destination;                          // group address

  destination=SAP2group(SAP);                // find group address
  t_sdu=encode(data, Read_Group_Res);           // code data+PDU
  T_Group.req(destination, class, t_sdu);          // multicast
}
```

Finally, this group telegram should arrive at the original sender. The local application layer accepts it with a corresponding handler. As soon as the response has been decoded and the relevant communication object identified, the data can be transferred to the user via the A_Read_Group.con service primitive.

```
T_Group.ind.sender_handler(destination, class, t_sdu)
  address destination;                          // group address
  class_type class;                      // priority of the message
  char[MAXLEN] t_sdu;                         // layer 4 user data
{
  SAP_type SAP;         // reference to the communication object
  char[14] data;        // data for the communication object

  if (decode_pdu(t_sdu) == Read_Group_Res) {
      SAP=group2SAP(destination);             // find group address
      data=decode_data(t_sdu);                    // extract data
      A_Read_Group.con(SAP, data);
  }
}
```

3.6 Application Layer

If one requires write access to a communication object, then the A_Write_Group service is used which comprises the A_Write_Group.req request, A_Write_Group.ind indication and local confirmation A_Write_Group.con.

After an A_Write_Group.req request has been directed at the application layer of the sender, then the corresponding group address must be found in the association table according to the communication object. A Write_Group_Req-PDU is then generated and with the T_Group.req operation the local layer 4 is activated. The parameters at hand-over are composed of group address and data (again a maximum of 14 octets can be transmitted at one time).

```
A_Write_Group.req_handler(SAP, class, data)
  SAP_type SAP;                           // communication object
  class_type class;                       // priority of the message
  char[14] data;                // data of the communication object
{
  char[MAXLEN] t_sdu;                     // layer 4 user data
  address destination;                    // group address

  destination=SAP2group(SAP);             // find group address
  t_sdu=encode(data, Write_Group_Req);    // code data+PDU
  T_Group.req(destination, class, t_sdu); // multicast
}
```

On the receiver side, the T_Group.ind indication of layer 4 activates the application layer. It is the task of the remote layer 7 to inform all communication objects associated with the group address of the arrival of new data (in the program fragment below the `for` loop attempts to perform this). Notification itself is achieved via the A_Write_Group.ind operation.

```
T_Group.ind.receiver_handler(destination, class, t_sdu)
  address destination;                    // group address
  class_type class;                       // priority of the message
  char[MAXLEN] t_sdu;                     // layer 4 user data
{
  SAP_type SAP;                           // communication object
  char[14] data;                // data for the communication object

  switch (decode_pdu(t_sdu)) {
  case Read_Group_Req:
      SAP=group2SAP(destination);         // find group address
      A_Read_Group.ind(SAP, class);
      break;
  case Write_Group_Req:
      data=decode_data(t_sdu);     // extract data and find SAPs
      for (;SAP=group2SAP(destination);SAP != NO_MORE_SAPs)
          A_Write_Group.ind(SAP, class, data);   // hand over data
      break;
  }
}
```

To finish with, a confirmation must be generated on the sender side. Contrary to the A_Read_Group service, this confirmation is local and is therefore derived from the confirmation of the local layer 4.

3 EIB Protocol

```
T_Group.con.sender_handler(destination, status)
  address destination;                           // group address
  status_type_layer2 status;// status information from layers 2,3+4
{
  SAP_type SAP;                 // reference to communication object

  SAP=group2SAP(destination);          // find communication object
  A_Write_Group.con(SAP, status);                    // convey status
}
```

Figure 3.36 is a simplified representation of the interaction of the individual service primitives.

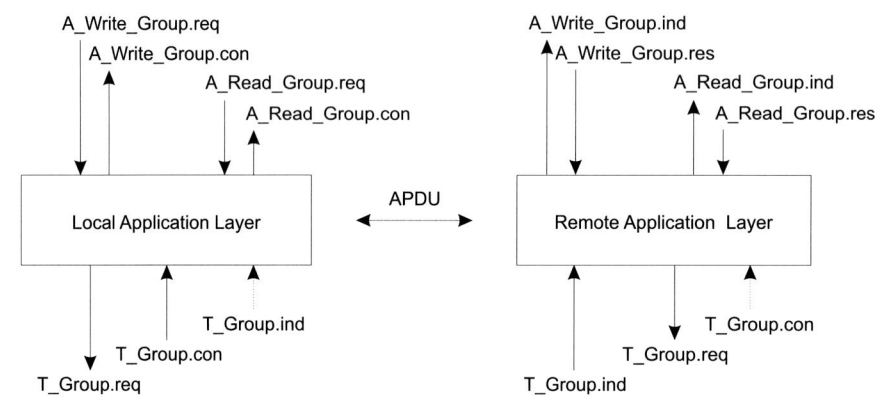

Figure 3.36: Services for Accessing Communication Objects

3.6.2 Device Configuration

The A_Set_PhysAddr service is called up by the user of the application layer when it becomes necessary to set or modify the physical address of an EIB device. The corresponding device must already have been switched to the programming mode. This is usually achieved by pressing a special button on the remote device.

This service consists of the A_Set_PhysAddr.req request, A_Set_PhysAddr.ind indication and local confirmation A_Set_PhysAddr.con. The request is mapped on to the T_Broadcast service of layer 4, which causes the sent message to reach all EIB devices. The telegram contains the user data of layer 7 which basically comprises the Set_PhysAddr_Req PDU and the new device address.

```
A_Set_PhysAddr.req_handler(newaddress)
  address newaddress;                         // new physical address
{
  char[MAXLEN] t_sdu;                           // layer 4 user data

  t_sdu=encode(newaddress, Set_PhysAddr_Req);           // code PDU
  T_Broadcast.req(system, t_sdu);                 // system message
}
```

3.6 Application Layer

In the devices connected to the EIB network the application layer is activated by the T_Broadcast.ind service primitive. After the PDU has been decoded, the prompt to reset the device address is called up via the A_Set_PhysAddr.ind operation.

It is the task of the remotely running application to check whether the programming key is pressed. If so, the device is set with the transmitted address.

```
T_Broadcast.ind.receiver_handler(source, class, t_sdu)
  address source;                        // physical address of sender
  class_type class;                      // priority of the message
  char[MAXLEN] t_sdu;                    // layer 4 user data
{
  address newaddress;                    // layer 7 user data

  switch (decode_pdu(t_sdu)) {                    // decode PDU
  case Set_PhysAddr_Req:                 // set device address PDU
      newaddress=decode_address(t_sdu);  // extract user data
      A_Set_PhysAddr.ind(newaddress);
      break;
  /* to be completed later */
  }
}
```

The application layer of the sender then passes on the incoming (local) confirmation originating from layer 4 to the user unchanged, via the A_Set_PhysAddr.con service primitive. The parameter type profile of this service primitive has the following structure:

A_Set_PhysAddr.con(status)
```
  status_type_layer2 status;    // status info from layers 2,3 and 4
```

Another possibility of setting device addresses is offered by the operations of the A_Set_PhysAddr_SerNr service primitive. There is no difference to the above mentioned A_Set_PhysAddr service primitive when it comes to the procedure of the underlying communication. This service primitive also consists of an A_Set_PhysAddr_SerNr.req request, local confirmation A_Set_PhysAddr_SerNr.con, as well as an A_Set_PhysAddr_SerNr.ind indication at the receiver. The only difference is that the communication partner is identified via a unique serial number composed of 6 bytes, which eliminates the pressing of the programming key. This serial number is transmitted throughout the EIB network together with the Set_PhysAddr_SerNr_Req PDU via broadcast. The parameter type profile of this service reads:

A_Set_PhysAddr_SerNr.req(serialnumber, newaddress)
```
  char[6] serialnumber;        // unique serial number of EIB device
  address newaddress;          // new physical address
```

A_Set_PhysAddr_SerNr.ind(newaddress)
```
  address newaddress;          // new physical address
```

A_Set_PhysAddr_SerNr.con(status)
```
  status_type status;          // status information from layer 4
```

A further service provides the user with the possibility of reading the physical address of a remote EIB device. This service consists of an A_Read_PhysAddr.req request, an A_Read_

3 EIB Protocol

PhysAddr.ind indication, an A_Read_PhysAddr.Res response and an A_Read_PhysAddr.con confirmation. We will now briefly outline the interaction of these operations.

To begin with, the user activates the application layer with the A_Read_PhysAddr.req request. As before, this request is mapped on to the T_Broadcast.req service primitive. The user data consists of a PDU which is now called Read_PhysAddr_Req. The message is sent as a telegram with the highest priority.

```
A_Read_PhysAddr.req_handler()
{
  char[MAXLEN] t_sdu;                          // layer 4 user data

  t_sdu=encode(Read_PhysAddr_Req);                        // code PDU
  T_Broadcast.req(system, t_sdu);                   // system message
}
```

On the receiver side, the application layer is activated by a T_Broadcast.ind indication. After successfully decoding the PDU, an indication is passed on to the user or application.

```
T_Broadcast.ind.receiver_handler(source, class, t_sdu)
  address source;                  // physical address of sender
  class_type class;                 // priority of the message
  char[MAXLEN] t_sdu;                   // layer 4 user data
{
  address newaddress;                   // layer 7 user data

  switch (decode_pdu(t_sdu)) {                    // decode PDU
  case Set_PhysAddr_Req:                   // set device address
      newaddress=decode_address(t_sdu);      // extract user data
      A_Set_PhysAddr.ind(newaddress);
      break;
  case Read_PhysAddr_Req:                  // set device address
      A_Read_PhysAddr.ind();
      break;
  }
}
```

If the programming key of the corresponding device is pressed, the device must react with the A_Read_PhysAddr.res response, which is intercepted on the receiver side with a corresponding handler. The task of this routine is simply to pack the transferred parameters (i.e. device address) into a broadcast telegram together with a new PDU, which is called Read_PhysAddr_Res. This telegram is transferred to the remote layer 4 with the T_Broadcast.req operation.

```
A_Read_PhysAddr.res_handler(addr)
  address addr;                                  // physical address
{
  char[MAXLEN] t_sdu;                          // layer 4 user data

  t_sdu=encode(addr, Read_PhysAddr_Res);      // code address+PDU
  T_Broadcast.req(system, t_sdu);                   // system message
}
```

On the sender side, the application layer is activated by T_Broadcast.ind. After decoding the PDU, this can then extract the device address from the user data, which is passed on to the

3.6 Application Layer

user via a confirmation. Note: Contrary to the confirmation of the A_Set_PhysAddr service primitive, this confirmation is caused as a result of the reaction of the remote communication partner!

```
T_Broadcast.ind.sender_handler(source, class, t_sdu)
  address source;                        // physical address of sender
  class_type class;                      // priority of the message
  char[MAXLEN] t_sdu;                    // layer 4 user data
{
  address addr;                          // layer 7 user data

  switch (decode_pdu(t_sdu)) {                       // decode PDU
  case Read_PhysAddr_Res:                // transmitted device address
      addr=decode_address(t_sdu);        // extract user data
      A_Read_PhysAddr.con(addr);
      break;
  }
}
```

As with setting physical addresses, there is also a service counterpart when reading that uses unique serial numbers to identify EIB devices. The corresponding service is called A_Read_PhysAddr_SerNr. In the course of communication, the Read_PhysAddr_SerNr_Req and Read_PhysAddr_SerNr_Res PDUs are exchanged.

As discussed, communication occurs between sender and receiver. It is enough therefore to list the parameter type profiles of the service primitives.

A_Read_PhysAddr_SerNr.req(serialnumber)
```
  char[6] serialnumber;    // unique serial number of the EIB device
```

A_Set_PhysAddr_SerNr.ind()

A_Set_PhysAddr_SerNr.res(addr)
```
  address addr;                                      // physical address
```

A_Set_PhysAddr_SerNr.con(addr)
```
  address addr;                                      // physical address
```

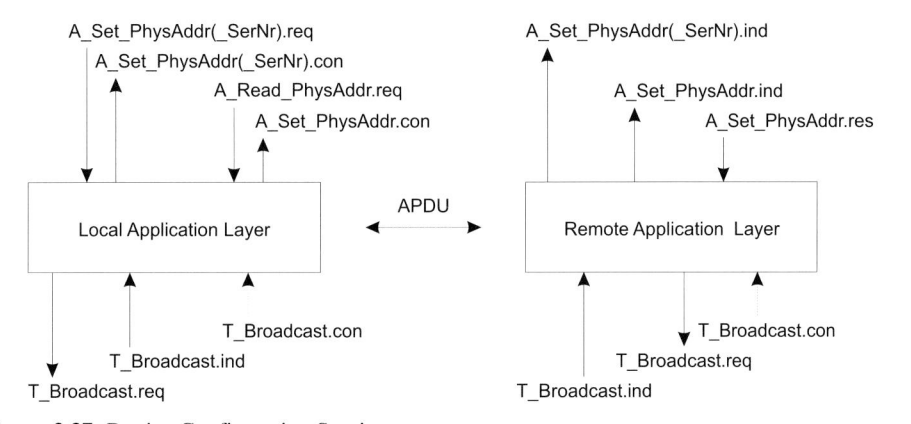

Figure 3.37: Device Configuration Services

133

3 EIB Protocol

Figure 3.37 illustrates the above-mentioned services of the application layer and the connection with the operations of the underlying transport layer.

Figure 3.37 does not show an additional service that can also be included in device configuration. This service allows service information to be read from a remote communication partner. Its name is A_Read_Service_Information. It too accesses the T_Broadcast service of layer 4. With its help, it is possible to check the current status of the remote user application, i.e. whether the application is currently active or not. Furthermore, this service can also be used to find physical addresses that exist more than once in the EIB network or to switch EIB devices to the monitoring mode.

3.6.3 Memory Access

With the A_Read_Memory or A_Write_Memory services it is possible to read or write up to 12 octets in the address area of a remote EIB device in one go. These acknowledged services are built on the connection-oriented communication of layer 4, which has been described in detail in chapter 3.5.1.

As soon as the user on the sender side has called up the A_Read_Memory.req or A_Write_Memory.req operation, the application layer is activated. The transmitted parameters firstly specify the physical address of the communication partner as well as the start address and number of data bytes to be read or written. If it is a new receiver, the application layer must now open a logical connection with this partner as the very first step. This is achieved with the T_Connect service of the transport layer.

If connection setup is successful, then with the help of the T_Data service, data can be exchanged between the two partners. For this, the application layer on the receiver side forms a Read_Memory_Req-PDU if reading or a Write_Memory_Req-PDU if writing.

On the receiver side, the user data is passed on to the application that is running there via indications, which if necessary cause the generation of corresponding responses. The responses are then packed into a Read_Memory_Res-PDU or Write_Memory_Res-PDU respectively. Once the sender has received all data and no further data transfer is required, it must terminate the connection in the proper manner. For this it uses the T_Disconnect service. Figure 3.38 highlights the situation in a clear and succinct way.

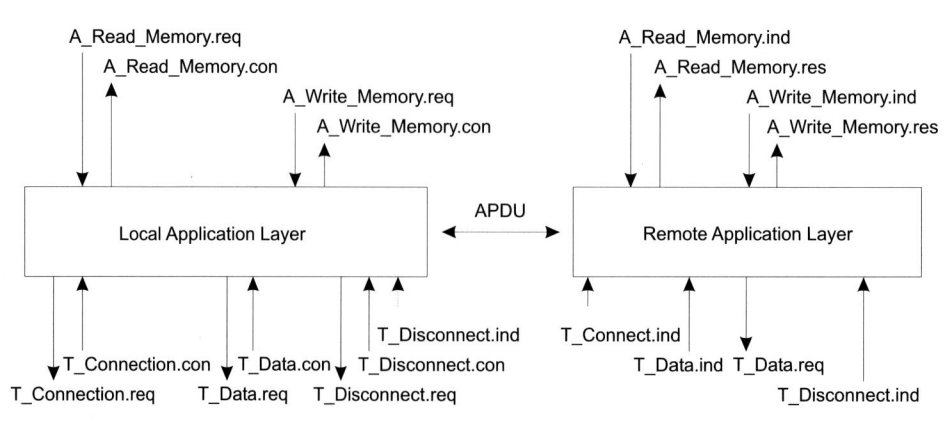

Figure 3.38: Services for Accessing the Memory

3.6 Application Layer

As we can see, calling up these services causes a series of other services to be activated in the underlying transport layer. This requires protocols, which must be implemented (transparent to the user) in the service primitives and corresponding handlers of the application layer. From time to time, these protocols can be quite complex. It is necessary for example to remember that on both the receiver and sender sides the connection can be interrupted without warning. We will therefore avoid going into a description of the implementation details and jump straight to the parameter type profiles of the service primitives.

The number parameter in the confirmation however requires a brief explanation. It usually specifies the quantity of read or written data. If memory access was not possible, this parameter is set to zero, which allows the application to detect that an error has occurred.

```
A_Read_Memory.req(destination, class, number, start)
   address destination;                    // address of receiver
   class_type class;                       // priority of the message
   char number;                            // number of data bytes to be read
   char[1] start;          // start address in address area of receiver

A_Read_Memory.ind(class, number, start)
   class_type class;                       // priority of the message
   char number;                            // number of data bytes to be read
   char[1] start;          // start address in address area of receiver

A_Read_Memory.res(class, number, start, data)
   class_type class;                       // priority of the message
   char number;                            // number of data bytes read
   char[1] start;          // start address in address area of receiver
   char[] data;                                      // read data bytes

A_Read_Memory.con(class, number, start, data)
   class_type class;                       // priority of the message
   char number;                            // number of data bytes read
   char[1] start;          // start address in address area of receiver
   char[] data;                                      // read data bytes

A_Write_Memory.req(destination, class, number, start, data)
   address destination;                    // address of receiver
   class_type class;                       // priority of the message
   char number;                  // number of data bytes to be written
   char[1] start;          // start address in address area of receiver
   char[] data;                              // data bytes to be read

A_Write_Memory.ind(class, number, start, data)
   class_type class;                       // priority of the message
   char number;                  // number of data bytes to be written
   char[1] start;          // start address in address area of receiver
   char[] data;                           // data bytes to be written

A_Write_Memory.res(class, number, start, data)
   class_type class;                       // priority of the message
   char number;                     // number of data bytes written
   char[1] start;          // start address in address area of receiver
   char[] data;                          // data bytes actually written

A_Write_Memory.con(class, number, start, data)
```

3 EIB Protocol

```
class_type class;           // priority of the message
char number;                // number of data bytes written
char[1] start;      // start address in address area of receiver
char[] data;                // data bytes actually written
```

3.6.4 Analog/Digital Converter Access

The application layer also offers a service for access to potential A/D converters in EIB devices. This service is called A_Read_Adc and is also based on the connection-oriented communication of layer 4.

After the application layer on the sender side has received a request via A_Read_Adc.req, which specifies the desired receiver, channel number of the A/D converter and the number of conversions to be carried out, the first step is to set up a connection (if not yet done so). This is achieved with the T_Connect service of the transport layer.

Once the user data has arrived on the receiver side, the application layer there generates the A_Read_Adc.ind indication, which causes the application to execute the conversions on the corresponding channel. If necessary, the application responds with the average value of the established A/D conversions and returns this with the A_Read_Adc.res response.

On the sender side, the established value is then passed on to the user of the application layer. If the receiver requires no further data, the connection is terminated with the T_Disconnect.req service primitive. If an error occurred during the conversion, the repeat parameter of the confirmation is set to zero.

```
A_Read_Adc.req(destination, class, channel_nr, repeat)
  address destination;                // address of receiver
  class_type class;                   // priority of the message
  char channel_nr;                    // required channel number
  char start;         // number of conversions to be carried out

A_Read_Adc.ind(class, channel_nr, repeat)
  class_type class;                   // priority of the message
  char channel_nr;                    // required channel number
  char start;         // number of conversions to be carried out

A_Read_Adc.res(class, channel_nr, repeat, sum)
  class_type class;                   // priority of the message
  char channel_nr;                    // channel number
  char start;         // number of conversions to be carried out
  char[1] sum;                        // averaged value

A_Read_Adc.con(class, channel_nr, repeat, sum)
  class_type class;                   // priority of the message
  char channel_nr;                    // channel number
  char start;             // number of conversions carried out
  char[1] sum;                        // averaged value
```

3.6.5 Property Access

In addition to the communication objects there are also so-called EIB objects that are of particular interest to network management. We will discuss this type of object in the following chapter in more detail. In the meantime, it is enough to say that we differentiate between

3.6 Application Layer

system and application objects. System objects include device objects, address table objects, association table objects and application program objects. EIB objects also have special attributes that are hereafter referred to as properties. In the case of a device object for example, the properties may be the serial number, manufacturer code and order code. For application objects the user is free to define the properties as required.

The application layer offers services that allow us to determine the structure of properties (A_Read_Property_Description), to read them (A_Read_Property_Value) and – if necessary – also to write them (A_Write_Property_Value). Access to properties is either connectionless or connection-oriented. This depends on the implementation. In order to show how access to such properties can be gained, we will highlight the example of the A_Read_Property_Value service and explain its functionality when messages are transported using connectionless communication.

The A_Read_Property_Value service consists of an A_Read_Property_Value.req request, an A_Read_Property_Value.ind indication, an A_Read_Property_Value.res response and an A_Read_Property_Value.con confirmation. As usual, the first step is the arrival of the A_Read_Property_Value.req request at the application layer. The request parameters comprise the physical address of the communication partner (it involves connectionless communication between one sender and one receiver) and the desired message priority with which the telegram is to be transmitted. In addition, the user must now specify which property he wishes to read as well as the start index and the number of elements of the corresponding property (until now we have failed to mention that a property can consist of a whole series of elements of the same data type).

The request must be packaged into a suitable telegram. For this, layer 7 must generate a special PDU that we have named Read_Property_Value_Req. Afterwards, layer 4 is activated with the T_Data_Unack connectionless service.

```
A_Read_Property_Value.req_handler(destination, class, object_id,
                                  property_id, number, start)
    address destination;                    // physical address
    class_type class;                       // priority of the message
    unsigned int object_id;                 // object identification
    unsigned int property_id;               // property identification
    char number;                            // number of values to be read
    char start;                             // start index
{
    char[MAXLEN] t_sdu;                     // layer 4 user data

    t_sdu=encode(object_id, Property_id, number, start,
            Read_Property_Value_Req);       // code PDU
    T_Data_Unack.req(destination, class, t_sdu);  // unicast
}
```

On the receiver side, the routine that is responsible for managing connectionless communication decodes the PDU and passes on the request to the application.

```
T_Data_Unack.ind(source, destination, class, t_sdu)
    address source;                         // physical address of sender
    address destination;                    // physical address of receiver
    class_type class;                       // priority of the message
    char{MAXLEN] t_sdu;                     // layer 4 user data
{
```

137

3 EIB Protocol

```
    unsigned int object_id;              // object identification
    unsigned int property_id;            // property identification
    char number;                  // number of values to be read
    char start;                                    // start index
    switch (decode_pdu(t_sdu))
    case Read_Property_Value_Req:
        object_id=decode_object_id(t_sdu);   // decode object number
        property_id=decode_property_id(t_sdu);    // decode property
        number=decode_number(t_sdu);                // decode number
        start=decode_start(t_sdu);             // decode start index
        A_Read_Property_Value.ind(source,destination, class,
                           object_id, property_id, number, start);
        break;
    }
}
```

The application must now generate a response that contains the desired data. If it is not possible to access the desired property (or if this is not permitted due to insufficient access rights), the `number` value that represents the quantity of returned data is set to zero. It is the task of the remote application layer to provide the parameter with a suitable PDU (Read_Property_Value_Res) and to pass it on to the transport layer. Transfer to the sender however occurs via connectionless communication.

Having arrived at the sender, the data pieces are extracted from the user data. The original requester is informed of the arrival of the message via a confirmation. The remaining service primitives have the following parameter type profiles:

```
A_Read_Property_Value.res(destination, class, object_id,
                     property_id, number, start, data)
    address destination;                        // physical address
    class_type class;                    // priority of the message
    unsigned int object_id;              // object identification
    unsigned int property_id;            // property identification
    char number;                         // number of read values
    char start;                                    // start index
    char data[];                                     // read data

A_Read_Property_Value.con(destination, class, object_id,
                     property_id, number, start, data)
    address destination;                        // physical address
    class_type class;                    // priority of the message
    unsigned int object_id;              // object identification
    unsigned int property_id;            // property identification
    char number;                         // number of read values
    char start;                                    // start index
    char data[];                                     // read data
```

At the beginning we mentioned that if necessary properties can also be written. The necessary communication basically follows the same procedure, except that it is the Write_Property_Value_Req and Write_Property_Value_Res PDUs that are exchanged between the partners.

With the A_Read_Property_Description service it is even possible to read entire property descriptions. This requires the object number and desired property index to be specified on the sender side. This service too can be mapped onto connectionless or connection-oriented communication.

Figure 3.39 shows all property services and their connections when accessing via connectionless communication.

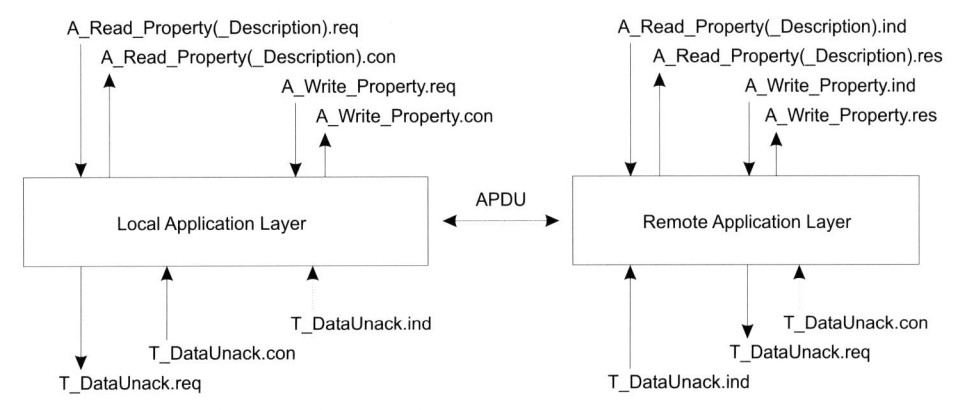

Figure 3.39: Services for Accessing Properties

3.6.6 Security

Security is becoming an increasingly important subject for fieldbus systems. The application layer should therefore offer a minimum amount of associated services. In the EIB system, security is considered in terms of the approach. The application layer does offer the possibility of setting a code for accessing an EIB device (or more accurately speaking its memory area or properties) and afterwards of carrying out authentication. However, this code is transmitted in its original form, i.e. it is not encoded. The reader should therefore be aware that such telegrams can easily be traced by listening in on the medium. This means that data areas which should be protected are easily accessed. In order to protect against such attacks, suitable coding algorithms should be used in the above-lying application.

After this short discussion on the subject of security, we would now like to look at the application services that are actually available. These are A_SetKey and A_Authorize. Both services use connection-oriented communication of layer 4. Every EIB device can manage up to 255 different keys each of which are associated with 255 different access levels (priorities). A key consists of 4 octets. The access code (key) and level must have previously been saved in the device.

With the A_SetKey service a user can newly select the access level with which a key is connected, whereby the new access level can only be less than or equal to the current priority. On the sender side, the application layer packs an A_SetKey.req request into a Set_Key_Req-PDU with the corresponding code and transmits this to the receiver after the connection has been set up successfully.

The application layer there generates the A_SetKey.ind indication. There is then a check to see whether the code is valid and whether the new access level can be set. If so, the remote application layer returns a Set_Key_Res-PDU after receiving the A_SetKey.res response, which contains the new access level.

3 EIB Protocol

On the sender side, the user is informed via an A_SetKey.con confirmation. If the receiver does not grant the new access level, the `level` parameter is set to FFh.

```
A_SetKey.req(destination, class, level, key)
    address destination;                    // address of the receiver
    class_type class;                       // priority of the message
    char level;                             // desired access level
    char[4] key;                            // access code

A_SetKey.ind(class, level, key)
    class_type class;                       // priority of the message
    char level;                             // desired access level
    char[4] key;                            // access code

A_SetKey.res(class, level, key)
    class_type class;                       // priority of the message
    char level;                             // actual access level
    char[4] key;                            // access code

A_SetKey.con(class, level, key)
    class_type class;                       // priority of the message
    char level;                             // granted access level
    char[4] key;                            // access code
```

To enquire of a remote communication partner whether a code is valid and which access level it grants, it is necessary to use the A_Authorize service. This service consists of the A_Authorize.req request, A_Authorize.ind indication, A_Authorize.res response and A_Authorize.con confirmation. The remote running application can therefore establish whether the sender is able to read data from the memory area and/or write data to it. On the basis of this, other services can be implemented which are then of interest to the network management. Within the course of the communication, the service primitives use PDUs known as Authorize_Req (in the direction of sender to receiver) and Authorize_Res (in the direction of receiver to sender). If the code is not known to the receiver, then it returns access level FFh in its response, which represents the lowest priority.

```
A_Authorize.req(destination, class, key)
    address destination;                    // address of the receiver
    class_type class;                       // priority of the message
    char[4] key;                            // access code

A_Authorize.ind(class, key)
    class_type class;                       // priority of the message
    char[4] key;                            // access code

A_Authorize.res(class, key)
    class_type class;                       // priority of the message
    char[4] key;                            // access code

A_Authorize.con(class, level)
    class_type class;                       // priority of the message
    char level;                             // granted access code
```

3.6.7 Other Services

It only remains to mention the remaining layer 7 services. These include a service to read the mask version in a communication partner (A_Read_Mask_Version). Another allows the communication controller to be reset in the partner (restart). Both services are connection-oriented. In the first case, Read_Mask_Req and Read_Mask_Res-PDUs are exchanged and in the second communication occurs via a Restart_Req-PDU.

There are also the optional A_UserData services. With their help it is possible to activate secondary device management operations. For example, it is possible to read and write remote, external memories, which means that it is possible to access the applications that are running there.

And last but not least, we should mention the services that allow the line couplers and backbone couplers to be programmed. This includes (among others) services to read (A_Read_Routing_Table) and write (A_Write_Routing_Table) the routing tables contained therein. They too use separate PDUs and the connection-oriented communication of the transport layer.

3.6.8 Application Layer Protocol Data Unit (APDU)

This concludes our discussion of the most important services of layer 7. In our deliberations we have taken care to incorporate the interaction of the service primitives with the possibilities of the underlying communication (multicast, broadcast and unicast). What we have as yet failed to do is present the *application layer protocol data units*, APDU, that are exchanged in the respective services.

Every protocol data unit consists of layer 7 specific control information (shortened to AL control field), which is coded after the control information of layer 4 and depending on the underlying service is either 4 or 10 bits long. This is followed by the actual data, which strictly speaking is the user data of the application layer (*application layer service data unit*, ASDU).

With a look at Figure 3.40 we would like to conclude our discussion of the EIB protocol. This diagram illustrates the different PDU formats and at the same time indicates how the communication methods are linked to the corresponding services. The abbreviations used in the diagram are defined as follows: B for broadcast, M for group addressing, CO for connection-oriented and CL for connectionless communication.

Those who have persevered this far have certainly earned a break before starting on the next chapter.

3 EIB Protocol

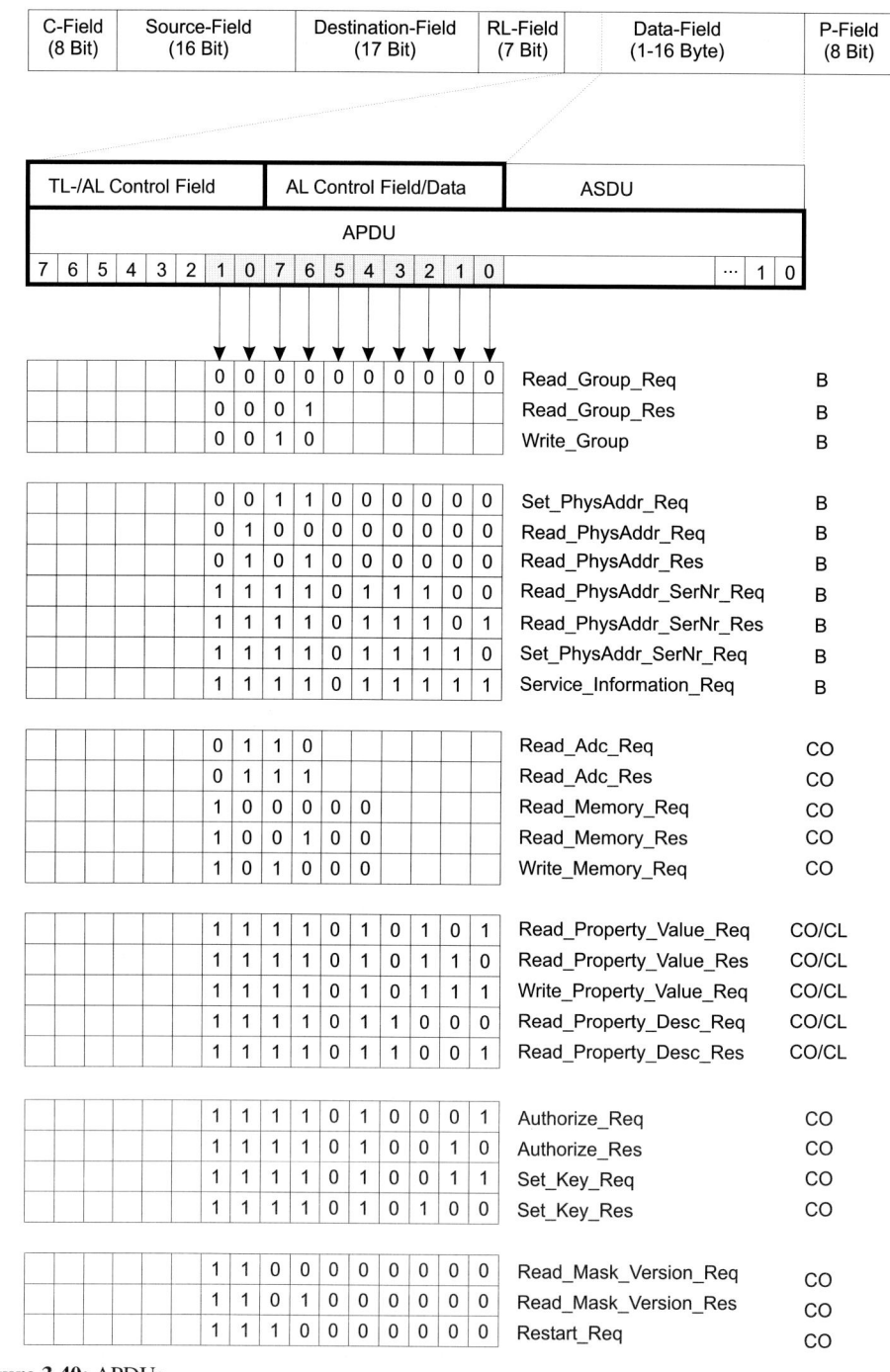

C-Field (8 Bit)	Source-Field (16 Bit)	Destination-Field (17 Bit)	RL-Field (7 Bit)	Data-Field (1-16 Byte)	P-Field (8 Bit)

TL-/AL Control Field | AL Control Field/Data | ASDU

APDU bits: 7 6 5 4 3 2 1 0 | 7 6 5 4 3 2 1 0 | ... 1 0

APDU bits	Service	Type
0 0 0 0 0 0 0 0 0 0	Read_Group_Req	B
0 0 0 1	Read_Group_Res	B
0 0 1 0	Write_Group	B
0 0 1 1 0 0 0 0 0 0	Set_PhysAddr_Req	B
0 1 0 0 0 0 0 0 0 0	Read_PhysAddr_Req	B
0 1 0 1 0 0 0 0 0 0	Read_PhysAddr_Res	B
1 1 1 1 0 1 1 1 0 0	Read_PhysAddr_SerNr_Req	B
1 1 1 1 0 1 1 1 0 1	Read_PhysAddr_SerNr_Res	B
1 1 1 1 0 1 1 1 1 0	Set_PhysAddr_SerNr_Req	B
1 1 1 1 0 1 1 1 1 1	Service_Information_Req	B
0 1 1 0	Read_Adc_Req	CO
0 1 1 1	Read_Adc_Res	CO
1 0 0 0 0 0	Read_Memory_Req	CO
1 0 0 1 0 0	Read_Memory_Res	CO
1 0 1 0 0 0	Write_Memory_Req	CO
1 1 1 1 0 1 0 1 0 1	Read_Property_Value_Req	CO/CL
1 1 1 1 0 1 0 1 1 0	Read_Property_Value_Res	CO/CL
1 1 1 1 0 1 0 1 1 1	Write_Property_Value_Req	CO/CL
1 1 1 1 0 1 1 0 0 0	Read_Property_Desc_Req	CO/CL
1 1 1 1 0 1 1 0 0 1	Read_Property_Desc_Res	CO/CL
1 1 1 1 0 1 0 0 0 1	Authorize_Req	CO
1 1 1 1 0 1 0 0 1 0	Authorize_Res	CO
1 1 1 1 0 1 0 0 1 1	Set_Key_Req	CO
1 1 1 1 0 1 0 1 0 0	Set_Key_Res	CO
1 1 0 0 0 0 0 0 0 0	Read_Mask_Version_Req	CO
1 1 0 1 0 0 0 0 0 0	Read_Mask_Version_Res	CO
1 1 1 0 0 0 0 0 0 0	Restart_Req	CO

Figure 3.40: APDUs

4 Application Environment and Network Management

The two topics of application environment and network management are so closely linked that we have included them in the same chapter. The first part of the chapter (application environment) describes how applications can access the EIB and the second part (network management) deals with the functioning of special applications that are used to commission, configure and in a further stage to manage EIB systems.

4.1 Application Environment

Communication objects form the basis for the communication between two or more applications within EIB. The network management is based on uniform structures called EIB objects. We will begin our deliberations with a discussion on which unit is ultimately responsible for the transmission of communication objects, followed by an outline of the set-up and transport of EIB objects. Once this preparatory work has been completed we will then examine the subject of network management [EIBA 95, EIBA 98].

4.1.1 Communication Objects

It is necessary to keep in mind that applications primarily communicate via an exchange of communication objects. In section 3.6.1 we have already indicated that communication objects are comparable with those that determine the object-based paradigm of information technology. They consist of a set of attributes and associated methods that facilitate access to the current values of the attributes (in the case of communication objects this is mainly their content).

In principle, we could view communication objects as part of the application, or part of the application layer. It is purely a question of taste. In order to solve this conflict, it is better to assign communication objects to neither the application nor the application layer but to a separate layer that we simply place above layer 7 of the EIB protocol and which is called the user layer (Figure 4.1).

In addition to a description and value, every communication object consists of a set of communication flags, which in EIB terminology (for reasons of implementation) are often referred to as RAM flags and which describe the current status of the object. How these RAM flags are actually implemented does not interest us at this point, a description is given in section 6.1.1. Here, we will concentrate on the abstract interface definition only. RAM flags can assume one of the following values:

4 Application Environment and Network Management

1. The `read-request` status indicates to the user layer that a communication object is to be read.
2. The `write-request` status indicates to the user layer that a communication object is to be written.
3. The `transmitting` status indicates to the application that a communication object is currently being transmitted.
4. The `update` status indicates to the application that a communication object has been received.
5. The `error` status indicates to the application that an error has occurred during the transmission of a communication object.

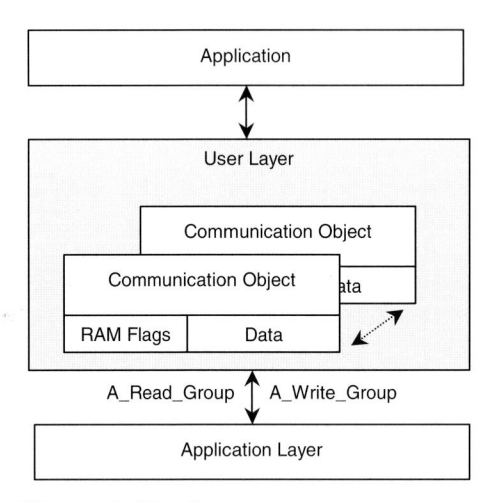

Figure 4.1: User-Layer

When we subsequently refer to the reading/writing/transmission of a communication object, we mean of course the reading/writing/transmission of the associated data.

Imagine that an application in end device 1 wishes to poll a remote communication object that is contained in the user layer of end device 2. By setting the RAM flag associated with the communication object, the application activates the local user layer, which then formulates a request with the services of the application layer (A_Read_Group, A_Write_Group). On the reception side, the remote user layer is informed of the request for a specific communication object by the corresponding service primitive. In the case of a read request, the object is packed into a message, which is then returned to the original service requester with the help of the application layer services. Note: This task is only undertaken by the remote user layer, the remote active application has nothing to do with it!

If we look at this situation from the point of view of the *client/server model*, oft-used within the field of information technology, then the applications that request access to communication objects always act as clients, and the user layer, in which the relevant communication objects exist, always acts as the server.

After this short explanation on the principal functioning of the user layer, it is important for later discussions to take a brief look at two possible EIB implementations in order to better

144

describe the interface between client and server. In this context we use the term implementation with regard to the hardware of an EIB end device and not the underlying medium. Other such implementations are examined in more detail in chapter 5. This is simply a taster.

EIB end devices may consist of a single processor only, which manages communication and has space for an (internal) application, or they can consist of two processors, of which one is responsible for communication (bus coupling unit), the other for the external application (application module).

We will begin with a look at the first variant in which the application and EIB protocol stack run on one and the same processor. It is possible in principle that both the application and protocol stack (or more accurately the application layer) access one and the same memory area, which also contains the communication objects. The definition of the user layer is useful in that it offers a consistent means of storing the communication objects and at the same time releases the application from the tasks of communication.

We will now take a more detailed look at the functioning of the user layer, for the two application cases of reading and writing a communication object.

If the application wants to read a communication object (Figure 4.2) it sets the corresponding RAM flags to the value `read-request` (1). On the transmission side, the user layer continuously monitors its status (2) – without wanting to make a statement on the implementation of the user layer, we will name this process polling. If the user layer sees that the current status has changed to `read-request`, it sets the corresponding RAM flags to `transmitting`. In doing so, it informs the application that it is now requesting the desired communication object via the network. For this, the local user layer uses the A_Read_Group service of the application layer and generates the A_Read_Group.req request (3). The application layer puts together a special layer 7 protocol data unit (APDU) and transmits this in a group telegram.

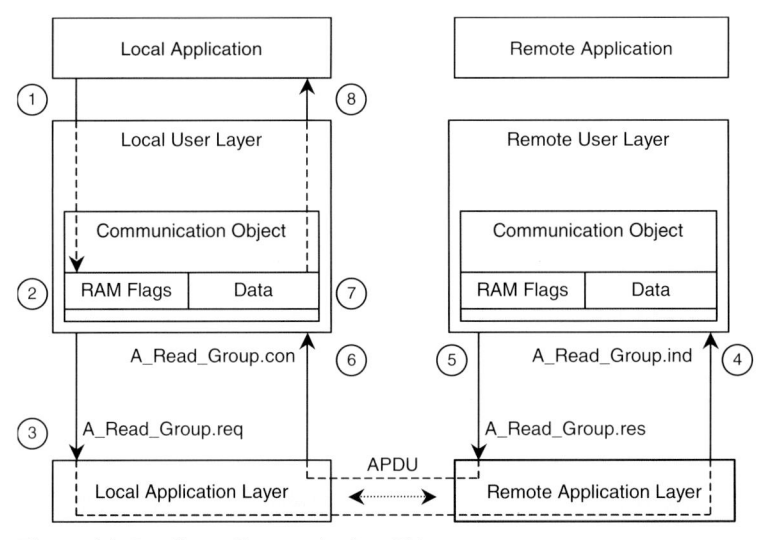

Figure 4.2: Reading a Communication Object

4 Application Environment and Network Management

On the reception side, the user layer is activated by the A_Read_Group.ind indication (4). With that, it grabs the desired communication object, packs it into a reply message and uses the A_Read_Group.res service primitive (5). The remote, active application layer then generates a corresponding APDU, which ultimately arrives at the transmission side and is then passed over to the local user layer via the A_Read_Group.con service primitive (6). This accepts the incoming communication object and signals its arrival to the application by changing the RAM flags to the value `update` (7). The application can now access the communication object (8). If an error occurred during access to the communication object (e.g. transmission failed), the user layer is informed by means of a negative acknowledgement, which causes it to set the RAM flags to the value `error`.

We will now turn our attention to the writing of a communication object (Figure 4.3). To begin with, the application must set the associated RAM flags to the value `write-request` (1). On the transmission side, the user layer detects the status change by continuous polling (2). It changes the RAM flags to the value `transmitting`, and with that informs the application that it is now attempting to gain write access to the desired communication object via the network. The A_Write_Group service of the application layer is now used and the A_Write_Group.req request is called up (3). The application layer puts together a special layer 7 protocol data unit (APDU) and transmits it via a group telegram. If it was possible to translate this service primitive into services of the underlying transport layer, the application layer transmits the A_Write_Group.con confirmation derived from layer 4 (we have referred to such acknowledgements as local confirmations in chapter 3). The user layer now resets the RAM flags (4). If the service request (local) could not be translated, the user layer receives a negative confirmation and it sets the RAM flags to the value `error`.

On the reception side, the user layer is activated via the A_Write_Group.ind indication (5). The remote user layer accepts the incoming data, adapts the relevant communication object and signals the arrival of new values by setting the RAM flags to the value `update` (6). During the next access to the communication object (or its RAM flags) this fact leads the application to conclude that new data has arrived (7).

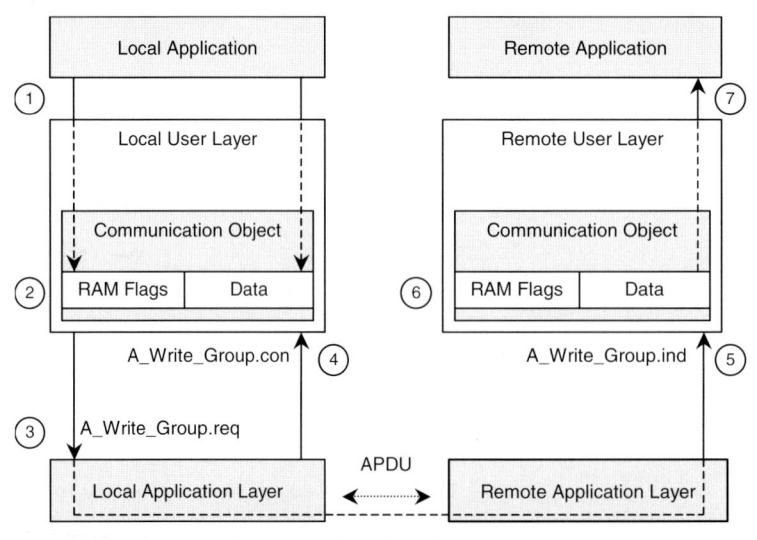

Figure 4.3: Writing a Communication Object

4.1 Application Environment

In the discussed hardware variant, the user layer is often implemented as part of the operating system. The application programmer can initiate the transmission of communication objects by setting the corresponding communication flags, which are located at specific points within the RAM. The necessary instructions can be formulated either directly using machine code commands or more comfortably via functions, which are summarized within an application programming interface, API. Listing the commands makes little sense as they always depend on the structure of the processor. The interested reader should consult the relevant literature. Chapter 6 contains a brief overview of APIs tailored to specific operating systems.

The introduction of a level that lies over the application layer and which is concerned with the transportation of communication objects is of course also beneficial to EIB end devices that consist of separate communication and application modules. With this hardware design it is not possible to assume a common memory area available to both processors. This means that the application, which is executed on the application module, must first contact the bus coupling unit on which the EIB protocol implementation is running as soon as it wants to access a communication object, before the desired communication object can be requested via the bus. The user layer now re-regulates the data flow originating from the application and converts it into service primitives that are understood by the application layer of the EIB protocol. This firstly relieves the load on the application as it need not concern itself with the correct access to the communication objects and on the other hand, does not overload the application layer of the EIB protocol. Again, the user layer acts as the communication object server!

To round off with, we will define the interface to this layer in accordance with the previously presented EIB protocol, followed up by a list of services (Figure 4.4).

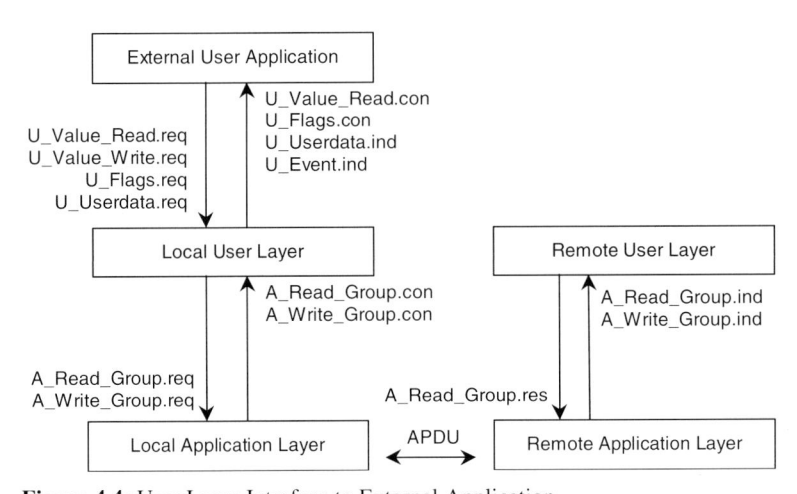

Figure 4.4: User Layer Interface to External Application

- The U_Value_Read service, consisting of the U_Value_Read.req request and the U_Value_Read.con reply, allows a communication object to be read.

- The U_Value_Write service, consisting of U_Value_Write.req request, allows a communication object to be written.

147

- The U_Flags_Read service, consisting of U_Flags_Read.req request and U_Flags_Read.con local confirmation, allows the status of the RAM flags to be polled.
- The U_Event.ind service primitive informs the application that the status of a communication object (or more accurately the associated RAM flag) has changed.
- U_Userdata services contain requests and indications that are not directed at communication objects but at a remote application module. They are passed on to the application layer unchanged by the user layer.

4.1.2 EIB Objects

As long as the mechanism is supported, EIB objects offer end devices a new method of communication. It is necessary to note that the exchange of EIB objects (or the data stored within) always occurs between two dedicated end devices whilst the transfer of communication objects always (under normal conditions) concerns groups of end devices. For this reason, services of the connectionless or connection-oriented communication are used when accessing EIB objects, which as we already know are used to control the data transport between two end devices, and when accessing communication objects, multicast services of the EIB protocol are used.

Subsequent references to the exchange of EIB objects indicate mechanisms for communicating with end devices that are to be configured. With that, EIB objects form one of the basic requirements for modern network management.

We will now turn our attention to the structure of EIB objects. They too, with certain restrictions, can be compared with "regular" information technology objects. Every EIB object typically consists of a series of so-called properties. A property is best thought of as an array, which contains descriptive elements in addition to the data. The descriptive elements comprise a unique property identification number, the data type of the array and an attribute that determines the access rights (access level) to the property. In the property itself it is possible to store not only a value but also a whole set of data of the previously defined type. For this reason, a further descriptive element is the maximum number of allowed values the property can contain. The current number of values momentarily stored in the array would also be a meaningful descriptive element. This value however is stored in the first element of the array. Figure 4.5 outlines the structure of an EIB object in a way that is easy to understand.

If an EIB end device supports EIB objects, there are certain objects that are stipulated, so-called EIB system objects. These objects are: device object, address table object, association table object and application program object. As we imagined the user layer as the server for communication objects, we must now imagine part of the network management as the server for EIB system objects. This server can also be partially implemented in the user layer. More on this in the next chapter.

It only remains to be noted that there can be other servers in the user layer, which control access to specific elements of an end device and react accordingly to special service primitives of the application layer. We presented the most important of these services in chapter 3.6 and concisely noted that with the arrival of an indication "the application must react with the desired response". With our current level of knowledge we can now elaborate on this. It need not be an actual application that returns the desired result, this task can actually be carried out by a special server of the user layer.

```
                    EIB Object
            ┌─────────────────────────┐
            │  Property               │
            │    Description          │
            │      char propertyID;   │
   Property │      char type;         │
            │      char access;       │
   Description       int max;         │
       char propertyID;               │
       char type;                     │
       char access;    type[0...max] value;
       int max;                       │
                                      │
   Value                              │
       type[0...max] value;           │
            └─────────────────────────┘
```

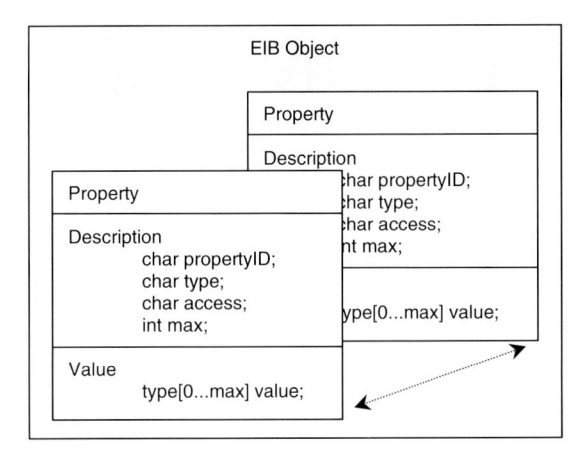

Figure 4.5: EIB Object

The mentioned services are those which facilitate access to memory (see chapter 3.6.3) or analog/digital converter (see chapter 3.6.4), or those that are capable of reading or writing properties or transporting their descriptions (see chapter 3.6.5). With this extended view of the user layer, this section is complete.

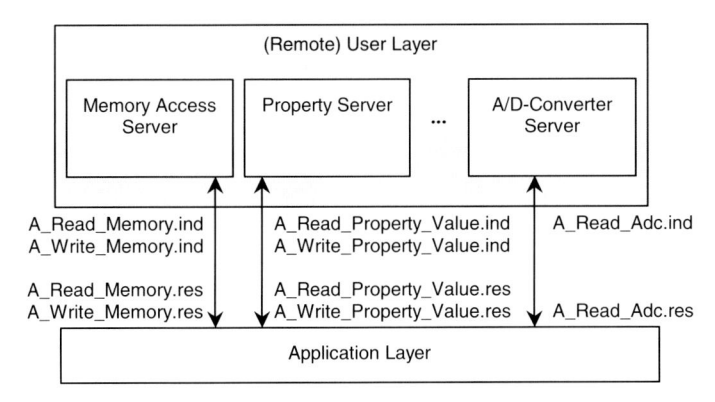

```
┌──────────────────────────────────────────────────────┐
│               (Remote) User Layer                    │
│  ┌──────────┐   ┌──────────┐      ┌──────────────┐   │
│  │ Memory   │   │ Property │ ...  │ A/D-Converter│   │
│  │ Access   │   │  Server  │      │   Server     │   │
│  │ Server   │   │          │      │              │   │
│  └──────────┘   └──────────┘      └──────────────┘   │
└──────────────────────────────────────────────────────┘
        ↑↓              ↑↓                   ↑↓
A_Read_Memory.ind    A_Read_Property_Value.ind    A_Read_Adc.ind
A_Write_Memory.ind   A_Write_Property_Value.ind

A_Read_Memory.res    A_Read_Property_Value.res
A_Write_Memory.res   A_Write_Property_Value.res   A_Read_Adc.res
┌──────────────────────────────────────────────────────┐
│                  Application Layer                    │
└──────────────────────────────────────────────────────┘
```

Figure 4.6: User Layer Server

4.2 Network Management

The tasks to be fulfilled by the network management are many and varied, e.g. incorporating new end devices into an existing network and combining them with other end devices to form functional units. In addition to carrying out these fundamental duties, a modern network management also offers mechanisms for the analysis and diagnosis of systems that are already up and running.

4.2.1 Introduction

Network management was not directly planned in the original ISO/OSI model. It is to be understood as existing in parallel to the OSI layers as it affects all of them:

- *Physical and data link layers:* All end devices must, for example, have the same channel configuration. This involves certain parameters that determine transmission on the underlying medium. In this connection, the applicable bit rate is of particular interest. Network management also concerns the repeaters, that operate on layer 1, and bridges, that are implemented in layer 2, because they, depending on their design, can also be used to link sub-networks with different channel configurations.

- *Network layer:* Every end device (and every group of end devices) must be provided with a unique address. With the help of intelligent couplers (routers), found in layer 3 of the ISO/OSI model, it is possible to form sub-networks and with that reduce the network load. These devices too can be contacted and then configured by the commands made available through the network management (the loading of new routing tables).

- *Transport and session layers:* These layers are mainly responsible for the service quality that is offered by the underlying network. Corresponding configuration possibilities, which can be carried out with the network management, are therefore closely linked to the term "quality".

- *Presentation layer:* Common syntax is a basic prerequisite for interoperability. The task of the network management in this level therefore, could be to inform two communicating end devices of the syntax of the matching data types.

- *Application layer:* Here, network management is concerned with the carrying out of tasks that relate to applications. It should be possible to load applications and define their specific configurations, but also to load, read and write tables that describe the communication relationships of applications.

EIB network management does not follow the approach of the ISO/OSI model, instead it is based on the *client/server model*. With the sending of network management messages, a client (e.g. a network management tool) generally prompts an end device to execute a particular administrative function. A server must be implemented in the addressed device, which is ready to receive this request in order to carry out the specified task. Network management messages can also, in special circumstances, be addressed to all end devices, in which case the entire EIB system is to be viewed as one large server. Generally however, the connection-oriented or connectionless communication of the EIB protocol is used to manage the transmission of data between a client and a server (or a dedicated end device).

We can imagine this network management server as a separate layer (over the application layer) or even as part of the user layer. Its task is to analyze incoming requests and to carry out the task contained therein. Similar to how the user layer encapsulates communication objects, the network management (or network management server) now manages configuration objects. In many cases, these configuration objects correspond to the briefly mentioned EIB system objects – device object, address table object, association table object and application program object. The information contained by these various EIB system objects is examined in the following chapter.

There are specific protocols to be observed on both the client and server sides that come into play after calling up a network management command. This means that after using a network management command on the client side, several messages are sent which on the server side need to be answered correctly in accordance with the given rules.

So much for the theory, now for the practice! The network management services provided in EIB can be roughly split into two categories:

1. Network configuration services facilitate the allocation of physical addresses or the scanning of the network for devices with specific properties.

2. Device configuration services are used to configure the parameters of a specific end device. However, it is only the form and not the content of the sent message that is established. How a specific end device is to be configured must first be made known to the network management tool.

There is another differentiation to point out with regard to the individual network management services. The selection of service must be made with regard to the medium that transmits the configuration message. On the one hand, network management messages can be transmitted via the EIB in order to contact connected devices, and on the other hand they can be passed on to the local bus coupling units via the external message interface (EMI). If a network management command is executed not locally but via the network, then either the broadcast addressing is used, or a node is contacted via its physical address. Group addressing is not supported by the network management, it is reserved for the communication of applications only. In the local situation, the messages arrive at the communication processor via the serial interface of the physical external interface (PEI). The services presented in the following chapters are available in all variants, it is only the names that differ.

We must again emphasize that network management tools are imperative, especially for systems that support complex configuration interfaces and high numbers of nodes. This has also been realized by EIB developers (if a little late). At the time of printing, the EIB management services were still in the specification phase. Therefore, we can only give a brief outline of their functionality in the subsequent chapters and again stress that the given service names are by no means fixed and may change in the final specification.

Before we take a look at the individual services, which we have combined into service groups, we need to clarify the databases they access and how these are organized. The associated structures are dealt with in the subsequent chapter.

4.2.2 Configuration Tables

Configuration data is stored in every end device. This data can be manipulated by network management services. It is managed in three different tables, which are closely linked to one another. We will now outline the structure of each of these configuration tables.

Address table

The address table records the physical address and the groups assigned to the end device (logical addresses) (Figure 4.7).

The network management is primarily interested in the physical address consisting of zone/line/device number. It uniquely identifies the corresponding end device in the network and is assigned during installation of the network and not, as is the case with other fieldbus systems, at the manufacturing stage.

4 Application Environment and Network Management

Address Table	
Length	(1 byte)
Physical Address	(2 bytes)
Group Address 1	(2 bytes)
...	...
Group Address n	(2 bytes)

Figure 4.7: Address Table

The remaining entries in the address table are used for group addresses. Depending on the hardware configuration of the end device, this can include up to 254 entries. This limits the number of groups to which an end device can be assigned.

Groups themselves can comprise any number of end devices. This organization results in the following properties for network management: New end devices can easily be inserted into an existing group – an entry in the address table of the corresponding end device is all that is required. It is not necessary to adapt the address tables of the other end devices! This is not obvious, as protocols with other group concepts require an exact knowledge of the group size in all group members. In these cases, it is necessary to reconfigure all existing members whenever a group is expanded.

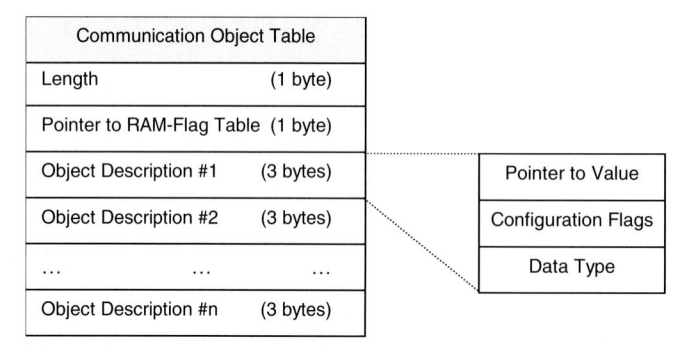

Communication object table

This table describes the layout and position of the communication objects, which the application of an end device can access. For every communication object there is one entry, which further classifies the corresponding object (object description). This records the data type of the communication object and specifies where it can be found within the address space of the device. The object description also contains a set of flags that are relevant to the network management (note: these flags have nothing to do with the RAM flags of the communication object. These are stored in a separate table!).

Figure 4.8 shows the principle layout of the communication object table.

Communication Object Table	
Length	(1 byte)
Pointer to RAM-Flag Table	(1 byte)
Object Description #1	(3 bytes)
Object Description #2	(3 bytes)
...	...
Object Description #n	(3 bytes)

Pointer to Value
Configuration Flags
Data Type

Figure 4.8: Communication Object Table

The data type of a communication object is always specified by the application and with that is omitted from the network management responsibility. When setting up connections it is only necessary to ensure that sender and receiver have matching semantics for the corresponding object (i.e. know which data type has been coded in the message), as otherwise the contents would be interpreted differently.

As can be seen from Table 4.1, communication objects can be simultaneously configured with read and write capabilities. This however is not recommended, as it weakens the boundaries between send and receive objects and the communication relationships become more complex and less clear.

Table 4.1: Communication Object Configuration Information

Flag	Value	Description
Transmit-Enable	0	No message sent after updating
	1	Message sent after updating (transmit object)
Value Memory Type	0	Value of the communication object stored in RAM
	1	Value of the communication object stored in EEPROM
Write-Enable	0	Network write access forbidden
	1	Network write access permitted (receive object)
Read-Enable	0	Network read access forbidden
	1	Network read access permitted
Communication-Enable	0	Communication object blocked for network
	1	Communication object released for network
Transmission Priority	00	System message
	01	Alarm message
	10	Message with high priority
	11	Message with low priority

Association table

We must now clarify how the group address is connected to the communication object. When the data of a communication object is transmitted, it is necessary to determine the allocation between communication object and corresponding group address. When a "group telegram" reaches an end device, it is then necessary to find the corresponding communication object. These relationships are recorded in the association table. It essentially consists of entries that explain the relationships. With that, it is possible to dissolve the link between group addresses and communication objects, as a communication object is not directly assigned an address but a communication reference. Changes to the group structure of a network therefore only affect the address table; the association and object tables remain untouched!

Figure 4.9 highlights the organization and interconnection of all configuration tables. To maintain clarity, only the most important entries have been listed.

When setting up the association table it is necessary to remember that a transmit object can only be linked to a single group address. Receive objects on the other hand can be linked to several group addresses, which allows them to react to messages originating from different

153

groups. In the above example, the association table consists of one transmit and four receive objects. The one transmit object (communication object 1) is – as stipulated – linked to a single group address. If we take a closer look at the receive objects, then it is clear that the arrival of a message with group address 3 affects communication objects 3 and 5. All other receive objects are linked to one group address only.

Address Table		Association Table		Communication Object Table
Length		Length		Length
Physical Address		Cref #1	Obj. #1	Obj Desc. #1
Group Address #1		Cref #2	Obj. #2	Obj Desc. #2
Group Address #2		Cref #3	Obj. #3	Obj Desc. #3
Group Address #3		Cref #3	Obj. #5	Obj Desc. #4
Group Address #4		Cref #4	Obj. #4	Obj Desc. #5

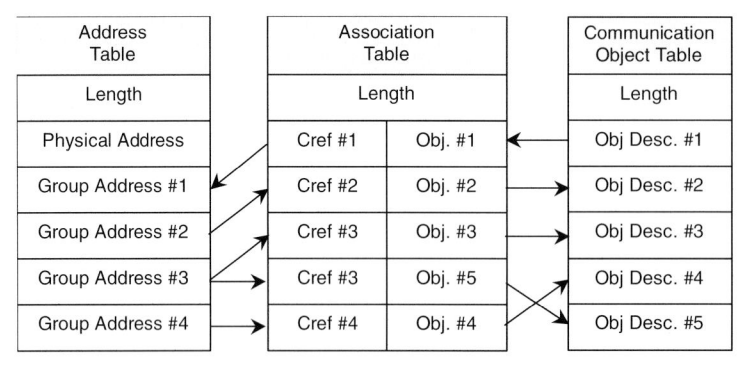

Figure 4.9: Association Table

4.2.3 Network Configuration

It is the task of the network configuration services to configure an EIB system so that it is then possible to obtain information about the connected devices. The individual services can be grouped as follows:

– *Management of the physical addresses*: These services are used to download and modify the physical addresses of one or several devices. Depending on the actual service used, only the devices in programming mode are contacted (implies the previous pressing of a programming button), or the corresponding devices are addressed via a unique (encoded in the message) serial number (no prior need to press the programming button). The corresponding services are: NM_PhysicalAddress_Read, NM_PhysicalAddress_Write, NM_PhysicalAddressSerialNumber_Read as well as NM_PhysicalAddressSerialNumber_Write.

– *Management of the domain addresses*: These services are used to download and write domain-IDs. Domain-IDs are used to split networks, which cannot be physically separated, into logical sub-networks. This logical subdivision is particularly attractive for open transmission media such as radio and powerline. Packet filtering is carried out on the basis of the domain-ID and must be done in the data link layer of the end device. The NM_DomainAddress_Read and NM_DomainAddress_Write services are provided for configuring a domain-ID.

– *Network scans:* These services provide a comfortable means of searching for end devices within an EIB system according to specific criteria. The search can be based on domain-ID, device, line or zone address. The most important services here are NM_DomainAddress_Scan, NM_PhysicalAddressScan and NM_Line_Scan.

4.2.4 Device Configuration

Device configuration services are used to configure individual devices. They depend on the respective device, which demands detailed knowledge concerning its specific implementation. Some of the adjustable parameters vary according to the hardware type of the device. As the communication occurs between a client (e.g. the network management tool) and a specific device, it is necessary to set up a connection to the device before executing the desired task. This is achieved using the DM_Connect and DM_Disconnect services which are mapped on to the corresponding services of the EIB protocol.

For many network management services it is also important that in addition to the ability to create and dissolve connections, there are also means of authentication available. These are used to poll and set access codes and access priorities in the end devices. For this reason, the DM_Authorize and DM_SetKey services have been defined for the network management, which are also mapped on to corresponding services of the EIB protocol. The service groups for device configuration are:

- *Reset and delay*: With the DM_Restart service it is possible to restart a device. An additional delay function, DM_Delay, allows the insertion of pauses between management operations.

- *Addressing*: Services in this group are used to read and write physical addresses or domain-IDs. Device addresses and domain-IDs are newly programmed locally and not via the network. The services are adapted from those of the network configuration and are as follows: DM_PhysicalAddressRead, DM_PhysicalAddressWrite, DM_DomainAddressRead and DM_DomainAddressWrite. If the configuration is to be defined via the EIB network, the corresponding device can be set to programming mode without pressing the programming button. The DM_ProgMode_Switch service has been defined for this purpose.

- *Physical external interface*: If an end device comprises communication and application modules, the PEI type can be ascertained with the DM_PeiTypeVerify and DM_PeiTypeRead services.

- *Memory access*: In the group of memory-related functions there are services for reading, describing and checking the communication and program memories of the connected devices. It is possible to transmit entire memory blocks and with that to load, for example, entire application programs. The functions in this group are provided in local and network-capable forms. They are DM_MemWrite, DM_MemVerify, DM_MemRead, DM_UserMemWrite DM_UserMemVerify and DM_UserMemRead.

- *Property access*: With the help of the network management, it is possible to read and write EIB object properties (see chapter 4.1.2). As with the memory-related functions, the services are again provided in local and network-capable forms. EIB objects are addressed via their type, and the properties contained therein via their index. It is also possible to poll entire object descriptions. The management server then transmits a list of its EIB objects with all contained properties. The specified services are DM_InterfaceObjectWrite, DM_InterfaceObjectVerify, DM_InterfaceObjectRead and DM_InterfaceObjectScan.

- *Access to EIB system objects*: A network management server can manage up to four EIB system objects, loaded using state machines. The EIB system objects include the address table, the association table, the application program and (in some hardware designs) the PEI program. The state machines reflect the respective status of the configuration. In the unloaded status no configuration has been loaded, the object is empty. In the loading status the configuration is in progress, and the loaded status indicates that the configuration process is complete. Transition between the statuses is triggered by network management services, with which the configuration process is managed.

In order to load a configuration table for example, the network management client (i.e. the network management tool) must set the corresponding state machines to the loading status. The table contents can then be transmitted. Afterwards, the state machine must be set to the loaded status, which signals to the remote end device that configuration is complete. The applicable services are DM_LoadStateMachineWrite, DM_LoadStateMachineVerify, DM_LoadStateMachineRead, DM_RunStateMachineWrite, DM_RunStateMachineVerify and DM_RunStateMachineRead.

– *Line coupler configuration:* Network management services can also be used to access line and backbone couplers, in order to read, write and check the memory contents. The configuring of routing tables is also supported. The corresponding services are DM_LCSlaveMemWrite, DM_LCSlaveMemVerify, DM_LCSlaveMemRead, DM_LCExtMemWrite, DM_LCExtMemVerify, DM_LCExtMemRead, DM_LCExtMemOpen, DM_LCRouteTableStateWrite, DM_LCRouteTableStateVerify and DM_LCRouteTableStateRead.

5 EIB Hardware

5.1 Topology

The EIB concept allows the system to be used in both small and large electrical installations [Seip 93]. Figure 5.1 illustrates the layout of an EIB system. The smallest functional unit is the line. If the bus line is a twisted pair (TP) then depending on the implementation it is possible to connect up to 64 (TP64) or 256 (TP256) communication units or end devices. Logically, the line has a bus topology and with regard to cable arrangement it has a free topology. In other words, the line can be laid with a star, tree or ring formation.

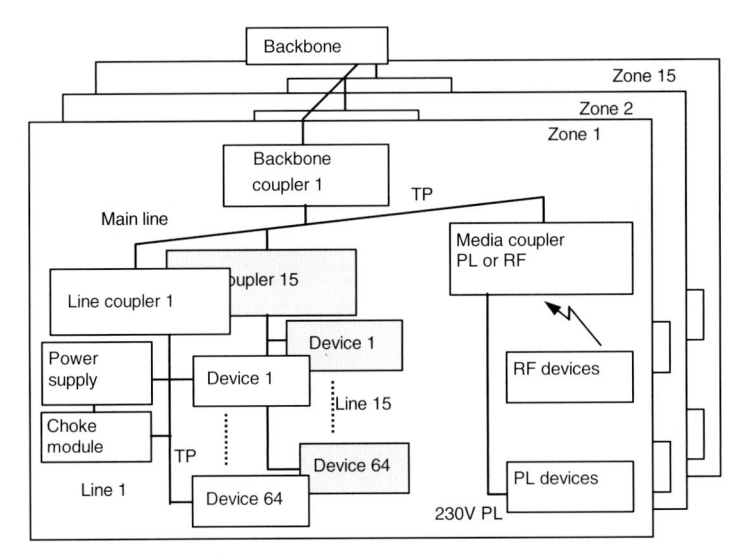

Figure 5.1: System Structure

The devices on a line are provided with a common low-voltage supply (30 V DC) from a remote source via the bus line. In the case of other media such as powerline (PL) or radio frequency (RF), each device is fitted with its own power supply (230 V, battery). With powerline, data transmission does not involve a separate line, it is conducted via the 230-V mains and with radio there are no lines at all. Depending on the supply conditions, it may be necessary to use phase couplers or band stops with powerline. If it is necessary to install more end devices than the maximum number permitted within a line, then couplers are used, which connect lines via main lines. If this coupling occurs between the same media, they are called line couplers and if the media are different, they are called media couplers. Basically,

5 EIB Hardware

the coupling of different EIB media always occurs in a certain hierarchy. In Figure 5.1 for example, coupling occurs on the "main line" hierarchy level with the TP medium. Depending on the system, it is possible to combine up to 15 lines within one zone. A maximum of 12 lines however is currently released. Three lines are reserved for future applications. If it is necessary to implement even bigger systems, up to 15 zones can be combined via backbone couplers. End devices can also be connected to main lines and backbones. The system configuration is defined via the bus. To achieve this, either a PC or a commissioning unit is attached to a data interface, and then the application is loaded into the individual devices or the corresponding parameters are set. The functions depicted as devices in Figure 5.1, such as power supply, choke module or line coupler need not be implemented as separate devices but can be combined into one.

Contrary to TP, the topology with PL is determined by the installed supply network. In general however, distribution follows a star structure, which is extended to a tree structure through branching. A further difference is the important role played by additional functions such as repeaters and phase couplers (see chapter 5.4.3), as electrical consumers do not create predictable signal attenuation or noise levels.

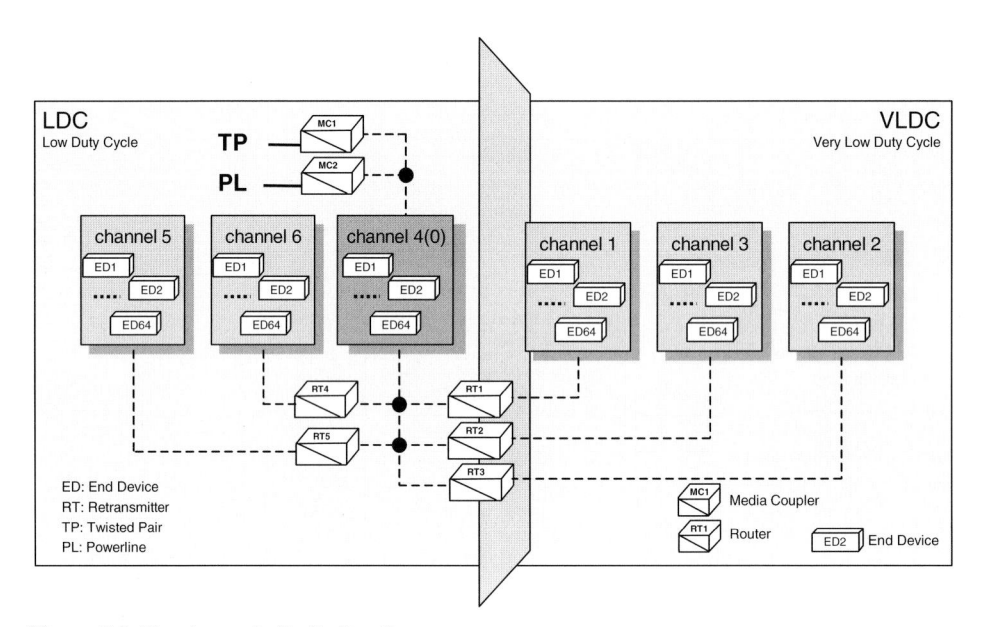

Figure 5.2: Topology of a Radio Bus System

The logical topology of a radio system or a radio sub-system (see Figure 5.2) corresponds in principle to that of a twisted pair system, in which the logical and physical topologies are identical. When using radio, the logical separation of the lines is achieved via domain addresses, the physical separation via different frequency channels. Devices on the same line therefore are distinguished by the same domain address and the same frequency channel. Within a radio bus system, it is possible to use up to three radio channels in each of the frequency bands, i.e. 868.0 MHz to 868.6 MHz for low duty cycle (LDC) operation and 868.7 MHz to 869.2 MHz for very low duty cycle (VLDC) operation. In the LDC band, a maximum channel occupancy of 1 % averaged over an hour is permitted and in the VLDC band, there is a maximum allowed channel occupancy of 0.1 % averaged over an hour. This means that there is sufficient time left over for the largely interference-free co-existence of various systems. Each channel represents exactly one sub-network with up to 64 end devices. Every sub-network has its own zone address, a 16-bit value which permits logical differentiation of the cells. Within a sub-network, up to three end devices can take on a retransmitter function in order to increase transmission security. Retransmitters are preferably end devices with mains supply (e.g. actuators). These sub-networks can be coupled via routers. Like the TP couplers (see chapter 5.4.1), the routers have a filter function in order to minimize channel occupancy in the sub-networks (lines).

5.2 Media

As we have seen in the last chapter, it is possible to use different media to set up an EIB system. The individual possibilities are examined in the following chapters.

5.2.1 Twisted Pair

The EIB standard line (e.g. YCYM 2x2x0.8) is a shielded and twisted plastic-sheathed cable with one or two wire pairs. The EIB only requires one wire pair for the remote supply of the devices as well as TP data transmission. The wire colors are defined in accordance with the polarity as follows:

+ red or dark blue
− black or blue

The second wire pair can be used as required. However, it is necessary to ensure that use of this pair does not affect the function or security of the EIB (regulations on the secure isolation of circuits, see EN 50090-2-2 or VDE 0829 part 2-2). Although not specifically stipulated, connection of the shield is however recommended if the EIB is to be used in an environment at a higher risk of lightning strikes or if expected interference levels are greater than normal. Further details are given in the EIB manual [EIBA 99]. Lines which pass a 2.5 kV AC voltage test, as specified in VDE 0829 part 2-2, can be laid in the direct vicinity of the 230/400-V lines.

The universal rules specified in Table 5.1 are valid when using a standard YCYM 2×2×0.8 EIB line. These rules also cover unfavorable conditions and combinations of end device and line arrangements.

Table 5.1: Global Rules for a Line

Parameter	Value
Transmission range	max. 700 m
Line length	max. 1000 m
Separation of power supplies (including choke) from end devices	max. 350 m
Separation of two power supplies including choke module	min. 200 m
Number of end devices	max. 64

5.2.1.1 Data Transmission Criteria

If it is necessary to stray from the rules listed in The universal rules specified in Table 5.1 are valid when using a standard YCYM 2×2×0.8 EIB line. These rules also cover unfavorable conditions and combinations of end device and line arrangements.

Table 5.1, e.g. a greater number of end devices are needed per line or an optimization is to be found for a particular configuration, then the following relationships must be observed. The most important parameters for the transmission are propagation time, attenuation and dc voltage level. Whilst the propagation time is almost exclusively dependent on the line length (when the same line is used), attenuation depends on the line length, the temperature as well as the number of end devices.

Propagation time

By virtue of the collision resolution method a maximum propagation time of 10 µs in one direction is permitted for a data transmission rate of 9.6 kbit/s. If the maximum transmission length is too large, collisions can no longer be resolved, telegrams are destroyed and the bus load increases due to the greater number of telegram repetitions.

Attenuation of the data signal

The data signals transmitted within EIB must be seen as ac voltage signals, generated by the transmitting end device with a specific internal resistance and then divided by the complex resistance and susceptance of the line and the parallel conductance values of the end devices. The divisor ratio of these resultant voltage distributors, in the case of a fully fitted line, should not exceed a level that causes negative violation of the input voltage threshold of the receiver (approx. 0.5 V). It is also necessary to remember that in practical use, an increase in the ambient temperature of the bus line (subsequent increase in line resistance) can lead to an increase of approx. 20 %.

Attenuation of the dc voltage for remote supply

As a result of the implementation, the transmission amplitude of the data signals can depend directly on the dc voltage that is available at the end device. This also specifies the dependency on the number of power supplies and the current consumption of the individual end devices. If two power supplies are used, including chokes, the best results are achieved when they are placed at each end of the line. Furthermore, the transmitting power of the end device that is necessary for the charge reversal processes during transmission creates a practical limit for the line length. A stretch of line that is not terminated forms a resonator. As the free topology of the EIB line is laid without termination, it represents a formation with many resonators of differing resonance frequencies. The line capacity determines the resonance properties. The maximum permitted capacitive load of a line is equal to 120 nF. Figure 5.3 illustrates the relationships between maximum number of end devices (ED), maximum transmission length, average line temperature and low dc bus voltage.

5.2.1.2 Global Power Supply

Every line requires a separate dc voltage for the remote power supply. Depending on the implementation, this supply is required for the transceivers, the communication modules or all end devices.

The power supply is voltage controlled and current limited. Short-term power failures can be bridged at full load for 100 ms. The power supply is connected to the EIB via a choke module. Power supply and choke module can be integrated into one device. If the EIB end devices require more than the power generated from a single supply, it is possible to use two power supplies in parallel. There should however be a minimum separation between these two supplies, including choke, of 200 m in order to guarantee the full system range. Furthermore, the power supplies should be installed near to the greatest concentration of end devices. These are, for example, the installation distributors. The power supply consists of two function blocks, the dc voltage generator and the choke. These two function blocks can be implemented separately or within the same unit.

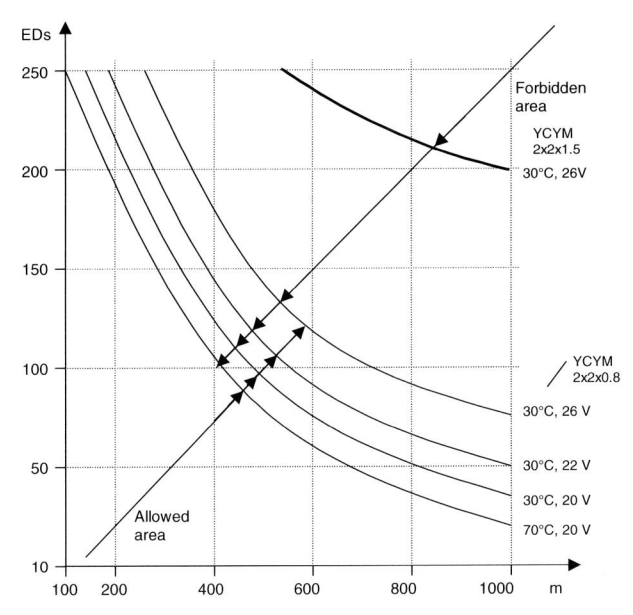

Figure 5.3: Characteristic Lines for Line Dimensioning

DC voltage generation

With EIB, protection against direct or indirect contact with components carrying bus voltage is guaranteed with the use of a safety extra low voltage (SELV). Safety extra low voltage means that there is a clear separation between the mains (230 V), earth and bus (e.g. via a safety transformer) and that a specific dc voltage level is not exceeded. The EIB output voltage of a maximum 30 V lies in the allowed region. Nevertheless, in order to guarantee a sure connection to earth of the otherwise floating EIB system, there is a link in the power supply via a protective impedance that maintains secure isolation. Due to the possible load of line arrangements and contacts, especially when using multiple power supplies, the short circuit current is limited to a maximum of 1.25 A. With nominal output currents, depending on the implementation 320 or 640 mA, the dc voltage generator guarantees a buffering time of at least 100 ms.

The generation of dc voltage is an essential component of the compound system which also influences the transmission properties of the system through voltage ripples and earth sym-

5 EIB Hardware

metry. The maximum output voltage ripple is equal to 100 mV$_{ss}$. The earth symmetry is essentially determined by the internal structure. When implementing the protective impedance to earth, it is necessary to guarantee the symmetric connection of the outputs. Figure 5.4 illustrates the typical current/voltage characteristics.

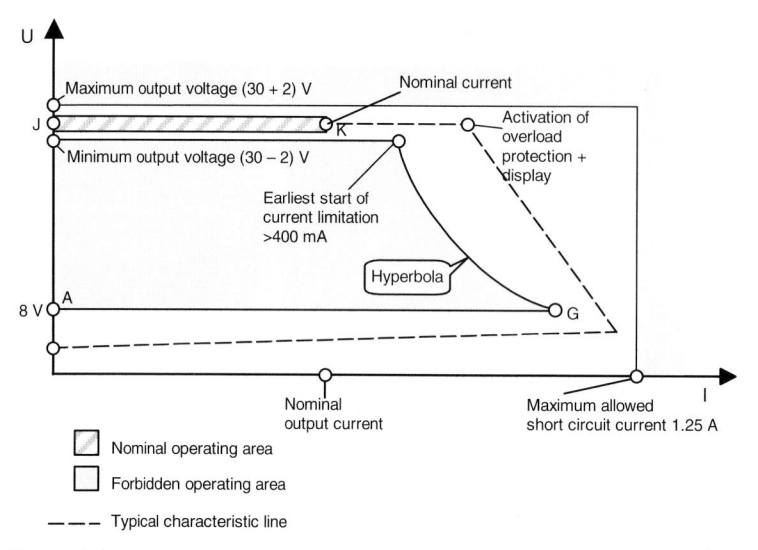

Figure 5.4: Typical Current/Voltage Characteristics

In the example characteristic line, point K corresponds to the nominal current for which the buffering time of the power supply has been defined. Through constant power output with increasing current, the hyperbola guarantees that the communication end devices work properly after the power is switched on, the power supply does not assume an operating point that lies outside the nominal operating area and that there are no undesirable oscillation effects. Point A specifies the minimum voltage for which the output level of the power supply must be calculated. For the communication devices supplied via the bus, this means that under 8 V they can only take low currents, as otherwise the current would increase sharply in accordance with the hyperbolic characteristic line (point G shifts to the right for smaller voltages).

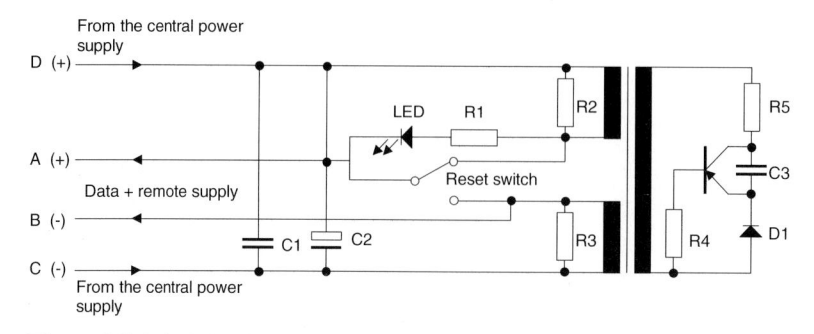

Figure 5.5: Principle Circuitry of the "Choke" Function Block

Choke

Because with EIB, the remote supply of the communication end devices and the data transmission occurs along the same wire pair, it is necessary to ensure that the connection of a dc voltage generator does not cause any feedback on the data transmission. It is therefore decoupled via a choke. Figure 5.5 illustrates its principle circuitry. The low impedance choke for the dc voltage represents a high impedance resistance for the ac "data" signal. There is also a compensation circuit integrated into the choke, which creates a certain bus terminal at this point. The choke also plays an important part in the generation of data pulses (see chapter 3.2). During the equalization pulse, it loads the energy stored in the active pulse back into the bus line. The maximum voltage drop across the choke is equal to 0.5 V DC with a nominal load.

5.2.1.3 Connection Possibilities

The implementation of electrical and mechanical interfaces to the bus has a significant effect on the reliability of the compound system, the installation work and the occurrence of installation errors [Grun 97]. Figure 5.6 shows the recommended and non-recommended connection possibilities of the bus line. Basically, the disconnection of a device must not lead to an interruption of the bus line. Examples of connection techniques include the data rail with spring-finger connectors in the devices for DIN rail mounted units (distributor installation site), the line/device connector with universal application possibilities (Figure 5.7) and the connector boxes for data networks (e.g. Western), which can also be used to connect portable units directly to EIB.

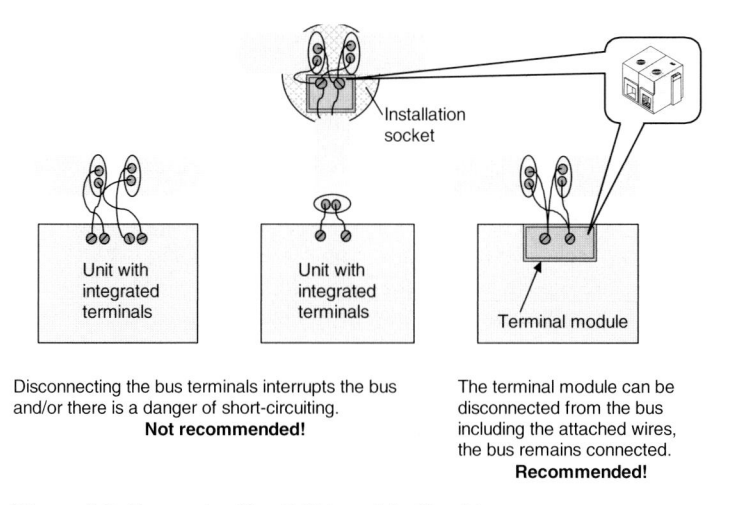

Figure 5.6: Connection Possibilities of the Bus Line

The data rail is a self-adhesive printed circuit board that is fixed into the DIN rail. The two inner conductors are used for the bus inclusive of the dc power supply and the outer ones are used, for example, purely for the dc power supply. When placed onto the DIN rail, rail-mounted units are then automatically connected to the bus via the integrated spring-finger contact system. Any part of the data rail not fitted with units should be covered to avoid

5 EIB Hardware

unnecessary pollution and also to guarantee the SELV (safety extra low voltage) conditions, even in the presence of other voltage supplies (e.g. 230 V).

The link between data rail and EIB line is achieved with DIN rail mounted connectors, which on one side make contact with the data rail via the contact pins, and on the other side has pins to which the line/device connectors can be attached. These line/device connectors are available with terminal screws or screwless terminals for the connection of the EIB line. The line/device connector is also provided for the connection of EIB lines, e.g. in installation sockets. The use of connection sockets for data networks has one major disadvantage – the link to the portable unit is directly connected to the bus and after separation of this link the device is no longer known to the bus and any damage to the connecting line can directly affect the bus. For this reason, portable units are generally coupled via a data interface, which contains a BAU/BCU and with that allows for decoupling. The following influencing factors should be taken into consideration when implementing a bus interface:

– Contact resistance, especially with regard to vibration and environmental influences
– Contact strength in combination with the used contact material
– Voltage and current ranges
– Number of insertion/withdrawal cycles
– Insertion/withdrawal force
– Reverse battery protection

The detailed requirements are listed in the EIB manual.

a) Data rail connection for DIN rail mounted devices

b) Line/device connector

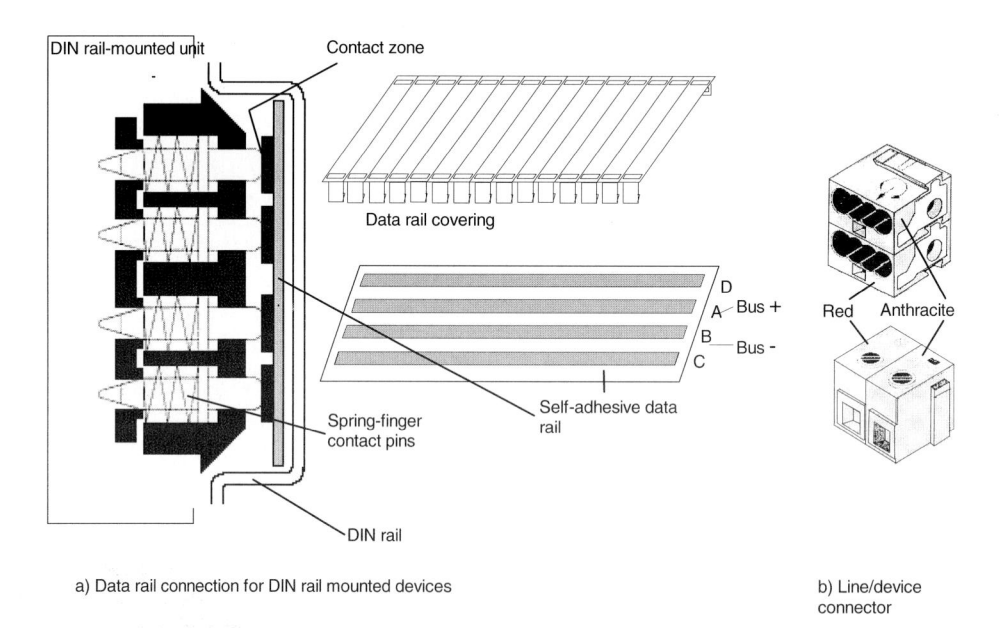

Figure 5.7: Examples of Connection Techniques

164

5.2.2 Radio Medium

With regard to a marketable and future-safe EIB radio system, particular attention has been paid to the selection of radio frequency. The main aspects for consideration were:

- High system reliability
- Feasible costs or prices
- Europe-wide availability of the corresponding radio frequency

Within the scope of extensive feasibility studies, several frequency bands were examined with regard to their suitability for a European bus system.

The examination was concentrated on bands in which low power radios, so-called short range devices (SRD) could be operated. They are as follows:

- 40 MHz
- 433 MHz
- 870 MHz
- 2440 MHz

The frequency range around 870 MHz was finally selected. The 40 MHz band was rejected on the basis of the large antennae it would require and the 2440-MHz band due to the current high cost and the high level of signal attenuation.

The 433 MHz frequency band

The 433 MHz frequency band is already highly used today. In addition to a number of SRDs, radio systems for industrial, scientific and medical purposes, so-called ISM radios, are operated on this band as well as amateur radio station transmitters (AF). Contrary to the SRDs, with a maximum transmitting power of 10 mW, ISM and AF systems work with higher powers. The transmitting power of AF devices can be anything up to 750 W.

It is not only the transmitting power but also the transmitting duration of all transmitters in this band that must be viewed as disturbing influences. For example, the 433-MHz band is used by a large number of audio units that are operated over periods of several hours. Depending on the local conditions, this can lead to interference, which significantly affects radio systems with low transmitting power or can even make them stop functioning altogether.

The 868 – 870 MHz frequency range

The 868–870 MHz frequency range has been accepted by the Council of European Postal Authorities (CEPT) within the scope of the coordinated assignment of frequencies and exclusively reserved as a special frequency range for short range devices. This frequency range therefore cannot be used by ISM or AF devices. In order to keep the mutual interference levels in this SRD frequency range as low as possible, certain essential restrictions have been defined for the radio units. These include:

- Limitation of the transmitting power for all users of this frequency range
- Limitation of the maximum transmitting duration (duty cycle)

The duty cycle specifies the total transmitting duration of a device summed up over the period of one hour (Table 5.2). The probability of interference increases with the increasing duty cycle.

The frequency range of 868 to 870 MHz is split into several bands each of which permit different duty cycles. This guarantees the co-existence of radio systems even with the expected increase in future use of this frequency range.

The messages associated with most applications of the EIB radio systems are so short that they can operate in the low-interference 'very low duty cycle band'. It is only necessary to fall back on the low duty cycle band in exceptional cases. Most European countries have since incorporated the license for this frequency band into national law, so that for the first time there is a European ISM band available within this frequency band.

Table 5.2: Duty Cycle of Universal SRD Bands in the 868–870 MHz Range

Duty Cycle	Abbreviation	Total Transmitting Duration	Frequency Band
Very low	VLDC	< 3.6 s / h	868.7 – 869.2 MHz
Low	LDC	< 36 s / h	868.0 – 868.6 MHz
High	HDC	< 360 s / h	869.4 – 869.65 MHz
Very high	VHDC	< 3600 s / h	869.7 – 870.0 MHz

The EIBA radio task force decided on this 868 - 870 MHz range for the reasons mentioned above. Groups of other radio users such as for example, automotive (radio operated keys) or telemetering (remote downloading of heating and energy consumption) have also recognized the advantages of this SRD radio area and will switch from 433 MHz to 868 MHz for future products as all these applications can co-exist thanks to the duty cycle regulation.

Depending on the interference situation or the propagation conditions for the radio waves, the range can lie anywhere between several tens of meters and one thousand meters. The changing interference conditions can cause the range of an installed network to change too. A media-specific retransmitter protocol guarantees that in spite of this, all devices on a line, even at distances of up to 100 m, are able to communicate with one another reliably.

5.2.3 Powerline

EIB powerline (PL), also known as "EIB power net", uses the available electrical network for two purposes: as a distribution network for electrical energy and as an information network. To achieve this, the EIB digital data packets are modulated onto the 50Hz sinusoidal oscillation of the 230V supply used as the carrier frequency. In itself, this technology is nothing new. Known as ripple control, it has been in use for more than one hundred years. Of note however, are the interference immunity and reliability of the system. With EIB-PL, the existing electrical installation in small to mid-sized buildings can be converted into a future-safe bus installation with the minimum of effort. For larger buildings the work required is slightly greater with the use of repeaters and/or media couplers (see description in chapter 5.4). The disadvantages include the poor transmission properties of the power network, which cannot be overcome without special transmission procedures.

With EIB-PL the bit coding is achieved using spread FSK. Contrary to regular FSK (frequency shift keying), which uses frequencies that are as close as possible for the transmission of 0 and 1 signals, spread FSK uses a greater separation between the individual frequency bands. A zero bit is represented by a signal with a frequency of 105.6 kHz, a one bit with 115.2 kHz. These frequencies lie within the lowest (95 to 125 kHz) of the three frequency bands released for the use of powerline transmission in Europe (the relevant standard is EN 50065-1). The maximum signal level is 1.26 V_{eff} (which corresponds to 116 dBµV). The bit duration with EIB-PL is a uniform 833.3 µs, which yields a data transmission rate of 1200 bit/s. In order to keep interference as low as possible, the frequency keying should occur on a continuous-phase basis, i.e. without jumps in the signal path.

As already mentioned in chapter 3, it is not only the definition of the physical protocol layer that differs from that of the twisted pair implementation, the data link layer is also different. Bus access in particular is controlled in a different way. Here there are no dominant and recessive signal forms with the modulation method used, as would have been made available by the CSMA/CA method common to the twisted pair configuration. Instead, a *time slot method* is used, in which each end device is only able to transmit at defined points in time. This avoids access conflicts.

In order to meet the special requirements of this medium, the data frame of the data link layer is also modified with EIB-PL. The data consists of the LPDU of the twisted-pair variant, but is also provided with a pre-amble, a checksum and a special system ID (Figure 5.8). This additional ID is used to identify logically connected EIB networks, which in an open medium cannot be electrically isolated from one another.

Sync 4 bits	Pre-amble 1 8 bits	Pre-amble 2 8 bits	Twisted pair LPDU	Checksum 8 + 4 bits	System-ID 8 + 4 bits

Figure 5.8: EIB-PL Telegram

The degree of miniaturization has developed rapidly in the field of EIB powerline. In addition to complete transceivers the size of a box of matches, there are a range of sensor and actuator products for flush-mounted sockets. This satisfies the requirements of the installer as regards upgradeable constructions and connection technology. The range of products for EIB-PL currently comprises more than 65 different units for all common applications. Coupling with EIB-TP (and with that RF) is also possible. For the users of EIB, EIB-PL offers an extra transmission layer in addition to the TP and RF media, whose special strengths lie in the existing infrastructure of the building.

5.3 Communication End Devices

The function of a communication end device is determined firstly by the contained system implementation and secondly by the application hardware and application program loaded at the time of commissioning. Each device essentially comprises all functions necessary for communication. This covers the transmission technology and bus access management as well as mechanisms that guarantee the interoperability of the application functions. This decentralization of the functions in the devices means it is possible to do without a communication center. Chapter 6 contains descriptions of various uses of communication software.

5.3.1 Components of an End Device

The complexity of the functionality contained within a communication end device varies greatly depending on the actual requirements [Anwe 97]. The availability of the EIB system on various micro-controller platforms, the definition of standardized interfaces, which are based on the layer definitions in the ISO/OSI model, and with that the ease of coupling with additional micro-controllers or other processing units, facilitate the functional scalability of the communication end devices that is required in practice.

The principle layout of a TP communication end device is depicted in Figure 5.9. Basically, a device consists of a communication module and an application module [Jahr 97]. In the basic layout, there is no difference between an RF or a PL bus coupling unit and the TP bus coupling unit. Coupling to the (radio) bus is achieved via an antenna that is connected to the transceiver. With PL, the transceiver is connected to the 230-V mains network.

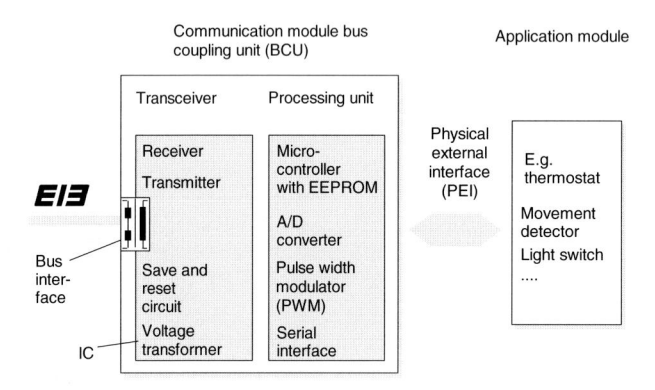

Figure 5.9: Layout of a TP Communication End Device

Communication module

The communication module consists of a transceiver and a processor, usually a micro-controller. The transceiver comprises transmitter and receiver, bus voltage monitoring circuits and usually a voltage transformer for creating controlled internal voltages from the bus voltage. This supplies the processor and the application module with power. The processor contains a micro-controller with ROM, RAM, EEPROM, A/D transformer, pulse width modulator (PWM) and serial interfaces. The ROM contains the system software with communication functions, which firstly serves to control the actual communication and secondly to connect communication with application.

Depending on the implementation, the communication module may also contain an application program that is loaded into a non-volatile memory (e.g. EEPROM). The use of the standardized physical external interface (PEI) between application and communication facilitates the modularization of the communication end devices, independent of manufacturer. Manufacturers are then free to concentrate on their own areas of competence. The PEI is a dynamically defined interface. The application hardware specifies required interface usage within a given, standardized set of possibilities. More details are given in chapter 5.2.4.5. When constructed as a stand-alone unit, the communication module is also referred to as a "bus access unit" (BAU). If the BAU is connected to the application module via the standardized PEI, this model is referred to as the "bus coupling unit" (BCU). The BAU or BCU contains a manufacturer code in the EEPROM that has been specified by EIBA.

Application module

With simple applications, such as for example, a light switch that can be used on the bus, the application module consists of a few parts only. The function of the application (light, switch, dim, open blind, etc.) is determined by the application program that is stored in the communication module. This program is loaded into the communication module via the bus during the commissioning stage. An application program is identified by a unique number. This consists of the manufacturer's code, a type code and a version number. Type and version are freely specified by the manufacturer. The manufacturer code of the certified application program must correspond to the manufacturer code of the communication module, otherwise the EIBA software tool (ETS) is unable to load the application program. This accesses the application module via the PEI. In the simple case of the light switch, the PEI is used as a simple input port for example. With more complex devices, which require their own processors, the application program is contained in the application module. Here, both modules communicate via a data interface; in the case of the bus coupling unit via the synchronous or asynchronous interface of the PEI.

In addition to the pure application function, the application program must also operate the interface to the communication system. Data exchange between the application and system programs occurs using communication objects. They represent a standardized interface between the application and communication, which is independent of the application. The application need not consider the technical communication details. The values contained in communication objects can, in principle, have different structures. It is only possible to evaluate the content when its structure, content and meaning are known. With EIB, it is assumed that data type and information type do not change during operation, which means that information relating to this is not transmitted in the telegram. This information is outlined in the corresponding device/application documentation. For standardized information to be transmitted via communication objects, it is necessary to use the corresponding EIB Interworking Standards (EIS) (see chapter 10.2). So-called EIB objects have also been defined to ensure that basic functions of different devices from various manufacturers are able to interact, where communication does not simply involve an exchange of individual values as with EIS. In principle, EIB objects are a standardized collection of descriptive elements, parameters, status information and communication objects. Chapter 6 contains details on communication objects and EIB objects.

5.3.2 Transceivers

Different transceivers are required depending on the transmission medium. The transceivers represented in the following chapter are implementation examples that are partially based on available ICs. The depicted circuits should clarify the principle functions.

5.3.2.1 Twisted Pair Transceiver

A number of different transceiver solutions are possible for the development of twisted pair devices. These solutions ensure optimization of the device in accordance with the area of application.

5 EIB Hardware

- Transceiver with transformer coupling and bit interface (IC FZE 1065, highly resistant standard solution with extensive functionality, certified)
- Transceiver with direct bus coupling and bit interface (IC FZE 1066, resistant solution with extensive functionality, facilitates miniaturization, certified)
- Transceiver with direct bus coupling and bit interface (discrete construction, low-cost but functionally restricted)
- Transceiver with direct bus coupling and UART interface (IC TPUART, miniaturized solution with few real-time requirements on the micro-controller, certified)

All TP transceiver solutions are designed for a data transmission rate of 9.6 kbit/s.

Twisted pair transceiver with the FZE 1065 IC

The FZE 1065 is a bipolar transceiver element for transformer coupling to the EIB. Two data signals are made available to the micro-controller, the "Receive" signal as a digitized bus signal and the "BitOK" signal that only outputs bit pulses of a defined minimum length. Transceivers with inbuilt FZE 1065 are distinguished by high symmetry. This is connected with a high immunity to common mode interference. The transceiver element contains a transmitter and receiver, a DC/DC transformer (output 5 V), a reset circuit and a low voltage detection mechanism. Figure 5.10 represents the block diagram. In order to be able to transmit a telegram, the Enable input must be "high" (TTL level) and Reset and Save must be inactive, i.e. the bus voltage must be greater than 18 V. "High" on the "Enable" input activates the transmitter and switches the bus separator circuit to low impedance. It is recommended that a capacitor is connected in series between the micro-controller and "Send", in order to protect the transmitter or bus transmission during periods of continuous drive caused by micro-controller error.

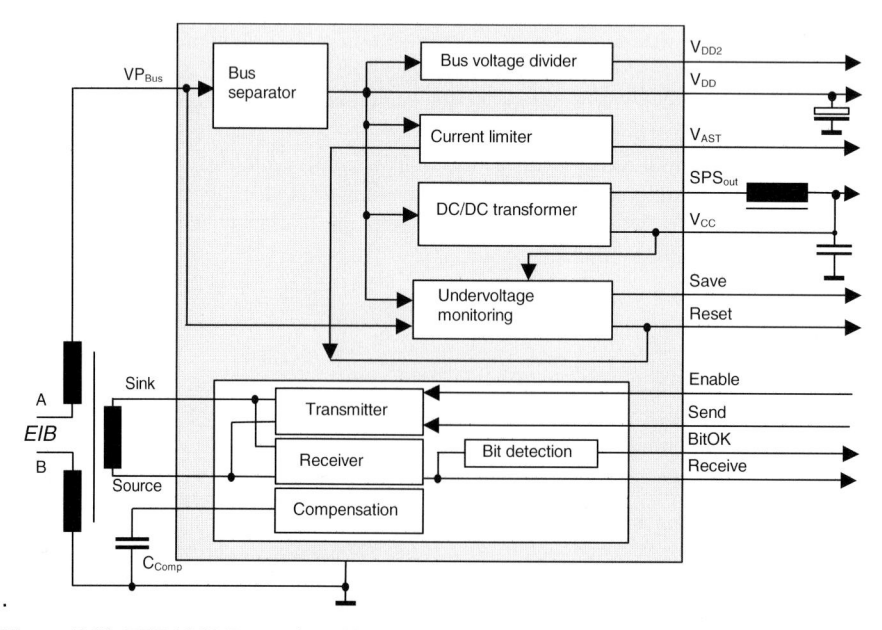

Figure 5.10: FZE 1065 Transceiver-IC

5.3 Communication End Devices

The receiver consists of a comparator that evaluates the differential voltage between source and sink. It has a typical hysteresis of 1.0 V. The purpose of C_{Comp} is to reduce the pulse tilts of an input pulse. The bus separator circuit contains reverse battery protection, which protects the device in case of incorrect connection and ensures that this does not affect the bus transmission of other devices. There is also a power management function which guarantees that when the bus power supply is switched on, the device capacitors are charged accordingly and that in the case of undervoltage (less than approx. 16 V) the bus device remains connected to the bus with high impedance to avoid any reverse discharging of the V_{DD} storage capacitor. The contained charge can be used to store important data in a non-volatile memory during a power failure. Current transceiver implementations guarantee this by buffering data volumes of up to 3 bytes for a period of at least 60 ms. If more data is to be protected, an additional capacitor must be connected to the 5V output to increase storage capacity.

Within the context of transmission technology, the transformer is an important link between bus and end device. On the bus side it has two coils through which both the dc current of the connected device flows as well as the ac current of the data signals. The inductance must have a certain tolerance (±5 mH) and must only change very slightly as a result of the load-dependent dc current (max. 12 mA) on the bus side.

The selection of the core material is extremely important. The magnetization curve must have a shallow S-shaped form in order to avoid too sharp an increase in the magnetization current during transmission and with that any shortening of successive bit pulses. Core materials with a hard saturation bend are unsuitable. Figure 5.11 illustrates a typical circuit for a transceiver with the FZE 1065 IC.

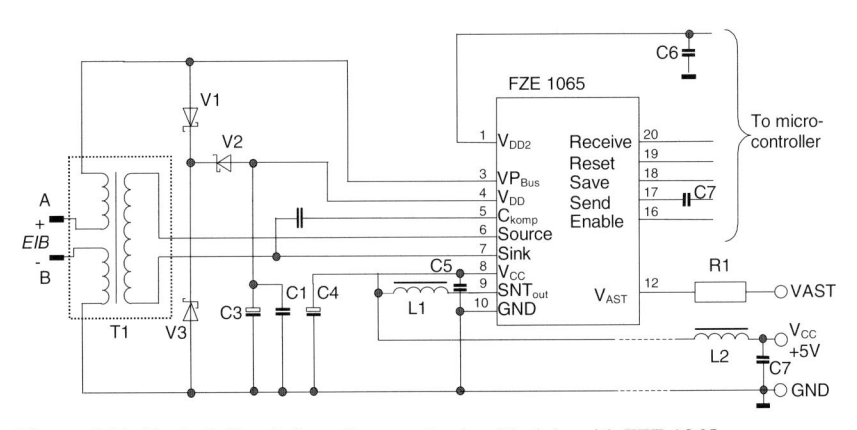

Figure 5.11: Typical Circuit for a Communication Module with FZE 1065

Twisted pair transceiver with FZE 1066

The FZE 1066 represented in Figure 5.12 is a transceiver element for direct bus coupling without transformer. This facilitates the further miniaturization of a communication module. As the FZE 1066 does not have a symmetrical layout, it is necessary when designing the circuit to pay particular attention to the symmetric arrangement of the conductors and sufficient decoupling of circuit parts. Figure 5.13 illustrates a typical circuit with this IC.

5 EIB Hardware

In addition to the functions of the FZE 1065, the IC also contains a temperature monitoring function which guarantees that the IC is not destroyed by overheating. Shortly before the IC shuts down as a result of high temperature, the "TEMP" output switches from low to high. This allows the micro-controller of the communication module to end its current transmission and, as long as this signal is active, prevents it from starting any new transmissions. A band filter is connected in front of the receiver to attenuate any interference signals. In special situations, it is also possible to increase immunity to interference by installing the L2 choke in series with the input.

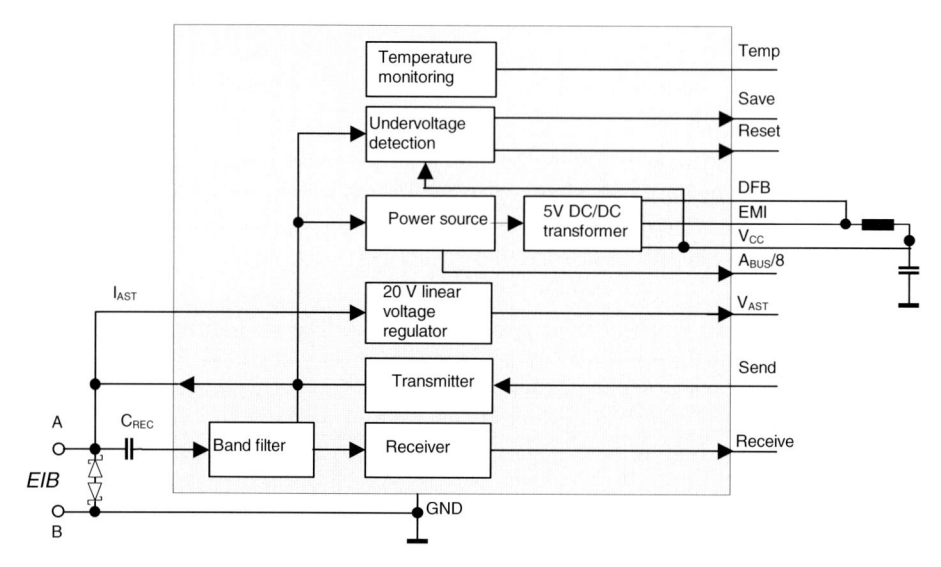

Figure 5.12: FZE 1066 Transceiver-IC

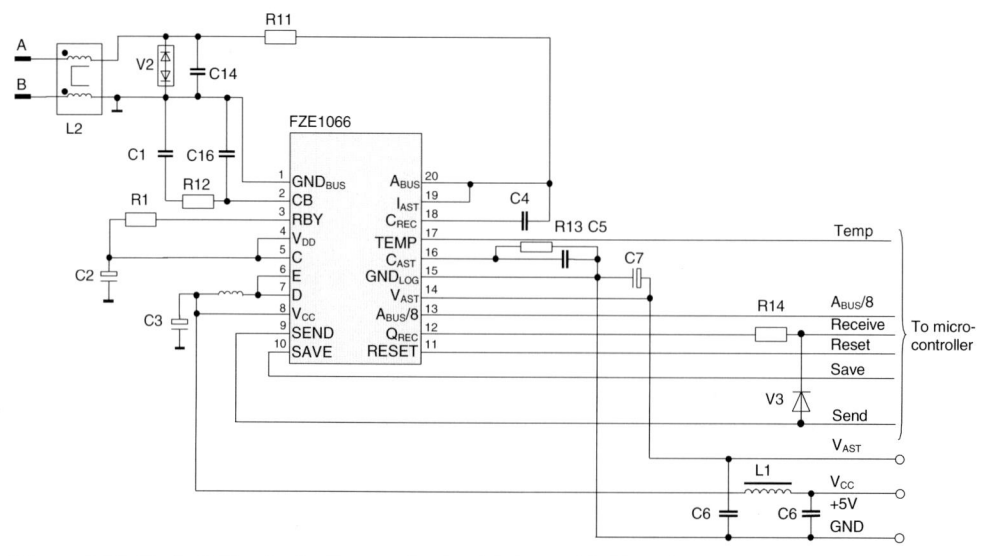

Figure 5.13: Typical Circuit of a Communication Module with FZE 1066

5.3 Communication End Devices

Discrete twisted pair transceiver

Discrete transceivers represent a low-cost alternative to transceiver-ICs with reduced functionality. Thanks to its discrete design, functions can be added or omitted in accordance with the actual requirements. Figure 5.14 illustrates a basic circuit. Compared with the transceiver elements, the represented solution does not have a save function or any 20V generation. The power supply is implemented as a linear regulator with a defined minimum current consumption. The maximum current draw is less than that of the other TP transceivers.

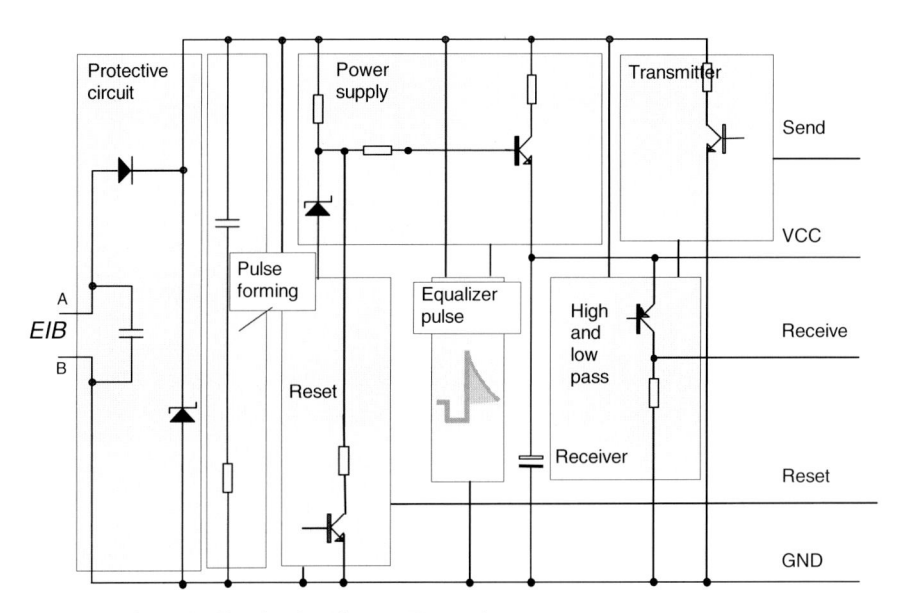

Figure 5.14: Basic Circuit of a Discrete Transceiver

Twisted pair transceiver, TPUART

The previously described transceivers demand short reaction times from the micro-controller with regard to the processing of bit pulses. The TPUART eliminates this problem thanks to integrated bit evaluation and generation. In this situation, transceiver and micro-controller communicate via an asynchronous interface (UART). This is a two-line interface (TxD, RxD) and with that facilitates low cost electrical isolation. Transmission to the micro-controller can occur at a rate of 9.6 or 19.2 kbit/s. As the communication no longer places time-critical demands on the micro-controller it can also be used for applications that place higher demands on the reaction times. The corresponding block diagram is represented in Figure 5.15. The analog part consists of a transmitter, a receiver with pre-connected band pass and a power supply. The power supply is a linear regulator with a controlled minimum current consumption of 3 mA. The linear regulator is supplied by a power source with a maximum di/dt of 1 mA/ms. The power supply also contains undervoltage monitoring for the generation of a reset signal and temperature monitoring. The transmitter regulates the transmission amplitude to a typical value of 7.5 V. The oscillator inputs of the digital part are designed so that it is possible to connect a quartz or a clock signal of the same frequency (4.9152 MHz).

The digital part of the TPUART is responsible for media access, layer 2 reply (acknowledgement frame or immediate acknowledge; see chapter 3.3.3.2) and, if necessary, telegram

173

repetitions. The micro-controller evaluates address and telegram length. The received data signals are sampled 128 times, interference suppressed via a digital spike filter and then transmitted to the micro-controller in a byte-wise and transparent manner. The telegram end is automatically detected by the micro-controller via a timeout of 2–2.5 ms. If addressed, the micro-controller must reply to the TPUART with an acknowledgement within 1.7 ms of receiving the address type/length byte. Telegram transmission from the micro-controller to the TPUART begins with a start byte that precedes the EIB control byte. A control byte is then placed in front of every other byte, which also contains the sequential number of the byte. The maximum number of data bytes is limited to 52. The TPUART buffers the bytes to be transmitted. The telegram is completed by the micro-controller with an end byte with subsequent check sum.

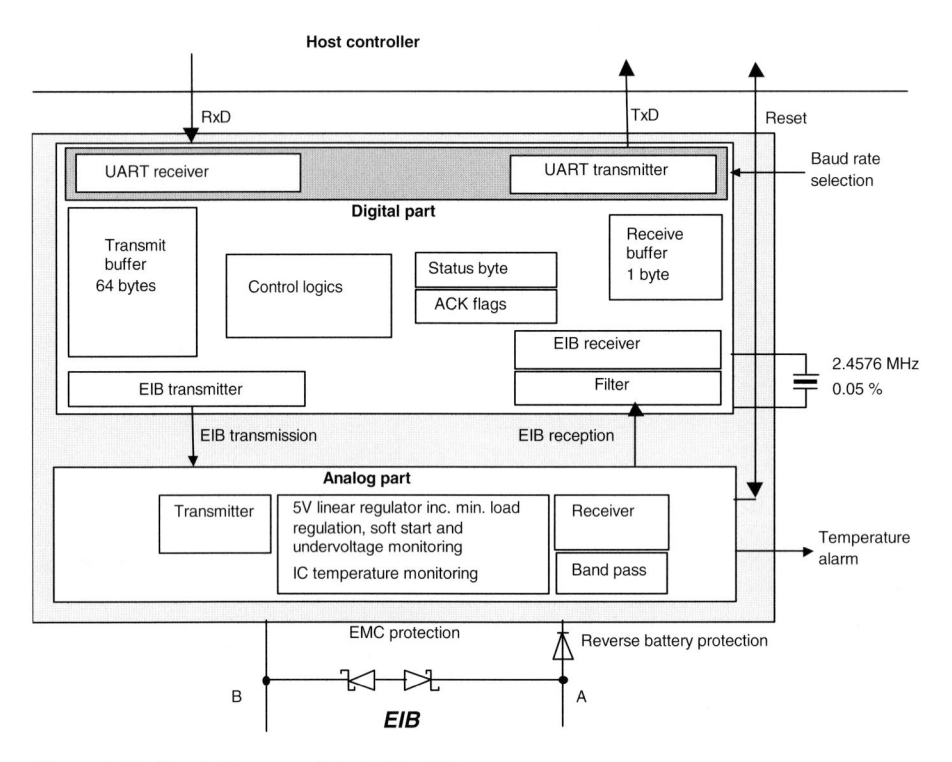

Figure 5.15: Block Diagram of the TPUART

5.3.2.2 Powerline Transceiver

Despite the poor transmission properties of the electrical mains installation, the PL transceiver facilitates fast and secure transmission for control purposes. The transmission procedures of the SNB1 semi-duplex transceiver (Figure 5.16) are based on spread FSK as discussed in chapter 5.2.3. The reception part of the transceiver is designed as an incoherent digital matched filter receiver, in order to guarantee the increased requirements of the EIB as regards interference free transmission in a medium subject to interference. This means the correlation coefficient is calculated between the stored 0 or 1 signals (references) and the input signals from which a decision is made. The transceiver is connected to the EIB-PL medium via the analog "IN" input and the analog "OUT" output. Signal processing on the chip is purely digital, the IC has been manufactured in 0.8-µm-CMOS technology.

The transmission part makes use of the principles of digital signal synthesis, whereby use is made of the reference sample values required for the matched filter receiver. The data to be transmitted is transferred to the "SHL" input. Switchover between transmit and receive modes is achieved with the "S/E" signal. It is possible to set a data rate of between 300 bit/s and 2400 bit/s via the "DR1/DR2" inputs. With "CHSEL1/CHSEL2" it is possible to select various transmission signals within the frequency range released for powerline transmissions (in Europe there are three bands from 95 kHz to 148.5 kHz).

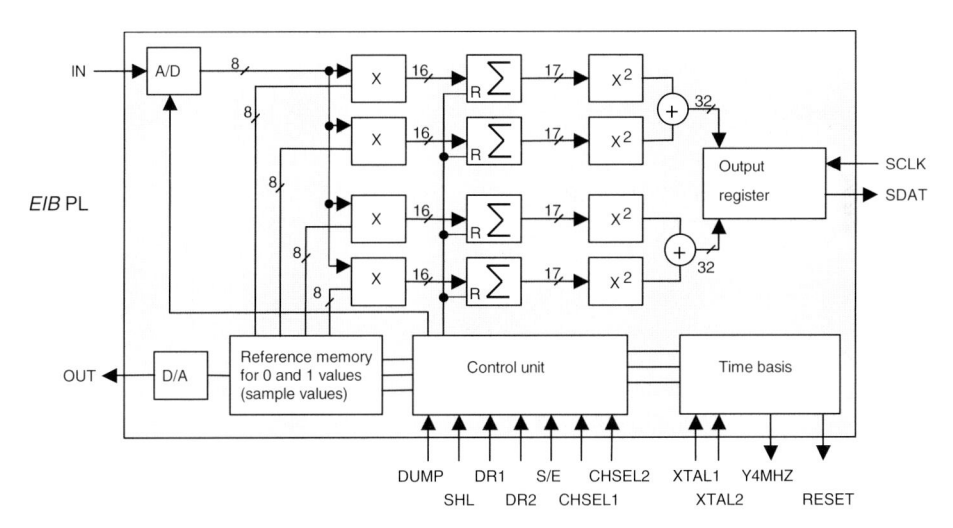

Figure 5.16: PL-Transceiver-IC SNB1

An external quartz (typically 4 MHz) is connected to "XTAL1/XTAL2". An integrated time basis provides the necessary control signals and generates the internal clock frequency of 24 MHz. The clock frequency of the quartz is buffered on the "Y4MHz" output. This frequency can also be used for the micro-controller of the communication module. The analog input signal from the mains supply is fed to the "IN" input. After setting the "DUMP" signal, the calculated correlator values are stored in the output registers and can be taken from the "SCLK" / "SDAT" serial interface as a 32 bit word. Hardware and software evaluations are possible to calculate the correlation coefficients. To simplify the transmission filter that is generally required for EMC reasons, the sinusoidal transmit signals of the transceiver are generated by a low-glitch 8 bit D/A transformer. The block diagram for the entire transceiver is illustrated in Figure 5.17.

The transceiver is completed with a circuit that generates synchronization signals and a power supply. The analog signals from the mains supply that are to be detected are fed to the transceiver via a high-frequency coupling network in combination with a pre-amp with automatic gain control (AGC). In transmission mode the high frequency signals are amplified with a driver stage and impressed onto the mains supply as a differential signal between phase and neutral conductors.

5 EIB Hardware

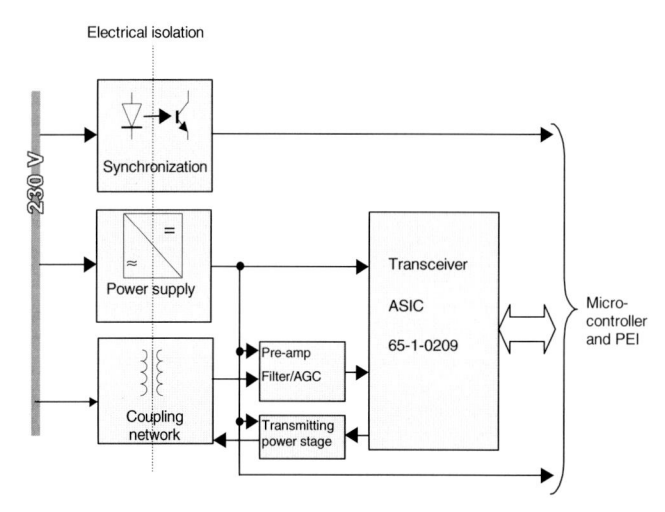

Figure 5.17: PL-Transceiver Block Diagram

5.3.2.3 Radio Transceiver

Radio transmission occurs in the frequency range between 868 MHz and 870 MHz. This range is reserved for short range devices and for the end user there is neither charge nor license requirements. It is characterized by the fact that the maximum transmitting power of the devices is limited and the maximum transmission time per hour is also restricted via the so-called duty cycle control. This results in reliable, interference-free operation with short reaction times.

In order to minimize transmission time and with that channel access, the system works with a relatively high transmission rate of 19.2 kbit/s when compared with EIB-TP and EIB-PL. Frequency shift keying (FSK) is used as the modulation method.

The radio transceiver (Figure 5.18) comprises a phase locked loop (PLL) for programming the actual frequency, a modulator with transmitter, a pre-amp with demodulator for the reception circuit, a high frequency commutator for switching between send and receive as well as a mains or battery module for the power supply.

In order to get the best use from the battery and with that to prolong its life, the radio transceiver works with voltages down to less than 2V. As the application module is also supplied with power from the battery via the PEI, the application circuit needs to be modified with regard to current consumption. In general, a differentiation is made between standby and active statuses.

The battery-supplied RF transceiver additionally contains intelligent activation logics (see Figure 5.19). The PEI is also supplied from the battery in standby mode. The transceiver itself (up to the activation circuit) and the micro-controller of the processor are switched off. Upon detection of a change in the current consumption of the application circuit (e.g. by pressing an operating key) the operating voltage is switched on, the DC/DC transformer activated and the application module continues to be supplied with 5V via the PEI. After successful transmission, the activation logics are returned to standby operation by the micro-controller via the "sleep" signal.

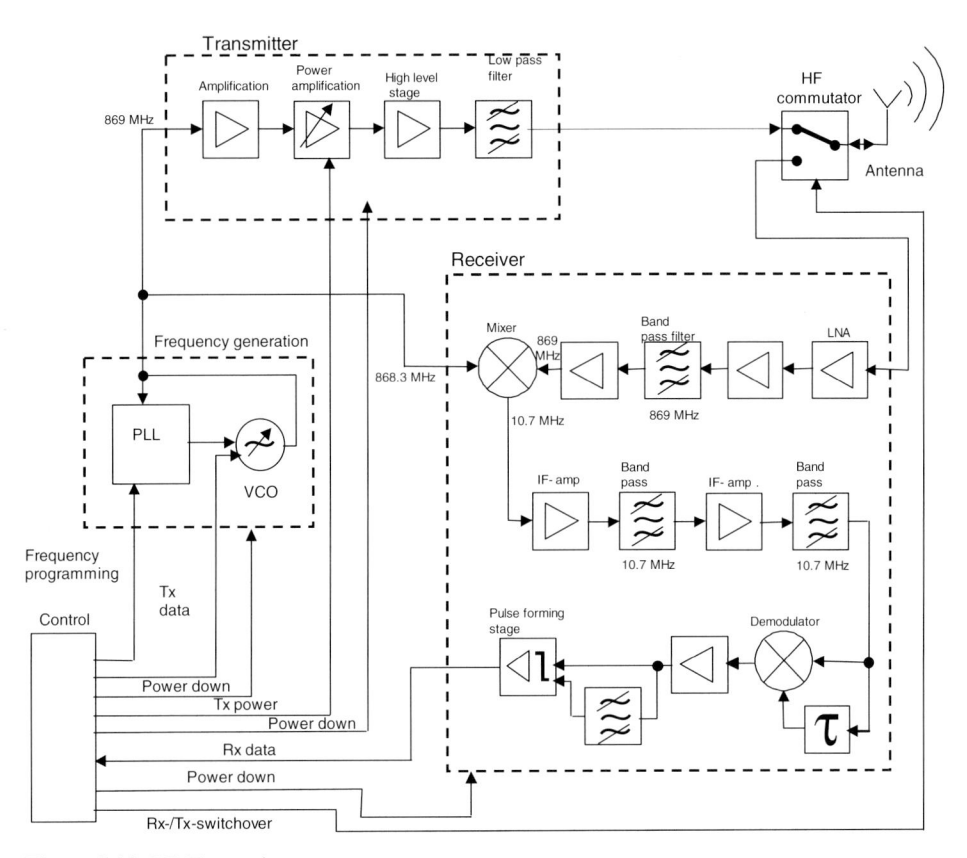

Figure 5.18: RF-Transceiver

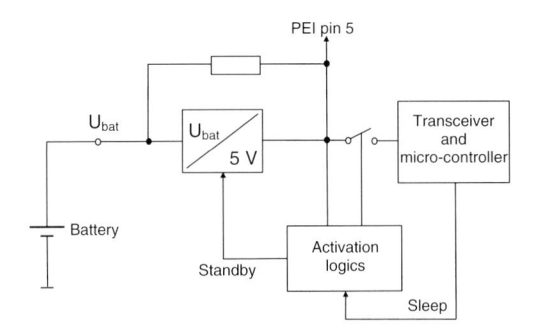

Figure 5.19: Activation Circuit of the Battery-Supplied HF Bus Coupling

5.3.2.4 Internal Power Supply

With TP, the internal power supply can either be derived from a third-party supply (e.g. 230 V, 24 V) or from the bus. If a device is to be supplied from a third-party, this supply must be isolated from the mains supply in accordance with EN 50090-9-2. If the power supply is to be taken from the bus, it is necessary to make sure that the current consumption is as low as possible and that load swings in the device do not lead to voltage differences on the bus that equal a signal to be transmitted. This means that under certain conditions it is necessary to use a buffer capacitor in the application circuit. In order to be able to save data in the case of power failure, the communication modules buffer the supply for a minimum of 60 ms.

With PL applications the internal device supply is derived from the 230 V supply and with RF applications either from the 230 V supply or from a battery. The battery supply of sensors in particular results in the flexible use of RF sensors as there are no cables to be laid. A 3V lithium cell battery is used. If the RF transceivers are used in actuators, the 230V mains supply is also generally available.

With regard to the 230V supply, a differentiation is made between a 2-line connection and a 3-line connection. At the installation sites of regular light switches, there is generally only the switched phase in addition to the phase (2 line). With the 2-line connection, the supply current for the transceiver is supplied via the consumer connected in series. This supply can only be used if the transceiver works with low current and the consumer also permits this current flow in the "switched off" status (e.g. not possible for fluorescent tubes). This supply option is available to RF transceivers. PL transceivers always require a 3-line connection, which means it may be necessary to lay a neutral conductor within the scope of a retrofit. If however, phase, switched phase and neutral conductor already exist (3-line connection) the above-mentioned restrictions do not apply.

5.3.3 Processing Unit

As already mentioned in chapter 5.3.1, the communication module is formed by the processing unit in combination with the transceiver. The implementations represented in the following chapters use Motorola micro-controllers exclusively.

5.3.3.1 Twisted Pair Implementations

It has already been stressed on several occasions that the twisted pair represents the "classic" EIB medium. With the passage of time, the increase in the performance of the processing unit can be largely attributed to the improvements in existing micro-controllers. The individual implementations are generally distinguished by the respective operating systems.

Operating system version 1.x

The 1.x implementation is currently based on the 68HC05B6 standard micro-controller. Communication modules with this solution are distinguished by the following functional units and properties:

- 8 bit CPU
- 6 kByte ROM
- 256 byte EEPROM
- 176 byte RAM
- Watchdog
- Serial, asynchronous interface

- Serial, synchronous interface with additional shift register on port A for reception
- 5 inputs/outputs (digital via port C or analog via port D)
- 1 analog input for type recognition
- 1 digital input for the programming key (port B)
- 1 digital output for the programming LED (port B)
- Up to 2 pulse width modulation outputs (PWM)

The micro-controller is operated with 4 MHz externally and 2 MHz internally. This means that even with low current requirements (typically 3.5 mA), supported by the 16 bit timer register and the capture and output compare register, it is possible to execute the software generation and sampling of the bit pulses. With this variant programs can be executed in the EEPROM. This is an important prerequisite for the remote loadability of application programs that is implemented with EIB. Figure 5.20 illustrates a basic circuit. The BCU and BIM implementations differ mainly in their use of port A. With the BCU this port is used for the synchronous, serial reception via a shift register. With BIM, this port is available for digital inputs/outputs.

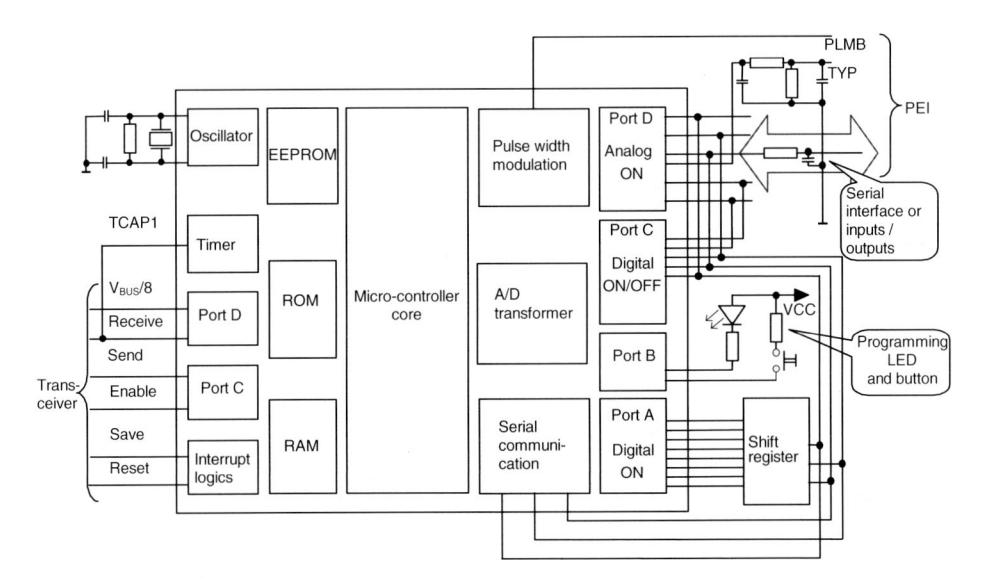

Figure 5.20: Basic Circuit with 68HC05B6

Operating system 2.0

The implementation is based on the 68HC05BE12 micro-controller. The 68HC05BE12 has been specially developed for EIB use. Compared with the 68HC05B6, it is characterized by a 12 kByte ROM, 1024 byte EEPROM, 384 byte RAM, multiplexed outputs and a programmable I/O controller (bit engine). By virtue of the multiplexed outputs the micro-controller is available in a smaller housing which facilitates a greater level of miniaturization. The bit engine relieves the 68HC05 core from the generation and sampling of the bits and takes responsibility for the additional time-critical parts of the medium-specific part of the EIB protocol. The micro-controller is usually operated with 4.9152 MHz. This allows operation of the asynchronous interface with 19.2 kbit/s in addition to the 9.6 kbit/s as with the 68HC05B6. The bit engine is a co-processor with a separate program and data memory. It has two program counters that work independently with a time offset. This means it is easy to implement serial EIB communication with the two main tasks of "send" and "receive". The maximum possible program size of the bit engine is 2×1 kByte. As a further specialty, the 68HC05BE12 also has an 8-bit port (port C) at which the timer inputs/outputs, pulse width modulator, A/D transformer and serial interfaces in combination with digital outputs can be multiplexed. The selection of the four per-pin possibilities is achieved via the software using two special registers. The switch is made between input and output functionality via the data direction register. Figure 5.21 illustrates this using the example of port C pin 6 together with port D pin 3.

Mux 15	Mux 14	DD 6	Function
X	X	0	Digital/analog input
0	0	1	Digital output
0	1	1	PWM A
1	0	1	Reserved = 0
1	1	1	BE-PORT 1

DD = Data direction
PWM = Pulse width modulator

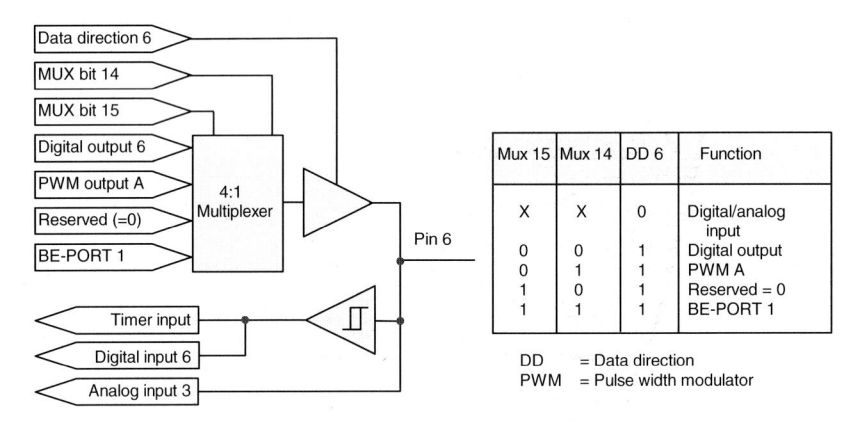

Figure 5.21: Multiplexed Outputs on Port C Pin 6

Operating system version 70.x

The implementation for operating system version 70.x is based on the 68HC11E9 micro-controller (Figure 5.22). By the externally available address and data buses it facilitates the installation of processing units with up to 32 kByte EEPROM and 16 kByte RAM. As the micro-controller does not have an internal pulse width modulator and due to its own current consumption cannot provide the required 10 mA for the application, it is not possible to operate a PEI without additional components.

Currently available BAUs therefore only offer the serial interface. As with the 68HC05BE12, the asynchronous interface can also be interrupt-controlled, as the controller is relieved from the task of bit sampling and evaluation thanks to the use of the "BitOK" signal of the FZE-1065 transceiver element (Figure 5.22). The maximum data rate to the application module is 19.2 kbit/s with asynchronous operation. With a correspondingly designed address decoder it is possible to operate extra inputs and outputs via latches. These should lie in the memory range of C000h–CFFFh. Other address ranges reserved for the system must not be used. The main applications for the 70.x implementation with 68HC11E9 are more extensive functions such as presence simulation, time and logic functions, interfaces and gateways.

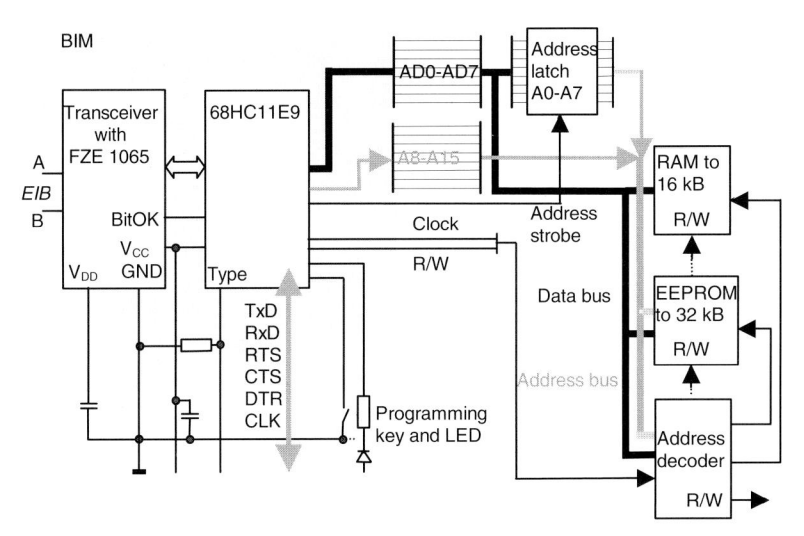

Figure 5.22: Block Diagram for a 68HC11E9 Implementation

5.3.3.2 PL Implementation (Operating System 1.x)

The processing unit for PL transceivers is based on the 68HC05B16 micro-controller. This differs from the 68HC05B6 due to larger ROM (16 kByte) and RAM memories (352 Byte). The example circuit depicted in Figure 5.23 shows the input and output settings of a power-line communication module.

Port B of the micro-controller is used almost exclusively to control the SNB1 transceiver ASIC. Capacitive or inductive networks can be used as coupling circuits to the mains supply.

Special attention must be paid to the power amp of a PL communication module. With low supply impedances, this must have the necessary driver capabilities. To prevent interfering with reception, it must be possible to switch the amp into a high resistance status (tristate).

5.3.3.3 RF Implementation (Operating System 2.x)

The 68HC05BE12 micro-controller is used for the RF variant of the processing unit. The radio implementation for this controller corresponds to the previously discussed TP implementation as far as the media-specific layers. The essential differences lie in the bit engine, which with radio must also guarantee synchronization among the end devices as well as managing the main tasks of send and receive.

5.3.4 Communication / Application Module Interface

The standardized interface between the communication module, bus coupling unit and application module, also advantageous for other product designs, is called the physical external interface (PEI). Mechanical data, electrical data and protocols are defined for the PEI. The protocols are discussed in detail in chapter 6.2.

With regard to the mechanical data, this includes in addition to the dimensions of the connectors, permissible values for the mating force (max. 3 N/pin), withdrawal force (min.

5 EIB Hardware

0.5 N/pin) as well as the maximum contact resistance (25 mΩ). A corresponding test is outlined in the EIB manual. For reasons of interference immunity, the PEI connection can be lengthened to a maximum of 100 mm. The PEI facilitates both parallel and serial communication. Figure 5.24 illustrates the occupancy of the 10-pin and 12-pin connectors.

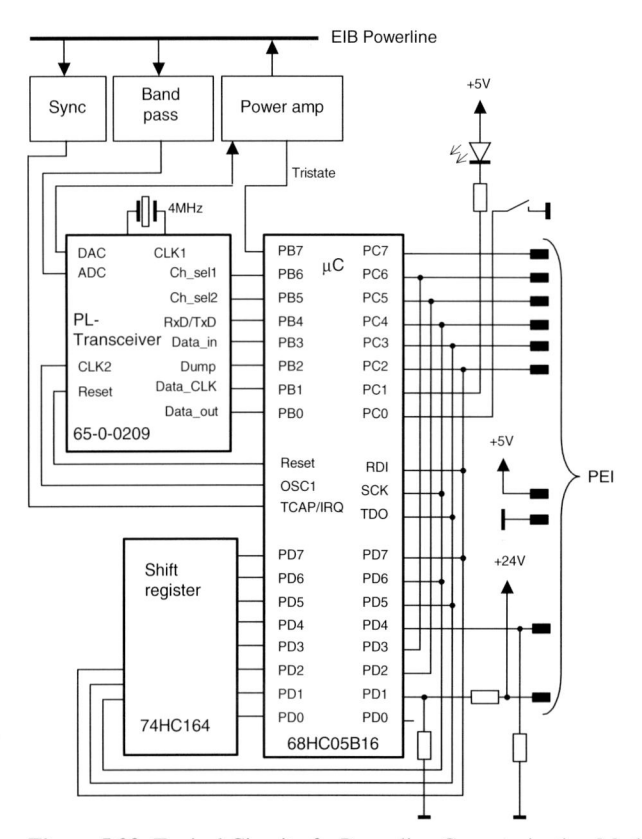

Figure 5.23: Typical Circuit of a Powerline Communication Module

Figure 5.24: Pin Configuration of the PEI

Figure 5.25 illustrates the basic functions of the PEI and examples of possible configurations. The 5V pin can take a maximum current of 10 mA and the 24V pin a maximum of 2 mA. The power drawn on both outputs must not exceed a typical value of 50 mW and under no circumstances must it lead to a bus side current of more than 12 mA with the smallest allowed power supply of the bus device (20 V). To suppress interference, a choke with an internal resistance of approx. 10 Ω is inserted into the V_{CCE} line to the application module. For the same reason, RC modules are connected in the data lines and the type line to the micro-controller. It should be noted that this reduces the edge steepness of the PEI output signals. These measures to suppress interference on the PEI inputs and outputs cannot prevent all possible interference. When designing a non-electrically isolated application module therefore it is necessary to ensure that the earth capacitance is kept as low as possible so that any occurring interference currents through the application, the PEI and the communication module are as low as possible.

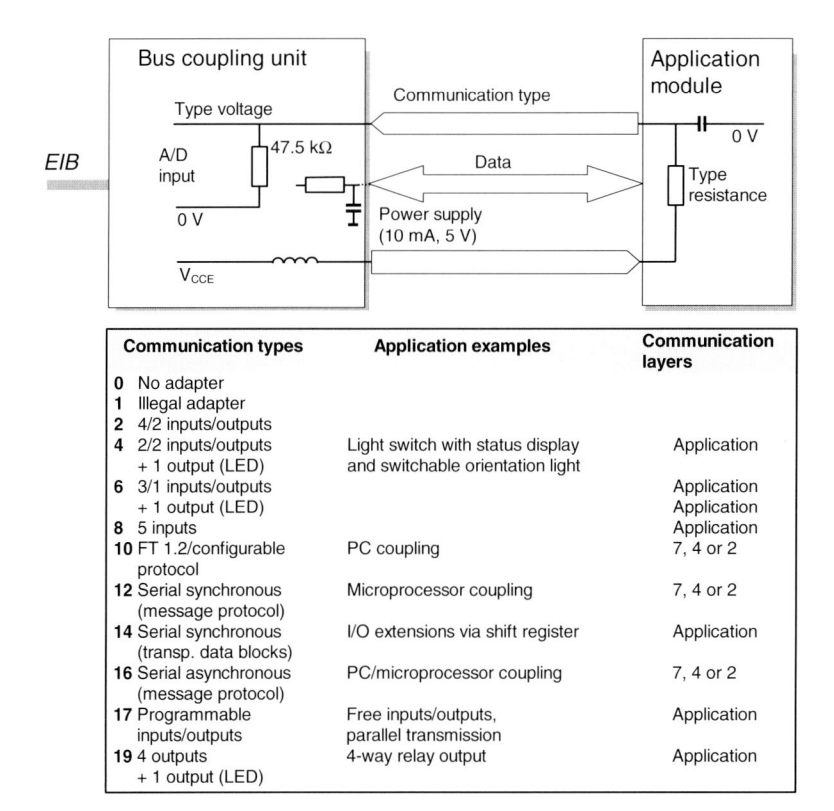

Communication types	Application examples	Communication layers
0 No adapter		
1 Illegal adapter		
2 4/2 inputs/outputs		
4 2/2 inputs/outputs + 1 output (LED)	Light switch with status display and switchable orientation light	Application
6 3/1 inputs/outputs + 1 output (LED)		Application Application
8 5 inputs		Application
10 FT 1.2/configurable protocol	PC coupling	7, 4 or 2
12 Serial synchronous (message protocol)	Microprocessor coupling	7, 4 or 2
14 Serial synchronous (transp. data blocks)	I/O extensions via shift register	Application
16 Serial asynchronous (message protocol)	PC/microprocessor coupling	7, 4 or 2
17 Programmable inputs/outputs	Free inputs/outputs, parallel transmission	Application
19 4 outputs + 1 output (LED)	4-way relay output	Application

Figure 5.25: Basic PEI Functions

Via the communication type, the application module informs the bus coupling unit of the required PEI configuration. The communication type is represented by an analog voltage value, which is determined in the application module via the type resistance (0-910 kΩ). Taking into account the tolerances of regular resistances, it is possible to define 21 different PEI configurations.

There are four different categories, described below:

Category 1 (PEI types 0 and 1)

PEI type 0 is always recognized when no application module or type resistance is connected. Type 1 must not be used by any application module. The setting of this type allows the application program to be stopped or deactivated. This application is always used where the application program has not been implemented in the communication module.

Category 2 (PEI types 3, 5, 7, 9, 11, 13, 15, 18 and 20)

These configurations are reserved and intended for future extensions.

Category 3 (PEI types 2, 4, 6, 8, 17 and 19)

These PEI types allow parallel I/O communication for both digital and analog signals.

Category 4 (PEI types 10, 12, 14 and 16)

The configurations in this category facilitate serial communication via the PEI. With PEI type 14 a serial protocol is activated that allows the transmission of application specific data blocks within the application. This means that transmission is completely transparent and with that it is possible to freely define the contents of the data blocks. A typical application is the I/O extension via a shift register in the application module. Contrary to this, the serial protocols with PEI types 12 and 16 allow messages to be exchanged at the layer boundaries between communication module and application module (see Figure 5.26). The desired layer can be selected via a configuration register. For message exchange, both the synchronous and asynchronous protocols use RTS and CTS lines for the hardware handshake in addition to the RxD (data input) and TxD (data output) signals.

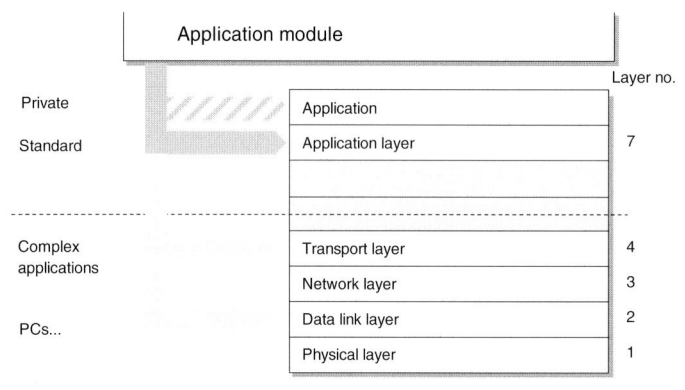

Figure 5.26: Inter-Layer Communication via Serial PEI Interfaces

Synchronous transmission can be achieved at a rate of between 300 bit/s and 41.66 kbit/s. Clock phase and polarity can be adjusted. With asynchronous transmission the data rate is fixed at 9.6 kbit/s. A data word consists of a start bit, 8 data bits and a stop bit. Parity information is not transmitted. Serial asynchronous transmission with PEI type 10 is supported by implementation version 1.2. The protocol is based on the international standards IEC 870-5-1 and 870-5-2 (DIN 19244). Only the RxD and TxD lines are required as the transmission is based on a pure software handshake. The data word additionally contains a parity bit (even parity). The transmission rate can be selected between 1200 bit/s and 19.2 kbit/s. The default standard value is 19.2 kbit/s.

5.3.5 Certified Communication Module

Depending on the implementation, the components of a communication end device previously outlined in this chapter can be combined in different ways to form a variety of units. Manufacturers are able to independently develop components and modules assuming they follow the specifications of the EIB manual [EIBA 99]. In order to guarantee the conformity of the device, certification costs and requirements [Kris 97] are relatively high in this case. In addition to the interworking, EMC and safety tests, it is also necessary to check the transmission technology, the communication protocol and the device management. This however is not necessary if using certified components such as the bus coupling unit or bus interface module (BIM). Figure 5.27 illustrates the standard modules.

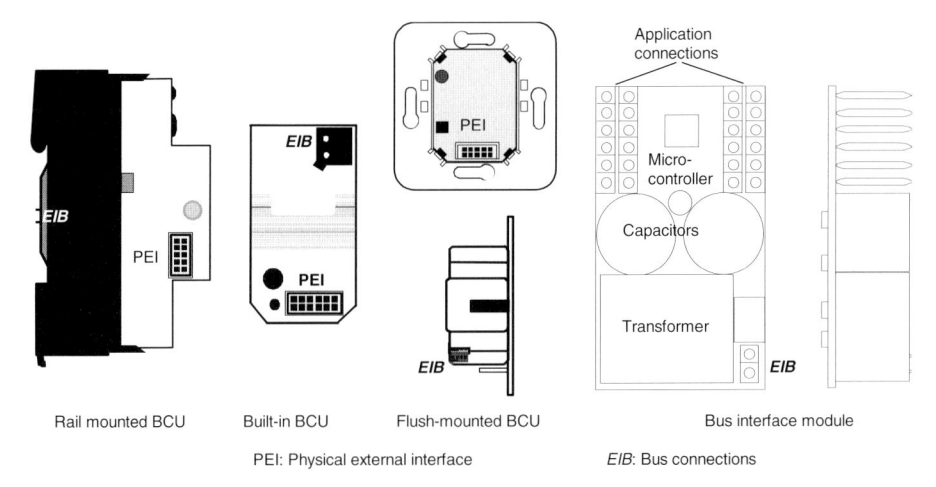

Figure 5.27: Certified Communication Modules

Bus interface modules are soldered or detachable communication modules, which with regard to communication technology, facilitate the low-cost development of integrated devices. A further advantage of the use of such modules is the ease of transferring an application to a different medium. Bus coupling units are available in a number of different designs suitable for various places of installation such as flush-mounted installation sockets, distributors or end devices. The use of available transceiver-ICs (FZE 1065/1066, TPUART) together with specified electronic components and micro-controllers with certified system software in the ROM or of tested and certified source code, allows the optimization of products with regard for example to design, size and cost. The cost of development and testing is therefore always less than that for a completely unique implementation.

5.4 Couplers and Interfaces

Couplers are used to connect EIB subsystems, which for electrical reasons cannot be connected directly. Couplers also allow a transition between different media. The classification of these functions, within the context of the ISO/OSI model, is examined in more detail in chapter 7. Here, we will limit our discussions to their concrete implementation. To finish

5 EIB Hardware

with we will then present the data interface that facilitates the transfer from the EIB system to a higher-order control (generally a PC).

5.4.1 Twisted Pair Coupler

The TP coupler, often referred to as simply the line coupler, is always required when several lines need to be linked together to form one larger system. As already explained in chapter 5.1, it is possible to combine up to 12 lines to form a functional zone and then to combine up to 15 functional zones using TP couplers. TP couplers used to combine lines are called line couplers and those used to combine zones, backbone couplers. Figure 5.28 represents a block diagram of these elements. Both devices are identical and differ only in the physical address that is assigned at commissioning. The physical address of the TP backbone coupler consists of zone "≠0", line "=0" and device number "=0"; with the TP line coupler the address comprises zone "≠0", line "≠0" and device "=0". The address assigned to the TP backbone and line couplers corresponds to the respective hierarchical stage.

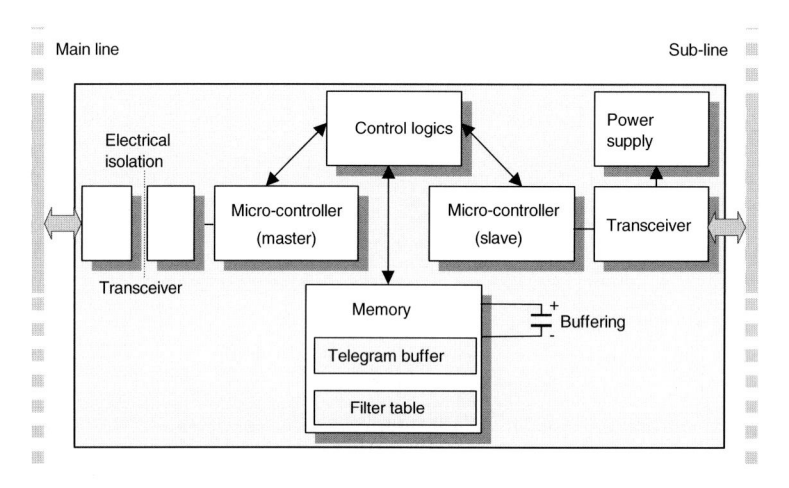

Figure 5.28: Block Diagram of the TP Coupler

When passing on telegrams it is necessary to distinguish between telegrams with physical addresses and those with group addresses. The passing on of telegrams with physical addresses is dependent on direction. In the direction of higher hierarchic levels (main line or backbone) all telegrams up to and including those of the current line are forwarded whilst in the opposite direction, only those that contain the actual line address (sub-line). As the group addressing is flat, i.e. independent of hierarchy or topology, it is not possible to read directly from the address whether or not forwarding is required. In order to avoid a situation in which all telegrams are passed on, which in the case of larger systems could lead to an overloading of the connecting main lines and backbones, configurable address filters (filter tables) are integrated into the TP couplers (see Figure 5.28). These address filters are generally configured automatically via the commissioning tool during the commissioning stage. For each group it is possible to specify whether or not the transferring of a telegram is necessary. Furthermore, the TP couplers must be able to buffer telegrams if the line is temporarily occupied on the other side.

In the current implementation this may involve up to 10 telegrams in each direction. If however the buffers are all full due to a high bus load then the TP couplers on the corresponding side acknowledge with "busy". When a telegram is passed through, the coupler reduces the routing counter contained in the telegram by one, if it is smaller than seven. If a negative value is obtained when the routing counter is reduced, the telegram is rejected. This prevents total breakdown in the bus communication due to continuously circulating telegrams in the case of a faulty direct link between two sub-lines (Figure 5.29). Once the routing counter reaches seven, the filter function of the TP coupler is deactivated for this telegram. This can be important for system diagnosis. The TP coupler electrically isolates the lines that it connects (600 V test voltage). This avoids any unwanted potential coupling.

The size of an EIB system attained with the use of TP line and backbone couplers is generally sufficient. In the case of extensions however it may prove impossible, e.g. on the basis of the topology, to locally insert extra line couplers. In such situations the TP coupler can also be used as a repeater. This facilitates the extension of the line by up to three segments, which must have a separate power supply in the same way as a single line and can only support the same number of end devices as a single line. As part of the physical address, a repeater must be assigned with a device number not equal to zero. If a telegram sent by a repeater is not acknowledged positively, it is possible to specify for the physical and group addresses separately how often transmission is to be repeated. When arranging the repeaters it is necessary to ensure that the serial consecutive connection can lead to an overflow of the routing counter of a telegram as otherwise the maximum number of 6 couplers could be exceeded (Figure 5.29).

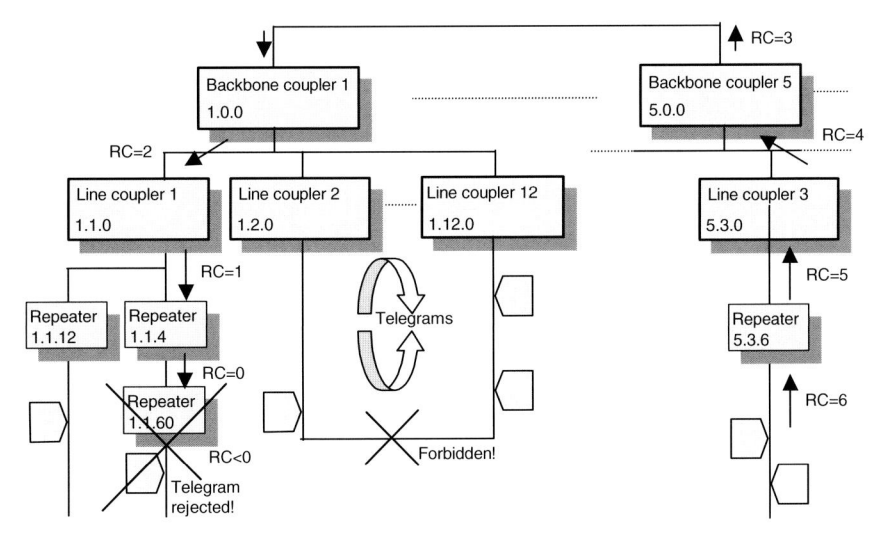

Figure 5.29: TP Couplers within a System

5.4.2 Powerline Band Stop

Band stops are used to separate the signals of PL systems from one another and from the official supply mains. This means that every system must be fitted with its own band stop. A PL band stop filters PL signals in the frequency range of 95 kHz to 125 kHz. The band stop

5 EIB Hardware

works in a direction dependent way directly on the physical layer of the EIB-PL medium. The applicable standard is prEN 50065-4.

It is necessary to remember that the "powerline" medium, like the radio medium, is an "open" medium. In other words, all devices connected to the mains network are, in terms of information technology, connected with one another whether this is desired or not. The band stop facilitates segmentation. Figure 5.30 illustrates the use of a single-phase band stop. In general however it is necessary to use a three-phase band stop, as the PL signals are coupled to all three external conductors due to the line capacities. The specified filter curve for a PL band stop is illustrated in Figure 5.31. Signal attenuation at the mid-frequency of the transmission band (110 kHz) must be greater than 30 dB. A minimum attenuation of 25 dB is required at the boundaries of the transmission band. The attenuation is measured from the PL signal area to the mains.

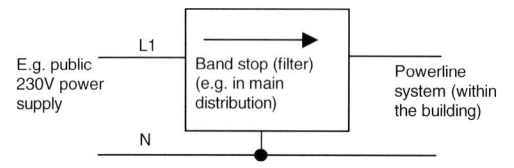

Figure 5.30: Using the EIB-PL Band Stop

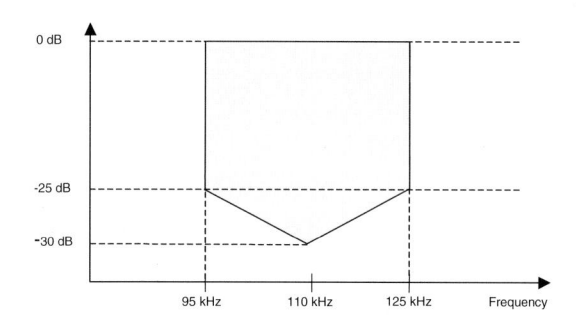

Figure 5.31: Requirements on the Frequency Response of a PL Band Stop

As already mentioned, the band stop is used for a number of different functions:

– The band stop facilitates the creation of closed signal areas in order to suppress the unwanted and problematic flow of signals from a system or system part. By virtue of its filter effect, the band stop suppresses the flow of interference signals into a PL system part.

– It is necessary to create communication islands for the defined coupling of different PL system parts via a media coupler. At the same time, the bus load in the individual islands is significantly reduced by the filter. PL band stops are inserted for every external conductor in the neutral point of a distribution or sub-distribution. In order to exclude any cross-talk effects, it is necessary to avoid the parallel laying of the incoming and outgoing lines. The maximum connected load of the band stop (e.g. 63 A) must be observed.

5.4 Couplers and Interfaces

5.4.3 Powerline Phase Coupler and Repeater

In the majority of cases, EIB-PL systems or end devices are designed for single-phase connections. However, communication in the powerline medium should cross all phases. Phase coupling of the signals is achieved through lines laid in parallel. For installation cables we can assume a unit length capacitance of between 20 and 100 pF/m, depending on the structure. The phase coupler is provided to allow phase coupling in PL networks. A phase coupler consists of defined coupling capacitors between the external conductors in a triangular connection. There must be no connection to the neutral conductor due to the signal attenuation effect. This type of passive phase coupling (Figure 5.32) is recommended for small to mid-sized systems.

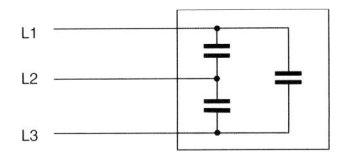

Figure 5.32: EIB-PL Phase Coupler

In the case of larger systems or where there are high levels of interference within an installation, active phase coupling via a PL repeater must be provided (Figure 5.33).

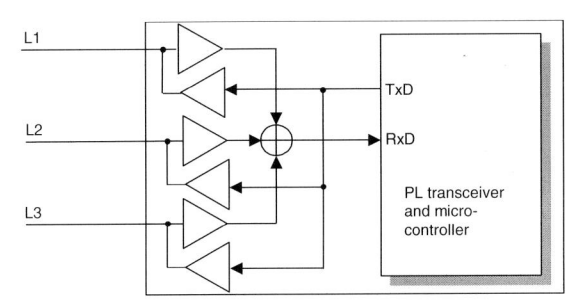

Figure 5.33: PL Repeater

The EIB-PL repeater consists of three identical transmitters and receivers. The device receives all telegrams of the three-phase network via the three RxD paths. Transmission errors lead to the absence of the acknowledgement telegram. For the EIB-PL repeater, this is the trigger for the three-phase repetition of the telegram. The advantage of this structure is that the repeater is only activated when required and the transmission times are not significantly prolonged during repeater operation. In order to utilize the full range, the repeaters should, if possible, be placed in the center or "neutral point" of a system. A system must be configured for repeater operation. Only one repeater is allowed in a system. Users should avoid the simultaneous use of repeaters and phase couplers.

5.4.4 Media Couplers

The hierarchy of an EIB system taking into account different media has already been outlined in chapter 5.1. The media coupler fulfils the functionality of a line coupler for different

5 EIB Hardware

media. Structurally, the media coupler consists of two media-dependent communication modules, a telegram buffer and a filter table (Figure 5.34). The telegram buffer is necessary to provide intermediate storage, necessitated by the different transmission speeds in the various media.

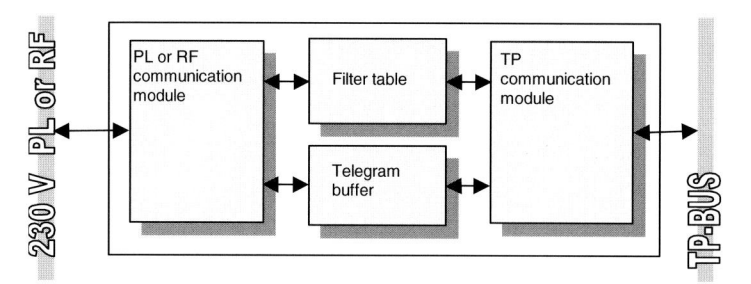

Figure 5.34: Media Coupler

The media coupler only passes on the group telegrams that are included in the filter table. The filter effect ensures the bus loads in the various media are significantly reduced. The PL part of the device is generally designed as a three-phase unit to ensure efficient coupling. With that, the PL/TP media coupler on the powerline side can also function as a repeater.

The system devices for the powerline technology must be arranged in a pre-defined order. A typical structure is illustrated in Figure 5.35. In order to achieve high signal attenuation between the PL lines, the use of band stops is absolutely essential. This doubles the signal attenuation of an individual stop as depicted in Figure 5.31.

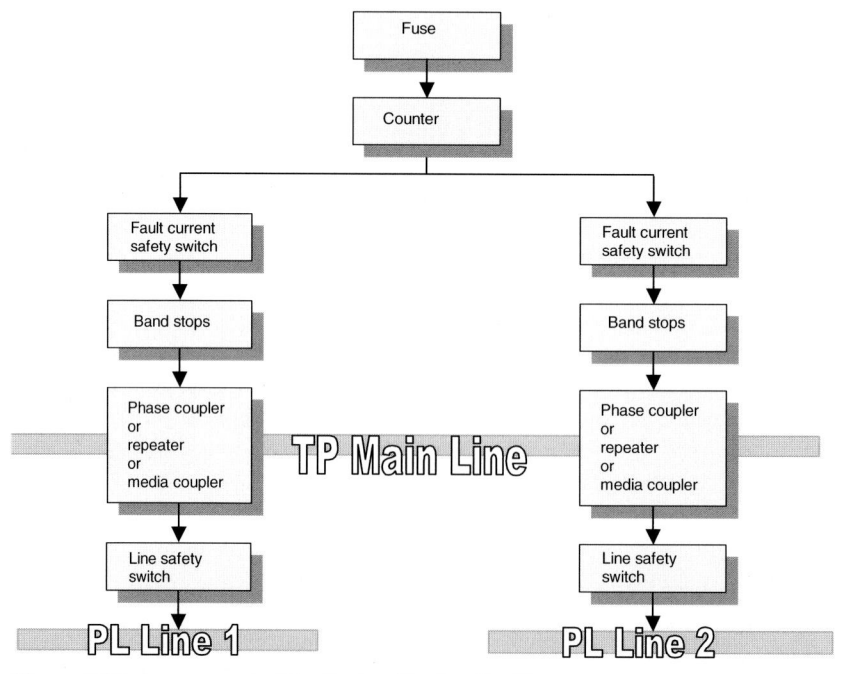

Figure 5.35: Arrangement of the System Devices for PL

5.4.5 Data Interface

The data interface is a device that facilitates data transmission between the EIB and a PC or other similar processing unit. The interface is available as a DIN rail mounted device for installation in the distributor or as a device for wall sockets. It consists of a standard bus coupling unit combined with an application module. The application module consists of electrical isolation, a level transformer and the Sub-D9 connection socket for the PC connecting line. The data interface works as Data Communication Equipment (DCE) and the PC as Data Terminal Equipment (DTE). Figure 5.36 illustrates the connections.

The data interface supports the RS 232-C (EIA-232-C, CCITT V.28, DIN 66256 T1) and RS 562 (EIA/TIA-562) standards with the following exceptions:

− The transmission rate must be at least 9.6 kbit/s

− DTR must be set to positive polarity

− After initialization there must be a time lag of 100 ms before transmission can begin.

RS 562 is used by some notebooks that work with low voltages and signal levels.

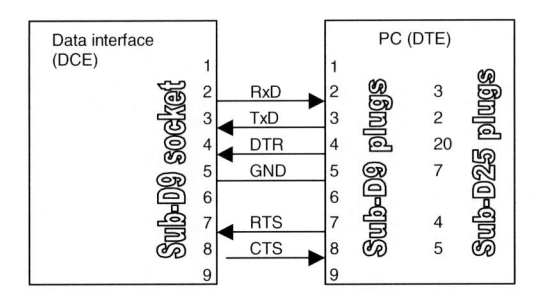

Figure 5.36: Connection Settings for the Data Interface/PC

The following data/conditions are true for the data interface:

− Minimum signal input level:	± 5 V
− Maximum signal input level:	± 15 V
− Minimum signal output level:	± 3 V
− Transmission speed:	Min. 9.6 kbit/s (requires high-speed opto-coupler for electrical isolation)
− Input resistance:	± 5 V - ± 9 V, $R_i \geq 2500\ \Omega$ $\geq \pm 9$ V, $R_i \geq 150\ \Omega$
− Electrical isolation:	4 kV AC for universal application

− DTR must be set to positive polarity

− After initialization there must be a time lag of 100 ms before transmission can begin

The line between the data interface and PC must not exceed 15 m with RS-232-C transmission and 5 m with RS-562 transmission.

6 EIB Software

EIB is a decentralized system in which the communication functions and to a large extent the application functions too are contained in the communication end devices [EIBP 97]. Whilst the communication software is permanently stored in the devices, the application software is generally only loaded into the devices at commissioning at which time the actual function of the device is determined.

Figure 6.1 represents the essential EIB software components, which fall into three parts. The communication module contains the system software and internal application, which are responsible for the following tasks:

1. The system software is responsible for the initialization, the cyclic polling of the application and communication via the network. It also regularly checks the communication module memory in order to detect any possible memory errors.

2. The internal application is generally responsible for communication with the application module via the physical external interface (PEI). This means that the internal application must be matched to the respective application.

The interface between the system software and internal application is implemented using communication flags (often referred to as RAM flags), which control access to the content of the communication objects.

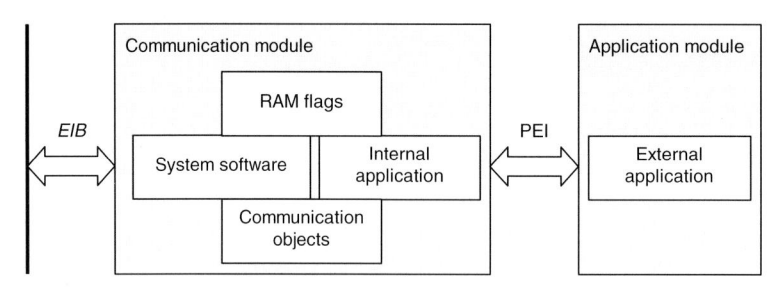

Figure 6.1: Components of the EIB Software

Now to the physical external interface (PEI): If we take a closer look at the application module, simple functions are often implemented here. Examples include the polling of a push-button or the switching of a lamp. For such situations there are predefined configurations for the PEI, which facilitate the direct connection of sensors and actuators. The internal application is solely responsible for the operation of these external units.

If more complex functions are to be implemented in the application module, which go beyond simple sensor or actuator mechanisms, the PEI can also be configured as a serial interface. In this case, the internal application, which operates one side of the interface, communicates with an external application which is executed in the application module. The serial protocols of the physical external interface therefore can be viewed as part of the EIB software. Figure 6.1 illustrates the situation.

The following sections are based on the previously described components of the EIB software. We will begin with a description of different uses of the system software. This is followed by a look at the interface between the system software and internal application. We then present the serial protocols of the PEI. As both the internal and external applications depend on the respective system and are not included in the EIB specification, we must limit our discussions to a description of the corresponding software interfaces.

All EIB implementations are based on the general software device model as depicted in Figure 6.2. Depending on the variant, only certain parts of the model may be implemented.

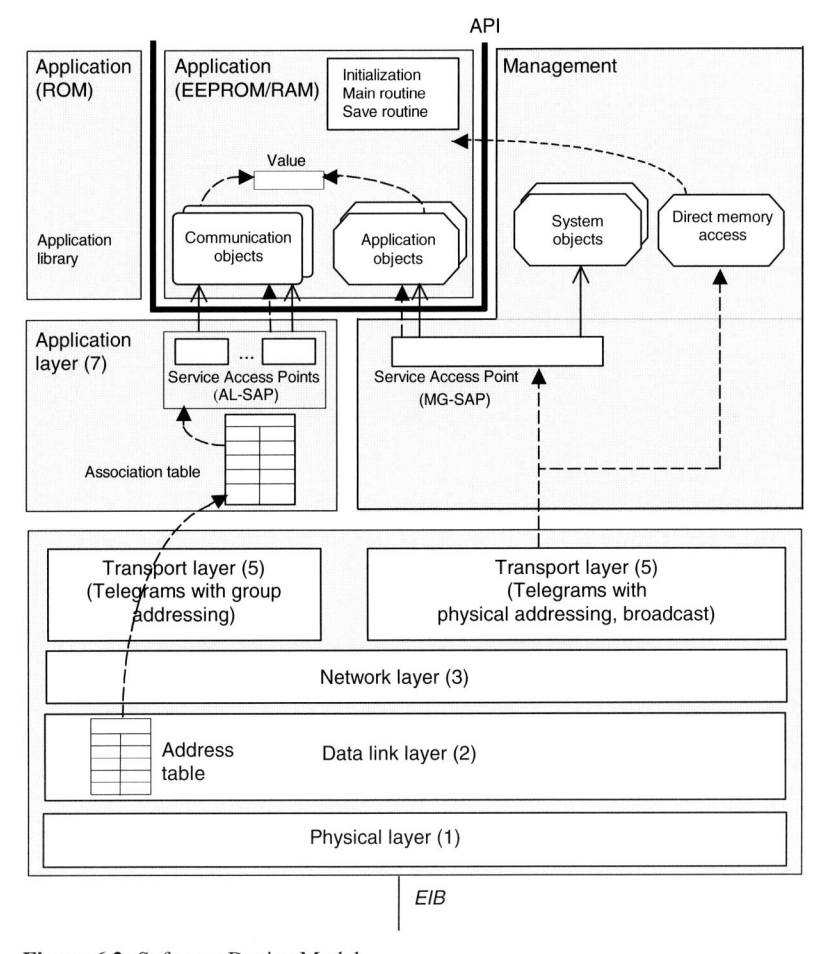

Figure 6.2: Software Device Model

193

6.1 System Software

The system software consists of two parts, a part that runs sequentially and an interrupt-controlled part.

The interrupt-controlled part is responsible for telegram reception from the bus and for the start of the save routine in the case of power failure. In the save routine, the application programmer determines which measures are to be taken after a loss of power (e.g. data backup).

The sequential part of the system software is, as shown in Figure 6.3, a large, cyclic loop. As a special feature of certain implementations, the few network layer functions are incorporated into the transport layer.

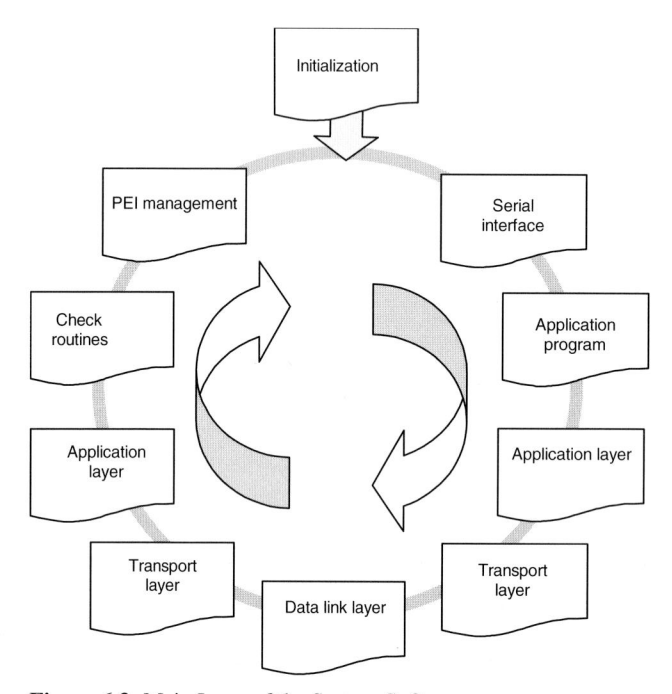

Figure 6.3: Main Loop of the System Software

6.1.1 Operating System Implementation 1.2

The implementation of the operating system in version 1.2 is based on the 68HC05B6 microcontroller.

Addressing

The three addressing mechanisms are supported – group addressing, broadcast as a special case of group addressing and physical addressing with direct memory access. Group addressing is a flat addressing method for the communication of operating functions via communication objects (see chapter 6.2). Due to the RAM/EEPROM memory resources, the maximum number of communication objects is limited to 12. Furthermore and in contrast to the general model (Figure 6.2), the communication objects are assigned to the application layer. It is not possible to program the end devices via group addresses. The physical addressing is a system wide addressing of the end devices that depends on the system topology and is reserved for management functions. Details are given in chapters 3 and 4.

The physical addresses are programmed as follows:

1. A check is made to establish whether the address already exists in the system. This involves an attempt to set up a connection with a device with the corresponding address and to then exchange data with this device. If no such device exists, there is no reply. If a reply is received, the programming process is interrupted.
2. A check is made to see whether a programming key has been pressed. For this, the Phys.Adr_Read service is sent via broadcast transmission. If several replies are received, i.e. several programming keys of communication modules have been pressed, or if no programming key has been pressed, the programming is cancelled.
3. Programming of the physical address.
4. Verification of the programming: An attempt is made to set up a connection with the device. If this is unsuccessful, the programming is cancelled.
5. Device is reset: A reset service is transmitted. With that, the programming LED on the programmed device is extinguished and the connection terminated.

Watchdog

The watchdog, as a component of the interrupt system of the 68HC05B6, is started automatically by the system software. Taking into consideration the required time for interrupt routines, the watchdog must be triggered every 1.5 ms. Concerning the application program, the watchdog is triggered by the system software before entry and after return.

Initialization

The initialization of a communication module comprises the following essential steps:

1. Wait until the save signal is no longer active
2. Initialize hardware
3. Initialize system software
4. Wait until the bus is free
5. Read PEI type
6. Check whether the programming key is pressed which means the application program should not be started
7. Check the checksum
8. Start the serial processing cycle (Figure 6.3)

After the initialization, the communication module is in the standard operating mode. Other important modes are the bus monitor and address allocation modes. In the bus monitor mode, the communication module is completely passive to the bus. Messages of the data link layer (acknowledge messages too) are passed directly onto the PEI. In the address allocation mode, the physical address can be read or programmed via broadcast transmission.

Application

The application is cyclically polled by the system software as a sub-routine. The time for this varies between 3 and 40 ms, depending on the possible communication via the bus or the serial interface. With this implementation it is only possible to operate "slow" running processes without additional preparation, such as those that occur in the vast majority of building automation and control applications. An application is only started if an application program has been loaded (i.e. the load control is in the "loaded" status), the application program is executable (e.g. there are no checksum errors) and the PEI types of application program and hardware match. If any of these conditions are not met, then even a running application program is stopped.

The start of an application program is always associated with the deletion of the application memory (RAM). The loading of an application program occurs via direct memory access. The save program is called up after a power failure (external interrupt), if the above-mentioned conditions for the application program are satisfied. After the save program has run through, the system executes a reset. The save program is not a component of the system software, as in this program the behavior of an end device is determined in relation to the application. A maximum dwell time of 10 ms is permitted in the application program. The maximum permitted duration of the save process is determined by the buffering time of the hardware. Use of the assembler commands *swi*, *cli* and *sci* is not allowed! In order to ensure compatibility with other variants or implementations, always avoid writing complete bytes to one port (e.g. *sta PortC, addr 02h, stx PortC*). This is necessary because parts of the port are used by the system itself, e.g. as transmission outputs.

Check routines

Using the EEPROM, the system periodically creates a checksum (EXOR). In case of error, the associated runtime error flag is set and the application program is stopped. A consequence of this is that it is no longer possible to access the communication objects. Other consistency checks such as length of the address table <116, PEI type in the EEPROM <20 and the address range of pointers supplement the checksum formation. The EEPROM content is only considered free from error when all these conditions are satisfied. The developer can decide how large the EEPROM area to be monitored should be. The area begins at 0108h and can end anywhere between 0108h and 01FEh. The checksum is recalculated after every write process to the range specified for monitoring.

Serial PEI function

Serial interface protocols with PEI configurations 12, 14 and 16 are supported. More details are given in chapter 6.3. Instead of a light switch for example, it is possible to connect an application module to the bus coupling unit that contains a serial data interface. If the bus coupling unit detects such an application module, the serial interface software is initialized. In this example, the application program remains inactive, as it is not assigned to a PEI configuration with a serial protocol.

The loading processes

The loading processes described below are used by tools such as the EIBA software tool (ETS). All loading processes are based on connection-oriented communication with direct memory access. The connection can be set up via the bus or via the PEI.

The steps are as follows to load the application program:

1. Set up the connection via the bus or the PEI,
2. Check whether the application is compatible with the communication module version,
3. Compile the EEPROM memory contents (data) for downloading into the communication module, including tables, parameters and user program,
4. Download the data into the communication module; the EEPROM memory areas up to the memory cells with write protection (e.g. manufacturer code) are loaded,
5. Set the length of the address table to the required value and reset the error flag. The program is in the "loaded" status.

Changing the address and association tables

The length of these tables can be modified within the scope of the reserved memory specified by the developer. The changes demand that the pointer to the association table may need to be changed in accordance with the newest status as the ETS is able to manage both tables as a common memory.

1. Set up the connection via the bus or the PEI,
2. Check whether the application is compatible with the communication module version,
3. Set all error flags in the communication module, set the length of the address table to 1,
4. Change the address table entries by direct memory access,
5. If necessary, change the pointer to the association table to the latest status,
6. Set the length of the association table to 0,
7. Change the association table entries by direct memory access,
8. Set the length of the association table to the required value,
9. Set the length of the address table to the required value and reset the error flags,
10. Terminate connection.

Application programming interface (API)

The following functions are implemented as the API:

- Manipulation of the communication objects:

U_flagsGet	Read RAM flags
U_flagsSet	Write RAM flags
U_testObj	Test RAM flags
U_transRequest	Transmit prompt

- EEPROM functions:

EEwrite	Write EEPROM
EEsetChecksum	Recalculate checksum

- Application functions:

U_debounce	Suppress bounce
U_delMsgs	Delete user messages

U_readAD	Read A/D converted value
U_map	Linearization function

- PEI functions:

U_ioAST	Binary Port_Access
S_AstShift / U_SerialShift	Byte exchange via serial PEI
S_LAstShift	Data block exchange via serial PEI

- Timer functions:

TM_Load	Start timer
TM_GetFlg	Read timer status
U_SetTM	Set user timer
U_GetTM	Read user timer status
U_Delay	Wait delay time

- Message management functions:

AllocBuf	Buffer reservation
FreeBuf	Release buffer
PopBuf	Retrieve message from buffer

- Arithmetic functions:

multDE_FG	Integer multiplication without sign
divDE_BC	Integer division without sign

- Various functions:

shlAn	Shift accumulator to the left
shrAn	Shift accumulator to the right
rolAn	Rotate accumulator to the left
U_SetBit	Write bit
U_GetBit	Read bit

- Table functions

AND_TAB	And table
OR_TAB	Or table

6.1.2 Operating System Implementation 2.x

Implementation 2.x basically guarantees that application programs and modules that have been developed for implementation 1.x and which only use documented features, can still be employed when using the EIBA software tool (ETS). The restrictions with regard to compatibility are essentially defined by the various transmission speeds of the synchronous interfaces, other pulse width modulation frequencies and by different access to the memory areas reserved by the system. The major extensions with regard to the system software are as follows:

- Serial protocol

- Support of EIB objects (see chapter 6.2.2, physical device communication with EIB objects)

- New management functions such as, for example, access protection
- Device polling mode

The main loop of the system software essentially corresponds to that of implementation 1.x (Figure 6.3).

Serial PEI function

This implementation additionally supports the serial protocol with PEI configuration type 10 (FT1.2). It is also possible to implement a separate serial protocol.

Access protection

Four access levels are supported. The meanings of the various levels are as follows:

1. Loading application programs
2. Loading the address and association tables
3. Changing application parameters
4. No access protection

These access levels are used for direct memory access and EIB object access. Access to communication objects is free. Memory areas such as address and association tables, application program, application parameters, protected user values, etc. can be assigned different access levels and passwords. Passwords have a length of 32 bits. Figure 6.4 illustrates the main procedure when accessing protected areas.

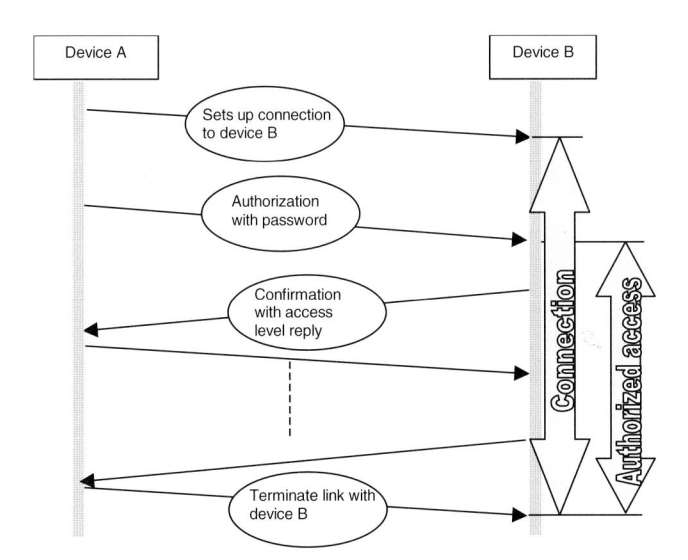

Figure 6.4: Access Protection

Serial number

Serial numbers are supported which are unique for each device. The European Installation Bus Association (EIBA) is responsible for assigning serial number ranges to manufacturers.

6 EIB Software

In addition to the methods described in chapter 6.1.1, programming with device serial numbers follows the procedure outlined below:

1. Check whether the physical address already exists in the system, if it does the process is cancelled.
2. Check whether a device with the given serial number exists, if not the process is cancelled.
3. Program the physical address using the broadcast service "Program physical address for device with serial number xy".
4. Verify programming by setting up a connection.
5. Reset the device.

Changing the address and association tables

With this implementation, the address and association tables as well as the application program are loaded via objects. The loading processes are controlled by loading state machines. The diagram represented in Figure 6.5 is valid for all state machines. The loading status of an object can be downloaded via a property assigned to it which also controls the loading process. The "not loaded" status does not simply mean that the corresponding data is invalid but also that associated resources such as memory for example, are released. Fundamentally, it is also possible to change tables/objects via direct memory access without having to reload. The length of the tables/objects however must not change.

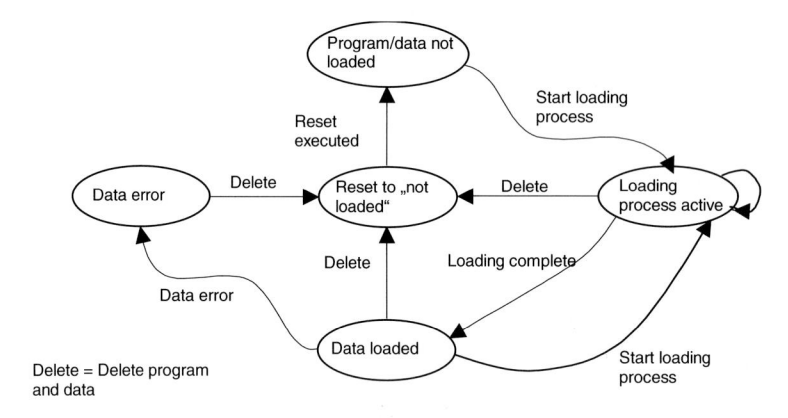

Figure 6.5: Loading State Machine

The following general procedure is valid for loading via a tool:

1. Set up connection via the bus or the PEI,
2. Authorization via password,
3. Check whether the application is compatible with the communication module version,
4. Delete program and data and wait till the status changes to "not loaded",
5. Set load control to "load" and wait till the status changes to "loading active",
6. Reserve necessary memory,

7. Load the data via direct memory access,

8. Load the segment control and information data blocks,

9. Set loading control to "loading complete",

10. Terminate connection.

Interface to the application

Operating system implementation 2.x involves a special form of the communication between system software and internal application. A message interface is implemented. The application communicates with the system software via a mailbox. Contrary to implementation 1.x, the communication objects belong to the application program and not to the application layer (Figure 6.6). This means that communication objects are only available when the application program is running.

The interface to the application is implemented as a callback function. The application program must install this at initialization thereby making it available to the system. Only then can the application layer access the communication objects. When it is ready to exchange data with the application, the application layer calls up this function asynchronously to the application program procedure.

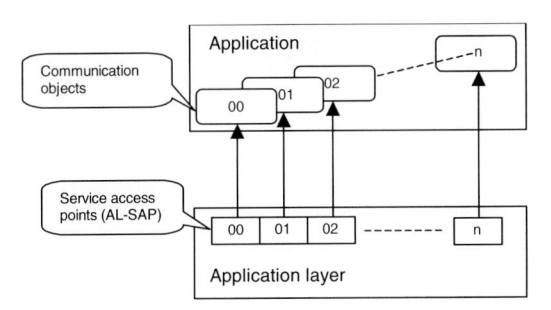

Figure 6.6: Position of the Communication Objects

Device group (polling)

This additional communication mechanism allows the status of a device group (maximum of 14) in a line (or segment when using repeaters) to be read with a telegram cycle. This mechanism can be used in the context of safety relevant applications, which require certain devices to be cyclically monitored in relatively short time intervals. Because there is no need to send individual telegrams for the polling, the overhead is relatively low and the bus load is not significantly increased. The polling mechanism is based on a master/slave principle. The master sends a status polling request to which the slaves respond with a single data byte in the time slots allocated to them. The polling group addresses have no connection with the otherwise used group addresses. Figure 6.7 illustrates the basic functioning of the polling mechanism. The master sends a status polling request to the AFFEh poll group. Device 1 of the polling group replies first with a status byte of 11h. Device 2 is faulty and fails to respond. In order to maintain the synchronization of the status byte, the master inserts a byte after a short wait with the contents FEh. This value is not permitted as a device status. Device 3 of the group then enters a status byte of 50h at the correct time.

6 EIB Software

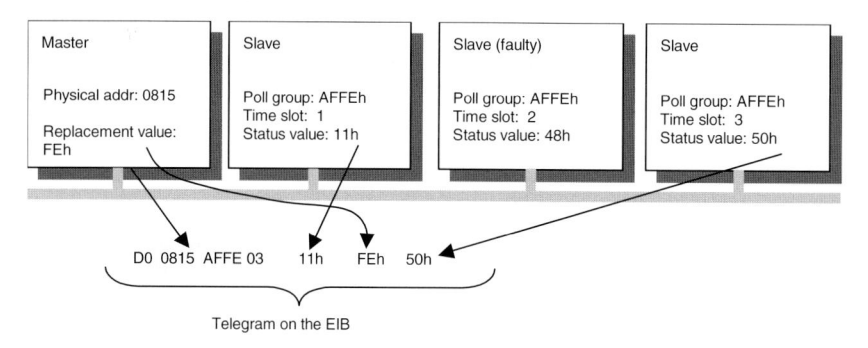

Figure 6.7: Example of Status Requests with the Polling Mechanism

Application programming interface (API)

Contrary to implementation 1.x, certain functions have been newly incorporated into the API.

- Timer functions:

U_TS_Set	Set timer
U_TS_Del	Delete timer
U_TS_Seti	Set timer within the interrupt routine

- Message system:

U_MS_Post	Send a message to the given message queue
U_MS_Switch	Divert a message from one queue to another

- Floating point conversion:

FP_Flt2Int	Convert IEEE floating point to an integer
FP_Int2Flt	Convert integer to an IEEE floating point

- Loadable PEI support:

U_FT12_Reset	Reset serial interface TYPE 10 (FT1.2)
U_FT12_GetStatus	Read FT1.2 protocol status
U_SCI_Init	Initialize asynchronous interface (SCI)
U_SPI_Init	Initialize serial synchronous interface (SPI)

- Various:

U_GetAcess	Read the current access level
U_EEWriteBlock	Write an EEPROM data block
U_SetPollingRsp	Set a value for polling status reply
AL_SAPcallback	Standard callback for communication objects

6.1.3 Operating System Implementation 70.x

Compared with the 68HC05 processors, the 68HC11 is characterized by a higher performance and greater memory areas. It is easily programmed in C or in Assembler. An extensive function library (in C) simplifies incorporation into the system environment. This implementation also supports multi-threading (see chapter 6.5.1). As a main application program, a thread can invoke two parallel running threads, which can also be quit at any time. Threads differ from tasks in that they must share the same resources. The general functionality of implementation 70.x essentially corresponds to that of implementation 2.x. Necessitated by the different resources and capacities there are a few major differences:

- Transmission and reception using the BitOK signal of the FZE-1065 transceiver are implemented using software as there is no hardware support for the media-specific layers (bit engine).
- A maximum of 254 group addresses and 255 associations and AL-SAPs are permitted.
- Access protection with up to 16 access levels is possible. This ensures that user software already loaded by the manufacturer, for example, is protected against overwriting by the EIBA software tool (ETS), whilst leaving open the possibility of modifying parameters.
- 16 bit pointers are used to a certain extent instead of 8 bit pointers.
- Each communication object is allocated 1 byte for RAM flags.
- The transmission status is extended to 3 bits. There is also an indication concerning whether it was possible to send a message or whether there is no association for the communication object.
- Multi-threading is supported.

We would like to take a closer look at the last point. A thread is a piece of program code, which runs independently of and parallel to a sequential program or other thread. Up to two additional threads can be active at the same time as the *main application* thread (Figure 6.8). The threads use the same resources (e.g. memory), but have their own stacks. Threads are executed in accordance with a time slot mechanism controlled by the system. If an application is quit, due for example to an EEPROM error, the system guarantees that all threads are quit. The *main application* thread is started by the system at the end of initialization. The main application is responsible for calling up subsequent threads.

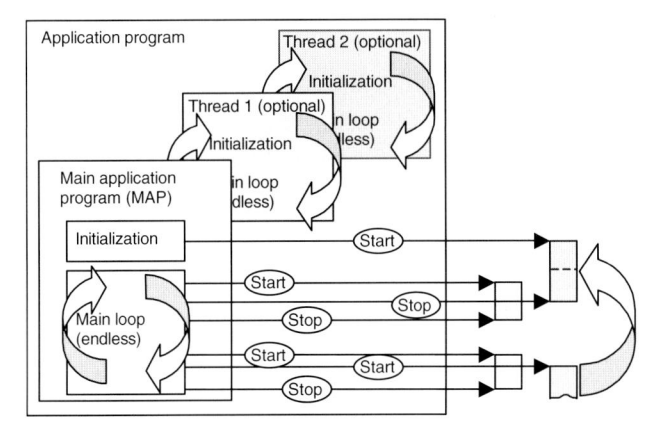

Figure 6.8: Use of Threads

6 EIB Software

6.1.4 PL-Implementation

System software 1.x of the EIB-PL bus coupling unit is implemented on the 68HC05B16 micro-controller. The structure is similar to that of implementation 1.x of the TP solution. The following section deals with the specific differences, in particular those in the media-dependent parts.

Figure 6.9 represents the main loop of the operating system software. A complete cycle takes approximately 8 ms with a processor frequency of 4 MHz. By pressing the programming key during the power-on reset, the system jumps past execution of the initialization and application program.

```
Initialization
   ↓
Check routines
   PEI type
   EEPROM
   Power supply
   ↓
Serial interface
   ↓
Application program
   ↓
Communication layers
   Application layer
   Transport layer
   Network layer
   Data link layer
```

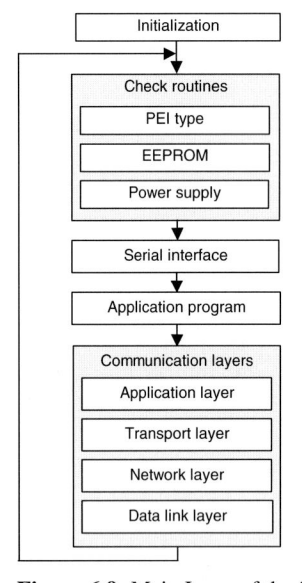

Figure 6.9: Main Loop of the System Software of 68HC05B16

The check routines include a power supply test in addition to the EEPROM and PEI type tests:

- The absence of the 50-Hz synchronization registers a failure of the mains supply (= bus voltage). The time between two crossovers of the mains ac voltage, T_{sync}, is established.
- T_{sync} = 20 ms: Normal operation
  T_{sync} > 60 ms: Suppress transmission, turn off driver stage
- T_{sync} > 200 ms: Start save routine, reset BCU
- The internal 24 V power supply is also checked. If the voltage drops below 15 V, transmission is suppressed.

There is an available memory of approximately 15 bytes for the system software. In the 68HC05B16 there are two user RAM areas of 176 bytes. RAM1 lies in the address range of 0050h to 00C0h, user RAM2 is addressable via 0250h to 0300h. There is a variable in storage location 0060h ("layer info"), which establishes the mode of the PL-BCU (Figure 6.10).

6.1 System Software

Bit #	7	6	5	4	3	2	1	0
Meaning	PARITY	DM	UE	SE	ALE	TLE	LLM	PROG

```
PROG    1 :   Programming mode
        0 :   Normal operation
LLM     1 :   Link layer in normal operation
        0 :   Link layer in bus monitor operation
TLE     1 :   Transport layer enabled
        0 :   Transport layer disabled
ALE     1 :   Application layer enabled
        0 :   Application layer disabled
SE      1 :   Serial PEI interface enabled
        0 :   Serial PEI interface disabled
UE      1 :   User program enabled
        0 :   User program disabled
DM      1 :   Download mode
        0 :   Normal operation
PARITY  Parity bit for this byte (even)
```

Figure 6.10: Powerline Bus Connection Mode

After a reset the application layer is active by default. Examples of allowed values of storage location 0060h are specified in Table 6.1.

Table 6.1: Possible Values of RAM Storage Location 0060h

Mode	Value
Bus monitor	90h
Link layer	12h
Transport layer	96h
Application layer	1Eh
Reset	C0h

The occupancy of the EEPROM memory of the 68HC05B16 implementation is identical to the occupancy of the 68HC05B6 memory. The differences only concern storage locations 0101h to 0103h.

0101h BaseConfig

The BaseConfig storage locations establish the operating modes of the powerline bus connection. Figure 6.11 illustrates the meaning of the individual bits. The default setting is FFh. Warning: Changes to this storage location can result in the inability of addressing the bus coupling unit via the bus!

0102h and 0103h SysID

For EIB applications using "open" media, such as radio or powerline, the SysID provides additional system identification. It is necessary for the reliable suppression of any unwanted crosstalk into neighboring areas. The SysID area ranges from 0001h to 00FFh. Figure 6.12 shows the content of these storage locations.

With the initial commissioning of a system it is necessary to guarantee that the selected SysID is also actually available. The programming tool therefore must be able to scan a

6 EIB Software

selected SysID. It is only possible to avoid scanning a SysID if the power supply network can be viewed as an "isolated network", e.g. if it is separated by filters or supplied via a separate transformer.

Bit #	7	6	5	4
Meaning	Must be 1	Frequency definition 1 1 = Channel A 1 0 = Reserved 0 1 = Reserved 0 0 = Channel B		Repeater exists 1 = Inactive 0 = Active

Bit #	3	2	1	0
Meaning	Bus monitor display mode 1 = Normal 0 = Extended	Repeater mode 1 = Inactive 0 = Active	Baud rate 1 1 = 1200 Baud 1 0 = Reserved 0 1 = Reserved 0 0 = Reserved	

Figure 6.11: Content of the BaseConfig Storage Location

Bit #	15	14	13	12	11	10	9	8
Meaning	must be 0	must be 0	must be 0	must be 0	must be 0	must be 0	must be 0	must be 0

Bit #	7	6	5	4	3	2	1	0
Meaning				SysID				

Figure 6.12: Format of the SysID

Figure 6.13 represents a block diagram of the compound structure of the system software with a connection to the application process in the implementation on the 68HC05B16.

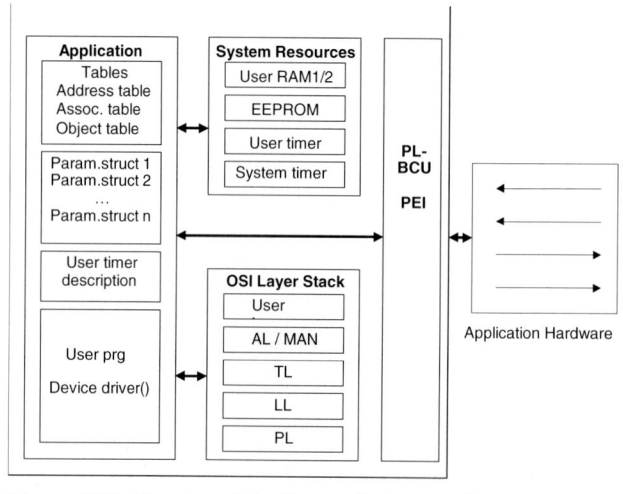

Figure 6.13: Structure of the System Software with Application Process

6.1.5 RF-Implementation

As the communication software is structured in accordance with the OSI standard, only layers 1 and 2 of the EIB protocol stack have been adapted for the "radio" transmission medium. Layers 3 to 7 are independent of media and remain unchanged. They correspond to the TP implementation for operating system 2.0.

Like powerline, radio transmission represents an open medium and for this reason, different systems are again distinguished from one another using system codes.

6.2 Interface to the Internal Application

Basically there are two mechanisms for transferring information as values. These mechanisms are represented in the general device model in Figure 6.2 and are split into:

– Group communication with communication objects (left path)

– Physical device communication (right path)

The physical device communication is split into communication with EIB objects and communication involving direct memory access. The subsequent chapters are concerned with the various methods of addressing.

6.2.1 Group Addressing with Communication Objects

Communication objects represent a neutral interface between communication and application. They exist in the form of data structures in the RAM and EEPROM of the communication module.

They basically comprise three parts (Figure 6.14):

1. Communication object description

2. Object value

3. Communication flags (often referred to as RAM flags)

The communication object description must contain at the very least the object type and transmission priority. The communication flags contain the status of a communication object. Communication objects are connected with the bus communication via group addresses. A communication object can receive data via several group addresses, but only transmit data via a single group address. Different communication objects of a device can be assigned the same group address. The group addresses are unique within the network and are assigned local connection numbers within the device. The application layer allocates service access points (AL-SAP) to these numbers, which permit access to the communication objects. These relationships are represented in Figure 6.15 for the case of reception and Figure 6.16 for transmission.

Both the address and communication object tables are subject to sorting. In addition to the functional representation in Figure 6.15 and Figure 6.16 the tables contain a specification of length at the start and the address table also contains the physical address in the second location. Access to the association table for transmission differs according to the implementation version. Implementations 1.x and 2.x access the association table directly via the index of the

6 EIB Software

communication object entry. This means that with these implementations, the communication objects used for transmission must be sorted in ascending order. Implementation 70.x looks for the relevant connection number in the association table. Figure 6.17 is a detailed representation of the interface between system and application software.

Communication object table

Number of objects	(1 byte)	Low memory
Pointer to RAM flag table	(1 byte)	
Description of object 0	(3 bytes)	
Description of object 1	(3 bytes)	
∘ ∘ ∘	∘ ∘ ∘	High memory

Object description

Pointer to value (1 byte)	Config. byte (1 byte)	Type byte (1 byte)

Communication flag table

			Low memory
Byte 0	Comm-flags 1	Comm-flags 0	
Byte 1	Comm-flags 3	Comm-flags 2	
∘ ∘ ∘	∘ ∘ ∘	3 2 1 0	High memory

Configuration byte

bit #	
7	Must be '1'
6	Transmit enable 0 = No transmission 1 = Transmission possible
5	Value memory type 0 = RAM 1 = EEPROM
4	Write enable 0 = Disabled 1 = Enabled
3	Read enable 0 = Disabled 1 = Enabled
2	Communication enable 0 = Disabled 1 = Enabled
1	Transmission priority 11 = Low operational priority 10 = High operational priority
0	01 = Alarm priority 00 = System priority

Type byte

bit # 7 6 5 4 3 2 1 0
       0 0   Code

Code	Symbol	Value size
0	UINT 1	1 bit
1	UINT 2	2 bits
2	UINT 3	3 bits
3	UINT 4	4 bits
4	UINT 5	5 bits
5	UINT 6	6 bits
6	UINT 7	7 bits
7	UINT 8	1 byte
8	UINT 16	2 bytes
9	BYTE 3	3 bytes
10	FLOAT	4 bytes
11	DATA6	6 bytes
12	DOUBLE	8 bytes
13	DATA10	10 bytes
14	MAXDATA	14 bytes

Communication flags

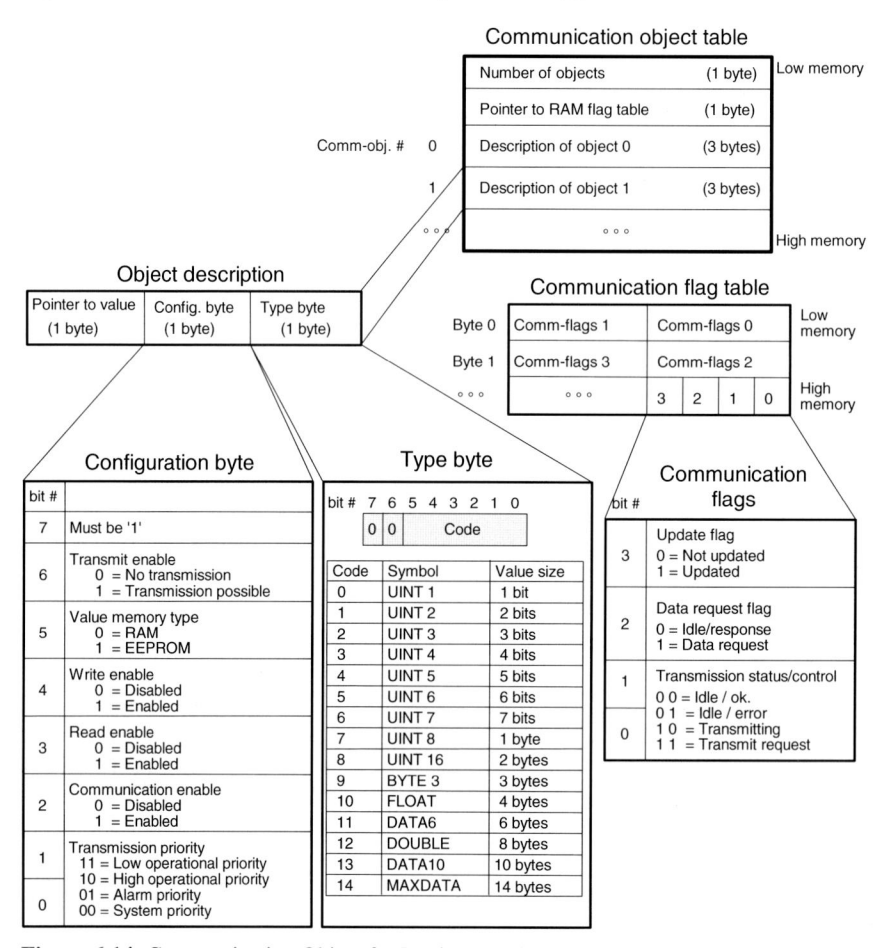

bit #	
3	Update flag 0 = Not updated 1 = Updated
2	Data request flag 0 = Idle/response 1 = Data request
1	Transmission status/control 0 0 = Idle / ok.
0	0 1 = Idle / error 1 0 = Transmitting 1 1 = Transmit request

Figure 6.14: Communication Object for Implementation with System Version 1.x

The update flag is set by the system software to inform the application software that a value has been received for the communication object. The application program polls these flags and copies the values of the marked communication objects for further processing. The application program sets the transmission status to "transmit request" when the value of a communication object needs to be transmitted. The application program need concern itself no further with the transmission of the value. Instead, it is informed of the status of transmission (see communication flags in Figure 6.14) by the system program and if successful is able to trigger the transmission of the next value by setting "transmit request". If the application program wants to read the value of a communication object of another device, the data request flag is set in addition to the transmit status flag. In this situation the system software sends a value request telegram, which does not contain any values for transmission. The addressed devices respond by returning the requested value of the communication object.

6.2 Interface to the Internal Application

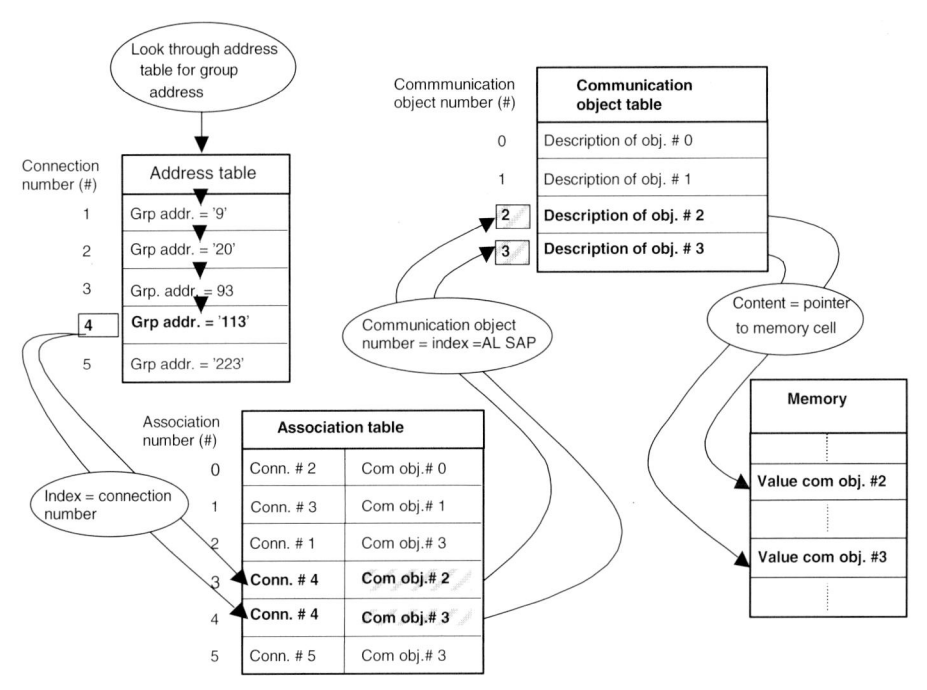

Figure 6.15: Telegram Reception with the Updating of Two Communication Objects

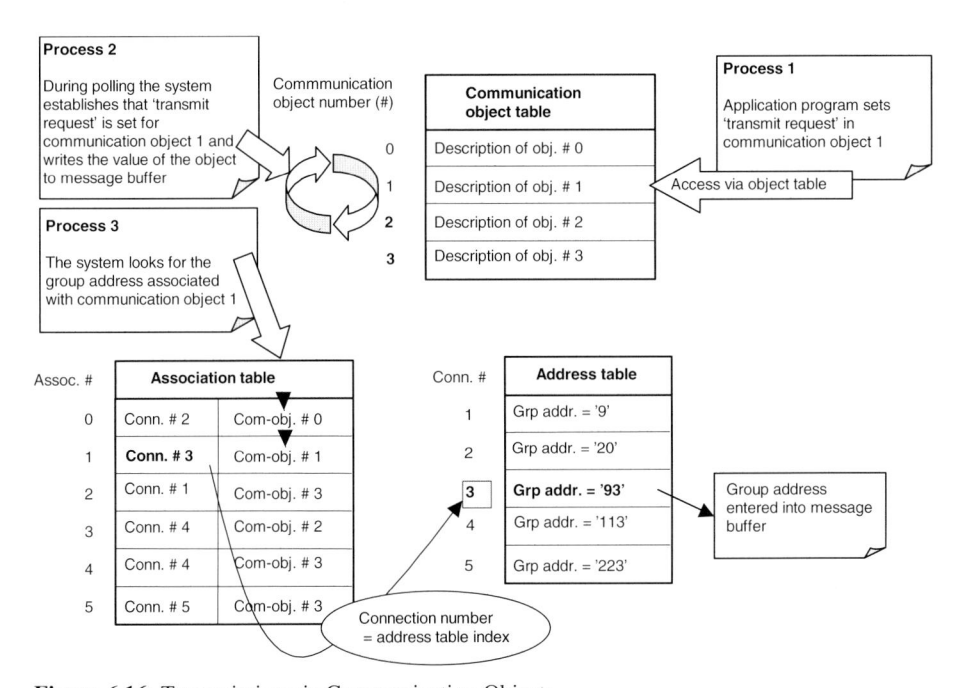

Figure 6.16: Transmission via Communication Objects

209

6 EIB Software

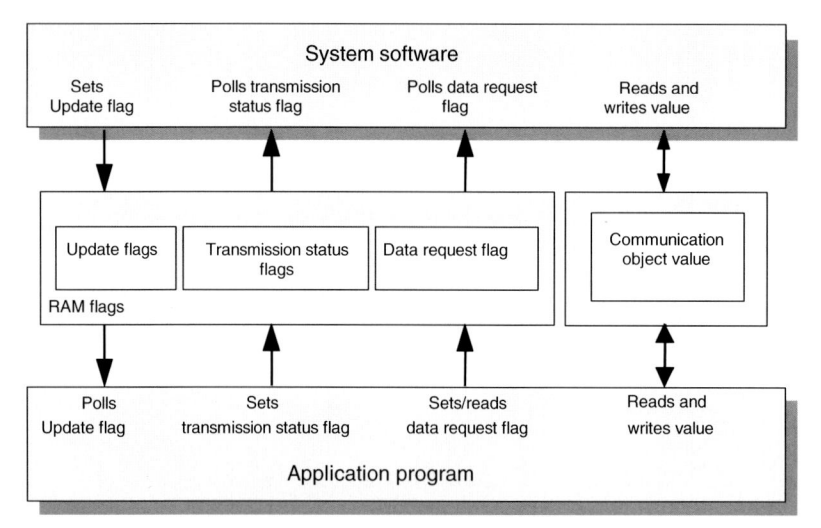

Figure 6.17: Interface between System and Application Programs

6.2.2 Physical Addressing with EIB Objects

EIB objects are data structures, which are assigned to device functions, and contain the information or properties (parameters and values) necessary for communication of the device function via the bus. Examples include time switches, dimmers or controllers. Communication with EIB objects is not so much used for the communication between application units such as sensors and actuators, but more for communication between such devices and central processing or display units.

Devices that use this method of communication support system and application objects (EIB objects) as shown in Figure 6.18. If a communication module supports EIB objects, then the system objects of device object, address table object, association table object and application program object always exist. The device object contains information such as serial number of the device, manufacturer code, order number, current PEI type, service control functions, in order for example to activate or deactivate layer functions, and polling address (see chapter 6.1.2).

In addition to the tables or program, the other system objects contain loading status information and the application program object also contains run time information and application program identification number. Every EIB object consists of at least one property, and every property consists of descriptive elements as well as a value or value field or an underlying function. The descriptive elements are the unique property identification number, the access level for read and write operations to the value or value field, the data type of the property, specifications that control the method of access to the value, value field or function (direct EEPROM, RAM/EEPROM value, value field, function) and if necessary the memory reserved for the value field (maximum length). The absolute maximum of the field length is 255. Element 0 of the value field contains the current number of valid field elements. This entry is increased automatically when an additional field element is entered after the last element.

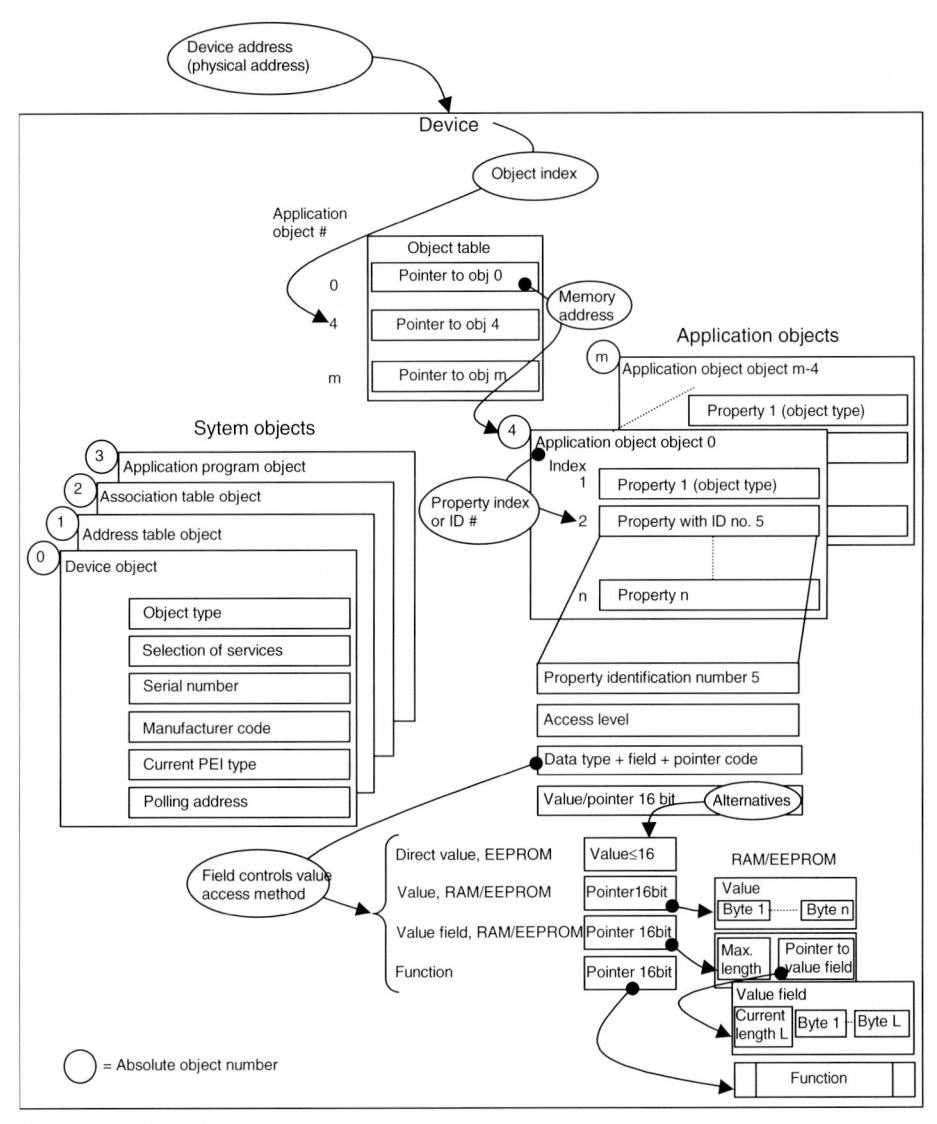

Figure 6.18: EIB Objects

Access to the property value of an object is highlighted in Figure 6.18. It occurs in a connectionless or connection-oriented manner via the physical address, object index and property index or property identification number. With connectionless access it is only possible to access the values for which there is no access protection. The first property of an object always contains the object type. Object types are uniquely defined for the system. Type codes up to 50,000 are reserved for standardized objects. In such standardized objects, the first 200 property identification numbers are not available for private, i.e. non-standard use. As shown in Figure 6.18, the property identification numbers do not correspond to the property index. They must be sorted in ascending order but not without gaps. This also permits the above-mentioned splitting into private and standardized properties.

6 EIB Software

The property index is always used when a functional, unknown device is to be downloaded for the first time, e.g. by a central operating station. In this situation, the objects and properties are systematically downloaded via index addressing until the "object or property does not exist" reply is received. In subsequent normal operation, the objects and properties of the device are known and the property identification number addressing is used instead of index addressing. It is also possible to directly call up an internal software processing or evaluation function via the addressing of a property (not to be confused with the device function that is assigned to the EIB objects!). The advantage of this possibility is, as shown in Figure 6.19, that initiated functions, e.g. a value that is derived from one or more measured values, is only calculated when needed i.e. when the function is called up, and for the bus, this read procedure does not differ in any way from that of a "normal" value.

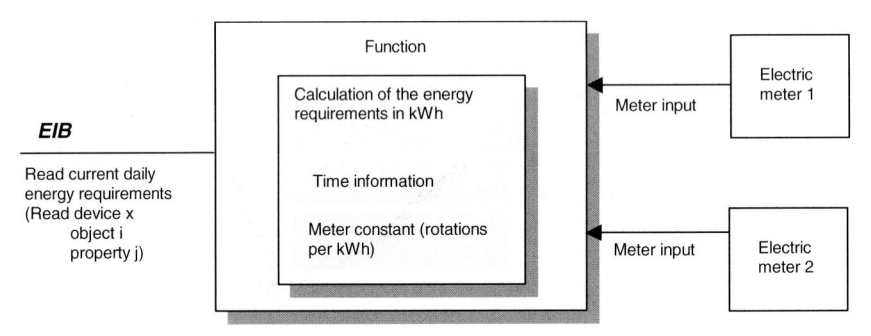

Figure 6.19: Example of Accessing a Function via a Property

6.2.3 Physical Addressing with Direct Memory Access

Physical device communication, which is always connection-oriented, is used for system management tasks (e.g. loading an application program) and can only ever take place between two devices. In order to be able to transfer data in this mode, a logical connection must be created between the communication partners using the connect service. The disconnect service can be used to terminate the connection again. The connection is also cancelled if specific response times are exceeded, or if the connection is no longer active. A sequential number identifies the data transmission process. This allows any necessary repetitions (maximum 3) to be uniquely recognized. This communication and associated services are described in detail in chapter 3.5.

6.3 Interface to External Applications

As already mentioned at the beginning of this chapter, external applications are executed in the application module. There are several serial protocols available via the physical external interface (PEI) for the communication between these applications and the communication module. A general, hardware-related explanation is given in chapter 5.3.4. Here, we will only outline the message formats and protocol procedure [EIBA 99].

6.3.1 Serial Protocols for PEI Configuration Types 12 and 16

These serial interfaces only differ in the physical layer (synchronous or asynchronous), the message formats are the same. A message consists of a length byte (length), a message code (meco), which contains the service primitives for the respective method of connection, and the data bytes (data). The length byte contains a parity bit (even) and 5 bits for the number of following data bytes. Message exchange comprises 4 phases:

1. Communication request
2. Establishment of data transmission direction by conveying the length byte
3. Data exchange
4. Pause

Communication module transmission and reception are represented in Figure 6.20. It should be noted that the interface signal pairs of RTS/CTS and TxD/RxD (=TD0/RDI with synchronous transmission) of the micro-controller are crossed. Communication is started by resetting the RTS line. The next step is to wait for a reply which involves polling the CTS line. As soon as CTS = low is detected, both micro-controllers involved in the communication send a length byte (fully duplex). If no data is to be transmitted, FFh is transmitted as the length byte. If both micro-controllers register a data transmission request and send a length byte not equal to FFh, the micro-controller of the communication module has priority. The application controller (host) must then reregister the data transmission request after the previous data transmission is complete. The direction for the subsequent data transmission is established upon completion of this process.

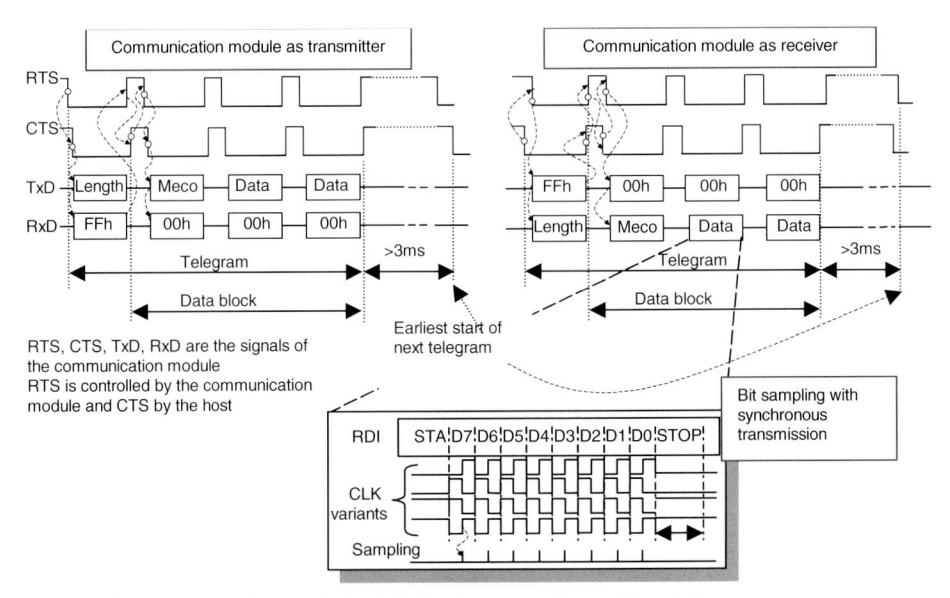

Figure 6.20: Serial Data Protocols for PEI Configuration Types 12 and 16

When exchanging data, a transmitted data byte is acknowledged by the receiver with a 00h byte. The reaction time for such a 00h byte typically lies between 0 und 3 ms for a communication module. After the transmission of a data block there must be a pause of at least 3 ms

(RTS, CTS = high). In case of error, e.g. if a timeout of 130 ms is exceeded for the transmission of an entire data block, the communication module resets the serial interface, sets RTS = high, waits until CTS = high is received and then waits at least 10 ms before reinitiating data transmission. If the communication module is reset, a length byte with length specification 0 is sent, in order to inform the application module controller of the resetting.

6.3.2 Serial Protocol for PEI Configuration Type 10 (FT1.2)

The serial point-to-point protocol type 10 is based on a pure software handshake. The communication module and the application module or connected PC can simultaneously transmit and receive a telegram. As both sides have the same access rights, there is no master/slave relationship. Telegrams of fixed or variable lengths can be transmitted. The telegrams are safeguarded by a checksum and must be activated by an acknowledge (one character). Figure 6.21 illustrates the telegram structures.

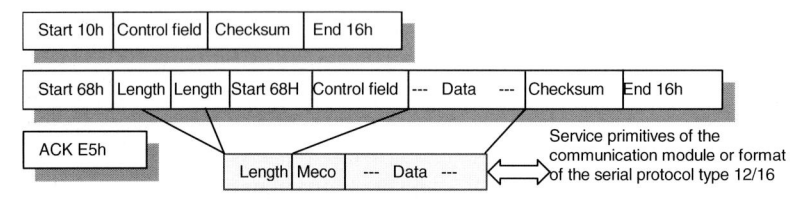

Figure 6.21: Telegram Structures of the Serial Protocol Type 10

The length specifies the number of data bytes plus the control field and can take on values between 2 and 23. How the length, message code (meco) and data of a communication module service primitive are mapped into serial protocol type 10 is also illustrated in Figure 6.21. The acknowledge character is used for positive acknowledgement. The control field contains information that identifies the telegram direction and initiating side, it contains control bits that support the management with regard to data flow control and specifies the service names such as Send_Reset, Send_Userdata, Request_Status, Confirm_Ack, Confirm_Nack or Respond_Status. Figure 6.22 illustrates an example of data exchange.

6.3.3 Message Exchange

Based on the previously described serial protocols for the PEI configuration types 10, 12 and 16, messages can be exchanged between the communication module and an external processing unit (micro-controller, PC, gateway etc.). The messages correspond to the service primitives of the OSI layer, on which the external communication is based. A unique message code is assigned to every service primitive. Whilst for implementation 1.x this message code is only used for external communication, implementation 70.x also uses it for internal message exchange. The available services on the PEI interface are given in Figure 6.23. The A_Value_Read and A_Value_Write messages allow, for example, read and write access to communication objects. With the activation of A_Event.ind all changes to the communication objects in the communication module are reported to the external application. Communication object parameters can be read via the A_Flag_Read message.

The A_User_Data messages facilitate the transparent transmission of data. In bus monitor mode, communication occurs with the data link layer via special service primitives, as in this situation it is also necessary to transmit faulty telegrams and acknowledgements to the external processing unit. In bus monitor mode the device is completely passive towards the bus.

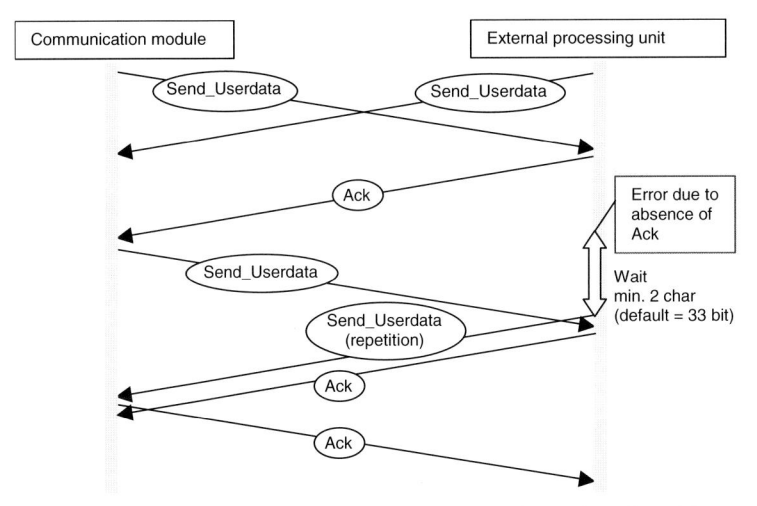

Figure 6.22: Example of Data Exchange via the Serial Interface Type 10

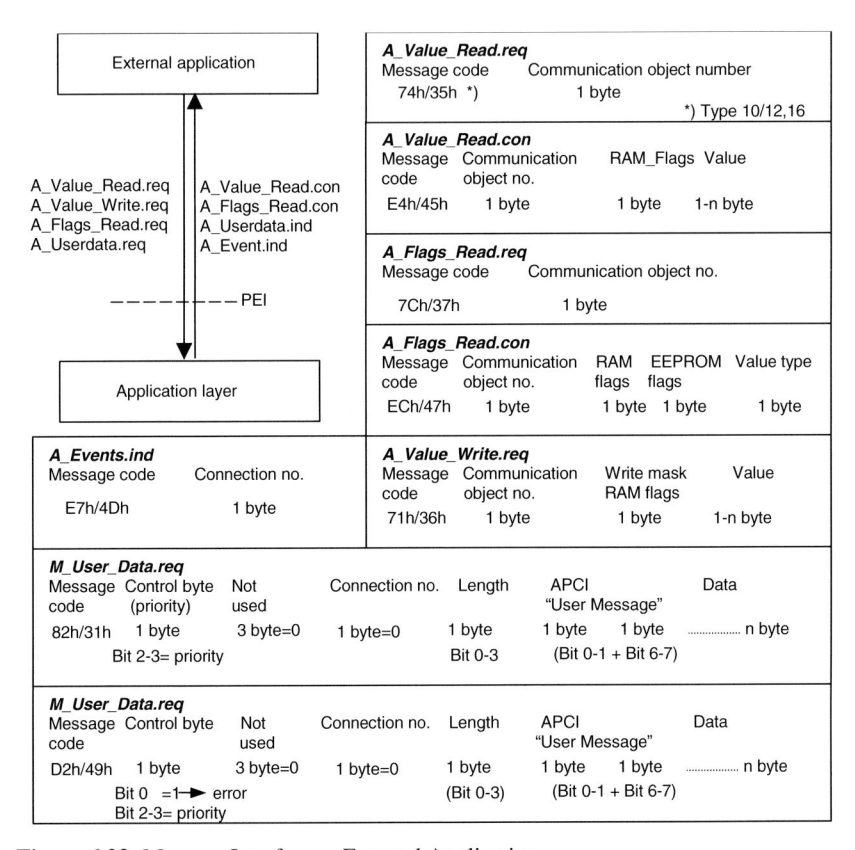

Figure 6.23: Message Interface to External Application

215

7 Integrating Building Networks

In the past, networks were developed for different areas largely independently of one another. In accordance with the technical possibilities, they were optimized for the respective area of application and with that seldom applicable to other tasks. The telephone network was developed for analog voice transmission, LANs for office communication and fieldbuses for automation tasks.

With the falling prices of the hardware and the improved performance of computers, it became possible to distribute the intelligence from a central location to the various sites wherever it was logically meaningful. With that, the foundation was laid for connections between previously isolated networks. The current situation is distinguished, in addition to increased networking at the lower level (via fieldbuses and LANs), by greater networking on a meta-level, i.e. by the interlinking of networks, inhomogeneous networks in particular. Because such interlinking of networks is applicable to the field of building automation and control, this chapter deals with the possibilities and problems of linking networks.

7.1 Types of Couplings

There are various possible ways of coupling networks. A reasonably meaningful classification is possible with the abstract OSI model, as discussed earlier. Here the decisive factor is the level to which the connected systems differ. With this division there would be a possible seven coupling methods, of which only four are commonly used [Tane 96]:

– Repeaters are simple signal regenerators, which work on the physical layer of a network. One application is the lengthening of network segments, beyond the limit of their physical specification (with regard to line length or number of nodes). On the other hand, repeaters can also be used to connect different media, if layer 2 remains the same. A classic example from the LAN field is the connection of twisted pair and thin-wire Ethernet segments. Repeaters do not filter or decode data.

– Bridges connect networks with different data link layers. A typical example is the connection of Ethernet with a token ring. In addition to signal regeneration, bridges also offer the possibility of filtering data packets. Not every packet that reaches the bridge from one side needs to be transferred to the other side. In this way, data traffic in two segments can be kept local and the overall bus load kept to a minimum. With EIB, this function is achieved with repeaters. Although they don't filter the data packets, they do acknowledge them on the data link layer.

– Routers work with different network layers. They are the essential components for large, structured networks and in connection with the Internet in particular, are now very well known. In networks based on TCP/IP routers separate different IP domains. With EIB, the functions of the backbone and line couplers are most closely related to those of a router.

- Gateways represent the most involved method of connection. They are used to connect networks that are completely different, i.e. those which either have different protocols right up to the application layer or those which do not correspond to the OSI model. As the protocols must be completely implemented for both sides of the gateway, it also by definition offers the greatest flexibility. The reproduction of the individual messages or commands between the two sides is achieved via a corresponding application.

In some respects this classification is theoretical, and in practice is often open to interpretation. In the LAN field in particular, there has been a wave of new expressions in recent times, mainly for reasons of marketing. A hub for example is actually a repeater with several connections. A switch fulfils the same purpose as a bridge. There are also hybrid forms, such as the layer-3 switch, representing a combination of bridge and router. A special form of the repeater is the remote repeater, which connects two spatially separated segments. The two halves of this repeater are typically connected by fiber optic cable and increase the extension of the network. Such types of constructions are also seen in fieldbus technology. There are also remote bridges, whose connections are transparently linked via a long distance network. Although the function of a gateway is undisputed, the term is often used inexactly.

Two other coupling elements have recently become popular, neither of which have anything to do with the given classification: proxies and firewalls. These two cannot be incorporated into the above classification, as they do not relate to the protocol level but to the underlying concept. With firewalls, the main issue is not protocol conversion, but security. The incoming PDUs are filtered in accordance with the security concept. Only the part that corresponds to the guidelines is passed on. Depending on the applied security policy, filtering occurs on different layers. With an Internet device for example, it is possible to filter according to IP addresses, but also in accordance with a specific higher protocol, such as HTTP (Hypertext Transfer Protocol). This allows a restriction in the use of protocols or special services in the areas in front of or behind the firewall [Str 97].

A proxy has a similar structure; the underlying strategy however is different. Proxies are well known to users of the World Wide Web. Their task is to buffer frequently requested data and with that to reduce the load on the connection between the LAN and the Internet. Proxies can also execute address conversions. Here, the filtering of the data traffic has nothing to do with security; it increases the throughput and optimizes the use of resources.

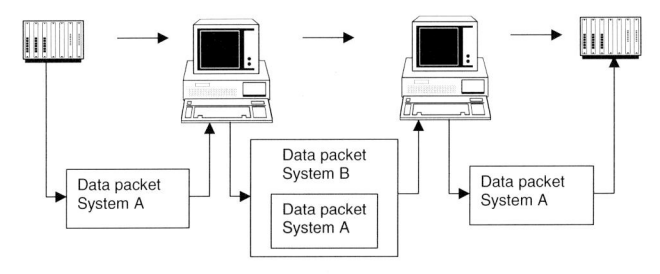

Figure 7.1: Tunnel Protocol

A special type of coupling between two networks is represented by the use of a tunnel protocol (Figure 7.1). In this situation, a communication system uses something other than a pure "transport vehicle", without any protocol conversion having been carried out. A complete data frame from system A is packed into a PDU (protocol data unit) of system B, sent to a target address in system B and there unpacked. System A, which essentially continues beyond this tunnel, can then use the data frame as if there had been a direct connection to the sender. The observant reader will object that the packaging of PDUs into other PDUs is already an essential characteristic of multi-layer protocols (see chapters 1 and 3). The difference to tunnel protocols is that with a usual protocol stack it is always a PDU of a higher layer that is transferred in a PDU of the underlying layer. However, we only talk of tunnels if a PDU of a *lower* protocol layer is packed in a protocol of a *higher* layer. In theory, we could tunnel EIB via TCP by putting a frame or packet (i.e. PDUs of layer 2 or 3) on the EIB side into a TCP message and then transmitting to a target computer via a TCP/IP network. There, the EIB-PDU is extracted from the TCP message and then passed on to the connected EIB without further processing. This illustrates two important properties of a tunnel protocol. Firstly, it always connects systems of the same type (in this case two EIB networks). Secondly, it is completely transparent; both end systems must in principle be able to communicate directly with one another without the tunnel path. In particular there is no routing, the tunnel simply splits a self-contained network into two spatially separated parts, which remain in logical terms connected [Tane 96]. We will discuss the tunnelling approach effectively defined for EIB below in section 7.4.1.

7.2 Using Couplings

After having outlined the various possible couplings, we will now turn our attention to the question of how these technologies are used with regard to EIB in particular. In the field of building automation and control, the coupling of networks can be meaningful for a number of reasons:

- The used system has restrictions that can be compensated by the connection to another network. An example of this could be a restricted extension, that could be rectified by coupling to a network better suited to greater extensions. This category also includes restrictions in the data transmission rate, which can be overcome with the use of a backbone.
- It is necessary to set up simple access to an automation system in order to poll data. With a special bus connection (e.g. a modem) it is possible to trigger actions or requests in the system via telephone. Another example is the data exchange between networks that have been installed in parallel for historic or legal reasons (e.g. fire alarm systems).
- Maintenance and installation costs should be reduced. Access via a global network allows the expert to undertake maintenance work or make configuration changes from afar. This access also opens up a multitude of possibilities for additional services.

Each of these main application areas is examined in more detail below.

7.2.1 Extending the Range

Networks that are optimized for a geographically restricted area require specific additions to cross greater distances or different technology that serves as a transmission aid and connects several networks of the same type. Even with the Internet, local networks have a different design than campus networks or intercontinental connections (LANs are preferably based on

Ethernet, the connection to other networks however is achieved via ATM). In fieldbus technology in particular, there are transmission media with limited range (e.g. with the common RS485 interface, a maximum extension of between 100 and 1200 m is permitted depending on the baud rate) and it is only possible to use a maximum of three repeaters in a transmission path. The situation is similar for EIB, where the line length of a segment must not exceed 1000 m and the number or repeaters or couplers is restricted to six. If however, it is necessary to bridge greater distances, in larger building complexes for example, then it is often possible to use the (TCP/IP) LAN that is usually already available as the transport medium, which allows simple tunnels to be created. There are also devices, so-called routers, that are capable of converting addresses. This becomes necessary when the same group addresses have been assigned for different purposes within two independent EIB networks, which are then to be connected. In order to avoid conflict, it is necessary to change the group addresses (which demands a reconfiguration) or to make the necessary conversion in the connecting node.

Links to other networks also offer a good solution when it is necessary to increase the data throughput. By virtue of the tree structure and the use of backbone and line couplers it is possible with EIB to keep part of the bus traffic local and only to allow the messages that are actually required in other zones or lines through to the higher levels. This filtering depends strongly on the coincidence of topology and data relationships. If communication mainly occurs between two different zones, then the main line in particular quickly becomes overloaded. This throughput problem can be solved by introducing a backbone with a larger bandwidth. This provides a higher capacity for the link between zones and overloading becomes less likely. The classic LANs are again suitable. There is no tunneling here as each tunnel only has two nodes (point-to-point connection). The coupling elements must have routing capabilities. Corresponding EIB/LAN routers are also available. For the future, EIB-Net has emerged as a suitable solution for the backbone problem.

7.2.2 Remote Control

It is often neither desired nor necessary to have all possibilities of a system available in the place where operation occurs. It is often sufficient, for example, to be able to press a button from a remote location. The translation of this control command into the fieldbus protocol occurs where the switch is located; the action itself is triggered by an external control signal. A typical example is the early means of remote control via telephone, which allowed the switching of relay contacts by dialing a specific combination of numbers (Figure 7.2). Infrared and radio remote controls also fall into this category. This technology also facilitates remote operation via the Internet. However, such a solution is inflexible, as the allocation of triggering event (control pulse) to reaction at the fieldbus is established by the configuration in the coupling element, where subsequently any modifications must be made. Moreover, this method is practically only suitable for binary commands; the transmission of additional parameters would require a great deal of extra work.

Strictly speaking, these simple remote controls are full gateways: A message (the control pulse), transmitted via a network (such as the analog telephone for example), is translated by a gateway application (in the form of a fixed allocation via the wiring or a configuration table) into a corresponding command on the fieldbus. For many simple applications, in the field of home automation in particular (e.g. to switch on the central heating), this has until now been sufficient, as well as easy to implement.

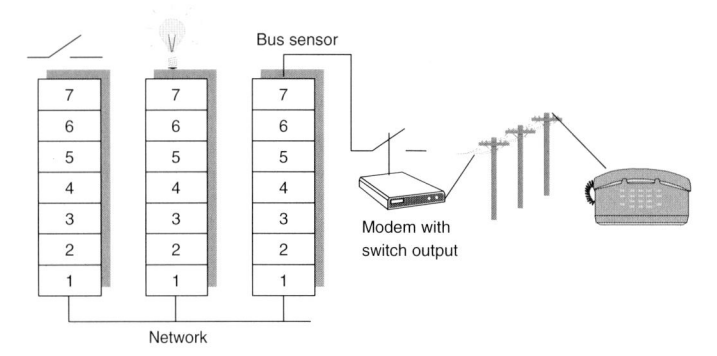

Figure 7.2: Remote Control via Telephone

A gateway is required for exchanging data between different fieldbus systems. There are a number of elements available to EIB for coupling to other bus systems, such as for example Profibus, M-Bus, BACnet or LonWorks. For accessing from further a field, there are methods of linking up to the analog telephone network, ISDN and Ethernet (with TCP/IP).

7.2.3 Remote Maintenance and Remote Services

With the improved technical possibilities and the greater networking, the more complex functions of remote monitoring and maintenance have developed from the simple remote control functionality. Here, the interaction between user and fieldbus system is associated with much more data and far exceeds the transmission of simple control commands. In addition to the control possibilities, it is possible to monitor the status of the automation system as well as its reaction to any change in the system parameters. Essential to maintenance is the possibility of accessing the component configurations.

The term "remote" in the case of remote access to an automation system must be viewed relatively. On the one hand, it can be used to denote the necessary bridging of large spatial separations. It is then possible for the creator of an industrial system to check any problems that may arise in the system from his office and make a corresponding diagnosis, all of which is achieved online, before dispatching the necessary experts. This means that many costly journeys can be avoided altogether. On the other hand, remote access can also denote simple data recording via the intranet of a company where the actual distances do not matter. It is much more about retrieving data from the self-contained environment of the fieldbus and making it available for automatic further processing in other tools. Visualization and general SCADA tools[27] belong to this group.

When it comes to coupling with the fieldbus, globally available networks are the only possibility. The Internet is ideal. It should be primarily viewed as a means of connecting stand-alone networks, all of which use a common set of protocols. The success of the Internet is all thanks to this uniformity. The actual transmission medium used is of secondary importance. Analog telephone lines are just as usable as ISDN, the TV infrastructure or Ethernet. It is also of no consequence whether access to the fieldbus occurs from the company intranet or the public Internet.

---

[27] SCADA: Supervisory Control and Data Acquisition

There are basically two possible methods for coupling fieldbuses with global networks [Kni 97]. The simplest method is to tunnel the fieldbus protocol to the user via the network and to carry out maintenance with the same fieldbus specific tools that would also be used locally. It would also be feasible to use proprietary protocols for data transmission between the coupling elements on the fieldbus and the maintenance tool. Although this variant is easily adapted for different applications, it relies on the user having a detailed knowledge of the fieldbus.

The second possibility involves the use of a gateway to convert the fieldbus protocol into a suitable equivalent on the global network. As with all gateways, this is associated with a certain loss of protocol functionality as it can seldom be fully converted. However, this protocol conversion does permit the use of tools that are completely independent of the fieldbus (e.g. web browsers) to monitor and control the fieldbus system. This variant is particularly advantageous in the field of home and building automation, where remote access is also required by the general user and not just trained experts.

The coupling of fieldbus systems with the Internet opens up a multitude of possibilities for services, assuming that the data exists in an abstract, Internet-compliant form. Many monitoring functions (break-in, fire, etc.) can be taken over by external service providers when the fieldbus allows access from the outside (see chapter 1). There are however certain security-related problems that need to be overcome when connecting fieldbus systems to the Internet. We will take a closer look at these problems later in the book.

From the EIB range of products, the various ISDN couplers and TCP/IP gateways are also suitable for remote access. The ISDN coupler could, for example, be used to set up a (tunnel) connection between two EIB networks via the telephone network or to link a computer via ISDN to the EIB, from where the system is then configured or monitored via a visualization tool. The TCP/IP gateway facilitates a direct connection to LANs. In addition, there are a multitude of coupling elements that offer serial interfaces which facilitate data exchange on the basis of a simple protocol. With a computer connected to this interface it would then also be possible to connect to a LAN.

7.3 Examples of Internet Access

The following sections concentrate on connecting EIB systems with the Internet, as this sometimes requires new approaches that go beyond familiar solutions. Even if the corresponding products are not yet available, it is worth highlighting the possibilities offered by the Internet for home and building automation [Rei 99].

7.3.1 Java-Oriented Approach

With the growing popularity of the Internet, it is the ideal medium for more and more applications. The major breakthrough came with the development of graphical interfaces (browsers) for representing and HTML (Hypertext Markup Language) for describing the documents distributed via the Internet. In this part of the Internet in particular, based on HTTP (Hypertext Transfer Protocol), the term WWW (World Wide Web) has become a household name. Java has become widely accepted as the programming language for Internet-based applications, a language which is not specifically aimed at the Internet and can be used quite independently. Its great advantage is its independence from any platform, one of the downsides of many other programming languages. In connection with Internet applications a differentiation is made between Java scripts and Java applets. It is the aim of both to extend the intrinsically static HTML pages and to allow user interaction. Java scripts are commands that are incorporated into a regular HTML page and then executed on the user's computer. They are comparable to CGI scripts[28]; the only difference being that these are executed on the server. Java applets are programs that are run on the computer on which the web browser also runs. They either run independently or they communicate with other computers. The latter is interesting in the context of a connection to EIB that is based on Java, where the form of the communication and the used protocol are not pre-specified.

The basic structure of a link consists of an HTML page that contains a Java applet. When this page is called up, the server downloads the applet which is then started on the initiating computer (client). It now attempts, via the Internet, to set up a link to the server to which the fieldbus is also connected (this configuration is not essential but is applicable in most situations). A Java servlet runs on this computer, which is another program that forms a counterpart to the applet and whose task it is to respond to incoming requests. The servlet is also matched to the application and forms the interface to the fieldbus.

This structure forms a comfortable user interface but is usually optimized for a particular application or fieldbus configuration and therefore only offers restricted flexibility. The applet could for example represent the ground plan of a building showing all sensors and actuators. If something changes within this configuration it would require a great deal of work to adapt the HTML page (or more accurately the contained applet) to match the new conditions. A more flexible way of linking the fieldbus to the Internet is examined in the next chapter.

7.3.2 SNMP-Based Approach

The Simple Network Management Protocol (SNMP) was developed for the maintenance of as well as the detection and removal of errors from devices in IP-based LANs. In accordance with the purpose for which it was intended, it was kept simple. The basic concept consists of a management station (Figure 7.3), that monitors other stations, so-called agents, and exchanges information with them [Sta 93].

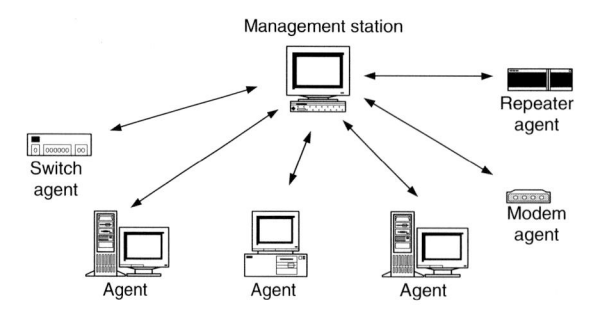

Figure 7.3: Manager-Agent Structure

---

[28] CGI: Common Gateway Interface

7.3 Examples of Internet Access

The good thing about SNMP is that the objects which the manager can access cannot be defined in any old way, but are established within a so-called management information base (MIB), see Figure 7.4. This MIB represents a hierarchic naming structure, the essential features of which are globally standardized. This guarantees that objects, which are available in different devices, always have the same names.

Not all objects that are defined in the MIB need necessarily be available in a device. There are certain objects that are compulsory for all SNMP devices, e.g. objects from the system sub-tree. Others depend on whether the associated service is implemented in the device, such as for example, the TCP group. A completely separate area is reserved for companies or private use. In a similar way to Internet addresses, the user can apply for an entry after which he is responsible for managing the associated sub-tree, e.g. iso.org.dod.internet.private.enterprises.companyXY. The company can then use and define the underlying names.

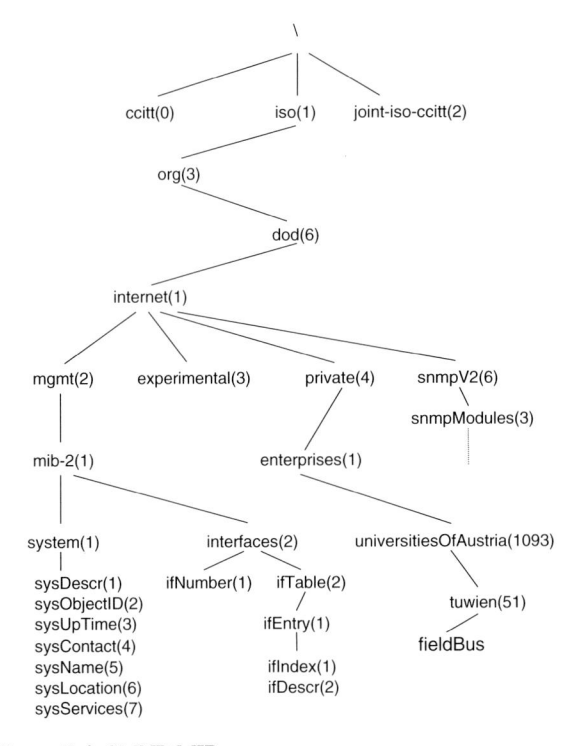

Figure 7.4: SMNP-MIB

In order to identify a specific object within this naming structure, simply follow a branch right to the end. With the name established in this way it is then possible to access the object value. With scalar values a '.0' must be added to the values and with tables the index that localizes the corresponding entry. Accessing iso.org.dod.internet.mgmt.mib-2.system.sysDescr.0 or 1.3.6.1.2.1.1.1.0 in Figure 7.4 for example provides a description of the system in all SNMP-compliant devices.

The protocol has a very simple structure and consists of only four PDUs.

- *Get* reads a single value from the MIB. From the above example: *Get (iso.org.dod.internet.mgmt.mib-2.system.sysDescr.0)* returns a description of the system. With a Windows NT computer, this has the following form: *Hardware: x86 Family 6 Model 3 Stepping 4 AT/AT COMPATIBLE - Software: Windows NT Version 4.0 (Build Number: 1381 Multiprocessor Free)*

- *GetNext* returns the value that immediately follows within the naming structure. In order to arrive at the same value as with the previous Get example, the command reads *GetNext (iso.org.dod.internet.mib-2.system.sysDescr)* or *GetNext (iso.org.dod.internet.mib-2)*, as both *system* and *sysDescr* are starting branches which are traced. *GetNext (iso.org.dod.internet.mgmt.mib-2.system.sysDescr.0)* supplies the value of iso.org.dod.internet.mib-2.system.sysObjectID, as this object has identifier 1.3.6.1.2.1.1.2. With the computer used in the above example, this yields the value *enterprises.311.1.1.3.1.1*. This command allows the user to scan a previously unknown MIB object by object.

- *Set* functions in a similar way to the Get command and permits write access to the value of an object. Here too the object must be accurately identified. If the object is denoted as read/write, its value is overwritten.

- *Trap* is the only command that works in the direction from the agent to the manager. In a similar way to an interrupt, it allows the agent to report unusual events to the manager.

There are two main factors in favor of using this protocol, also valid in connection with fieldbuses - firstly its availability and secondly its simple, clear layout. Its ready availability also means that there are a large number of programs for the administration of SNMP devices. Products range from simple public domain software right through to professional management tools. The simplicity of SNMP also permits simple implementations. Two steps are necessary to adapt this protocol for remote access to a fieldbus. These are the definition of a new MIB and a gateway that carries out the conversion between SNMP and the fieldbus protocol. In the gateway, the MIB must correspond to a data structure that allows manager access from the Internet. All data relevant to the fieldbus system must have a representation in this data structure.

There are of course also certain disadvantages associated with SNMP. One of the main ones being a lack of user-friendliness for the non-professional user. Now, however, in the field of building automation and control, Internet access should be made available to everyone in the home and this demands clear, graphical tools. A solution could be offered by the simple operability of a web browser in connection with Java. With that, the interface could have the "look and feel" of the WWW and the underlying communication could be based on the well-structured SNMP. This would also guarantee access for professional building management – without the web interface but with powerful tools. Before the connection between fieldbuses and the Internet can be used on a large scale however, there is one major problem that needs to be resolved and it is that of access protection. This area is discussed in the next section.

7.3.3 SOAP/HTTP-based approach

The Simple Object Access Protocol (SOAP) bundles the power of the HTTP communication protocol with rich, structured message content provided by XML. The result is a standardized, general-purpose Remote Method Call protocol for IP environments. The first standardization effort for high-level IP protocols tailored to EIB focuses on SOAP. SOAP is already intensively used by web-based services like eCommerce, or business-to-business applications, and is furthermore the intended run-time protocol for the Universal Plug and Play (UPnP) concept.

A SOAP request for opening an EIB connection may look as follows:

```
<SOAP:Envelope xmlns:SOAP="urn:schemas-xmlsoap-org:soap.v1">
  <SOAP:Body>
      <m:Open xmlns:m="www.falconserver.com">
          <eLayer>Linklayer</eLayer>
      <m:Open>
  </SOAP:Body>
</SOAP:Envelope>
```

An example for a request's answer may be:

```
<SOAP:Envelope xmlns:SOAP="urn:schemas-xmlsoap-org:soap.v1">
  <SOAP:Body>
      <m:OpenResponse xmlns:m="www.falconserver.com">
          <eError>DeviceOpenErrorNoError</eError>
      <m:OpenResponse>
  </SOAP:Body>
</SOAP:Envelope>
```

The XML code of a SOAP request or response is easy to read: The parts that have a fixed structure are defined within the SOAP namespace, while the SOAP body then contains the request or, respectively, response specific method name and, within that scope, the parameters or return values of the call.

For discovery within IP networks, the Simple Service Discovery Protocol (SSDP) has been defined. It is based on HTTP and also XML and allows searching for devices, optionally of specific type, over IP multicasts. The device description together with information on the device type and operations supported, are then retrieved via HTTP in XML format. This allows subsequent accessing of the device via SOAP. SSDP is also an essential part of the UPnP infrastructure.

7.4 IP Connectivity Models and Applications for EIB

The dream of a unique encompassing solution consisting solely of IP (Internet Protocol) components certainly has its merits and adepts. Gradually though, the realization is dawning upon engineers and marketeers alike, that probably the world isn't flat after all:

– Distributed control applications feel better at home on a dedicated, compact (with a stack footprint smaller than 5 kB and narrow band) and economical *Device* or *Component Network*[29], in other words, the bus as provided by EIB.

– An IP superstructure, whether local (intranet, LAN) or remote (extranet, Internet), complements the building's Device Network to provide all kinds of integration and service functionality: meet the Service Environment[30]!

---

[29] To keep the distinction between the "control bus" level and its "IP environment" (whether local or remote) clear, we will refer to the former explicitly as "Device Network" throughout this section.

7 Integrating Building Networks

Precisely this idea was the starting point for the EIB Association to trigger the development of *ANubis* (*Advanced Network for Unified Building Integration Services*) in 1999. ANubis aims to standardize the EIB adaptation of these general concepts, where necessary. The central design objectives for this initiative can be summarized as follows:

- To provide a Service Environment based on IP which transparently integrates with EIB Device Networks for Field Level distributed control.
- ANubis must be able to deal with the requirements of the traditional Automation and Management Level in a megabit IP setting.

Beyond these, the concepts and realisations to render any building equipped with EIB – residential as well as commercial – service-enabled, whether in a (local) intranet or a (remote) extranet environment. Figure 7.5 gives an idea of the scope of the ANubis program, in terms of integration and service connectivity.

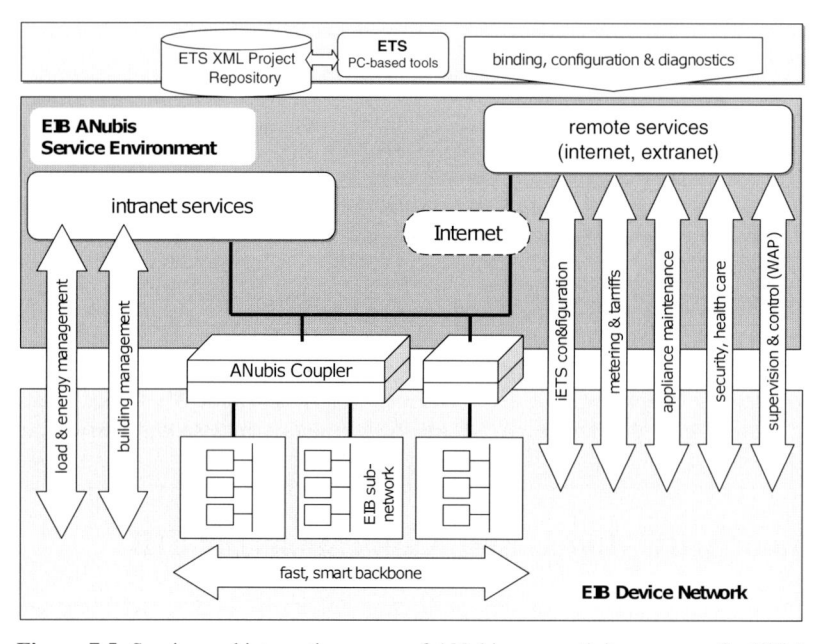

Figure 7.5: Service and integration scope of ANubis connectivity program for EIB buildings

At any rate, the whole idea hinges on IP. Basing their models and protocols on this universal approach, the ANubis specifications become largely independent the physical communication medium. The EIB ANubis protocols and middleware can indeed be operated on any IP network, with medium options ranging from 10 or 100 Mbit Ethernet (IEEE 802-2), Gbit Ethernet, over 11 Mbit Wireless Lan (IEEE 802.11) up to FireWire (IEEE 1394b) etc. It can

---

[30] Note that the 2-tier model Device Network / Service Environment collapses the traditional "Field Level / Automation Level / Management Level" model for process automation. At the same time, the concepts are extended, since an IP Service Environment can provide a variety of sophisticated and remote services, hitherto not envisaged.

also be run across point-to-point IP links, such as PPP dial-up connections (on V90/92, ISDN, ADSL, TV cable, etc.). For specific needs, USB and Bluetooth will be used.

The next paragraphs show that several needs have to be addressed, in direct relation with Building Control, when integrating IP environments and the control bus. This is one more illustration of the fact that on IP we have to and indeed can exploit the "mixed protocol" nature of this environment. Here, a critical design constraint was that to the largest extent possible, EIB ANubis solutions had to be based on established Internet Standards (IETF). The following digest shows how this was solved and demonstrates the pivotal role played by various standard IP protocols.

7.4.1 Device Network Protocol Transfer

The idea is to make the full EIB Device Network Protocol accessible over IP for specialized clients. Two alternatives are required:

Frame Tunnelling ("thin gateway")

As outlined in section 7.1 above, we want to transport essentially native binary EIB bus frames over IP. It is clear that in this case, the client must be aware of the binary encoding of bus frames.

In ANubis, this is covered by eFCP (e-Field Control Protocol), which defines a common Link Layer tunnelling content encoding for bus frames, based on EIB's standard External Message Interface (EMI). In fact, this encoding standard applies generally for EIB tunnelling, not just on IP but also on RS-232, USB etc. The IP version is labelled EIBnet/IP, and has been submitted as an update of the EIBnet official standard. Both a TCP (reliable) and UDP (compact) flavour are defined; the latter allows for competitive entry-level implementations.

The attentive reader will notice that the (Link Layer) EMI protocol differs strictly speaking from the EIB bus message LPDU. In fact, EMI is a light-weight client-server protocol, which allows an EMI client to perform bus communication through a proxy stack, acting as EMI server. So the tunnelled message consists of the PLDU, preceded by a request code to the (remote) proxy.

Tunnelling may be tricky, as it exports timing aspects of the wrapped protocol across the link, which may not always be able to sustain "real-time" responses.

Protocol Mapping ("thick gateway")

Here we go one step up in abstraction, hiding the binary details by mapping the Device Network Protocol to an abstract method API. The method calls are performed through SOAP (Simple Object Access Protocol). In this way, we simplify the life of the client, but now the gateway server has to deal with the mapping conversion, i.e. with composing and parsing frames. Even here though, the client or gateway user still gets to see representation of the actual bus protocol, though no longer binary. This approach also gets rid of the timing pitfalls posed by tunnelling: the gateway acts as a powerful local proxy of the remote system.

Concretely, ANubis selects the abstract method API of the eteC Falcon Component to realise this aim. Note however, that the specification as such is open and independent of the (Windows) implementation, which is provided by EIBA itself.

7.4.2 Common Datapoint Model

Whereas the Protocol Transfer is sufficiently powerful and low-level to permit transparent management and configuration access to the underlying Device Network, it seems too specific for an adequate run-time representation of the control applications running on it.

So we further climb the ladder of abstraction and encapsulate the Device Network Protocol for the client behind an Object Model with structured, abstract datapoints. On the IP-side, these are self-describing (through XML) and are accessed via a SOAP method calls. Observe that this is very close to the UPnP (Universal Plug and Play) approach.

On the EIB control bus side on the other hand, data points are mostly represented by a Group Address, sometimes by a property of an Interface Object. It is clear that the mapping to be performed by the gateway will be largely installation-specific. The key to the ANubis approach is that not just the run-time API is standardised, but also an XML meta-model needed for the configuration of the gateway engine, and that upcoming ETS versions will support this feature.

7.4.3 Applications in EIB Controlled Buildings

Let's have a look at how the protocols and models described above are used in practice, and the added value they bring to Home and Building Automation. Evidently, it is the EIBlib/IP tunnelling protocol, which was first used by the iETS ("Internet ETS") extension to the ETS tool, thus bringing support for remote configuration and maintenance.

The Falcon-based protocol mapping via SOAP essentially covers the same needs, but in a more sophisticated fashion. Its application is planned for the next generation releases of the EIB system tools, and is getting attention on the part of larger providers of Internet services.

For the Common Datapoint Model, applications are typically in the SCADA domain, i.e. for visualisation and supervision, or as gateway to datapoint-oriented partner networks. Indeed, the Common Datapoint Model may easily be:

- Mapped to OPC (OLE for Process Control), BACnet etc., or

- Combined with SSDP (Simple Service Discovery Protocol) discovery, and integrated into Service Environments like Universal Plug and Play.

In fact, all three methods discussed here may be used to achieve an IP fast backbone, integrating different EIB subnetworks. Detailed definitions for EIB were still under review by EIBA, at the time this book was published.

7.5 Security Aspects

Control tasks often involve sensitive information. The field of home and building automation is no exception. This is particularly valid for applications that involve links to global networks such as the Internet. Good examples include meters that can be read remotely or alarm systems whose statuses can be read or controlled via a global network. It is obvious that access to such sensitive data and system parts needs to be restricted. A consideration of the link between EIB and global networks is not complete without taking into account the question of security.

For this reason, the following sections of this chapter are dedicated to the most important security concepts in control technology. We will also look at elements for secure software architecture. For more details please refer to [Rank 99].

7.5.1 Data Security Concepts and Definitions

It is first necessary to define the term *security* in more detail. In the field of information technology, security is used to denote the minimization of the ability to attack information systems. In other words, all the weak points of a system that could be used to provide unauthorized access to either the system itself or its contents should be eliminated or at the very least reduced. These weak points are established by identifying possible threats and through risk analysis. Both are essential analysis techniques for the development of secure systems. In this context, threat is used to denote possible violation of the system security.

Each information source can be assigned various properties that can be the target of a threat. The main properties are defined below, damage to which can have a negative effect on the compound system:

- *Availability* means that the source can make data available whenever it is required.
- *Integrity* means that data cannot be changed in any unauthorized way or destroyed
- *Confidentiality* means that information is not made accessible to any unauthorized users, groups or processes.

For every system the first step is to define a security policy before specification and development are undertaken. This policy consists of an informal description of the security guidelines and applicable methods, usually in a natural language. These rules generally establish what is allowed in normal system operation and what is not allowed. The guidelines form the highest hierarchic stage of a security specification.

Concepts and methods for converting the security policy are looked at in the second part of [ISO 89]. The first part of this ISO standard defines the OSI model. The second part extends this with a definition of the security architecture for open systems. This specifies a series of security services that as an option can be made available within the OSI model. These services are outlined below.

- *Authentication* proves the association of the communication partner and data source to a group with specific rights.
- *Access control* protects against unauthorized access to system resources. This service can refer to various access methods, such as for example, the use of a communication channel, reading, writing or the deletion of data.
- *Data confidentiality* protects the data against being passed on without authorization. These services ensure the confidentiality of connections, individual data fields and the data flow.
- *Data integrity* is achieved by protecting the information flow from change, insertion, deletion or the repetition of data.
- *Non-repudiation* means that the receiver (or transmitter) can be certain of the origin (or reception) of the data. This is necessary to prevent a communication end device from denying the transmission or reception of data.

7 Integrating Building Networks

In practice, the services are called up in the suitable OSI layers and in relevant combinations in order to satisfy the requirements of the security strategy and the user. Authentication, for example, can be incorporated into layers 3, 4 or 7, in accordance with [ISO 89]. Actual implementation of the services uses safety mechanisms such as, for example, encoding or digital signatures. This type of mechanism is generally used in three different security stages [ITSEC 91].

There is another aspect worthy of note – the system user must have faith in the implemented security measures. An evaluation of system security – preferably by an independent adjudicator – should create this trust. Some evaluation criteria have been worked out, valid on a worldwide scale. A European variant [ITSEC 91] defines seven stages from E0 to E6 that state the degree of trustworthiness in the security services and mechanisms. Stage E0 indicates faults, E6 the highest level of trustworthiness.

The remaining sections in this chapter describe how these concepts can be applied to fieldbus - Internet links.

7.5.2 Attacking Fieldbuses

In order to be able to assess the security of links between fieldbuses and other networks, it is first necessary to identify possible points of attack. Figure 7.6 illustrates a relatively general architecture of such a connection to the Internet. There are three essential components. The first is the fieldbus itself with a number of nodes and specific network architecture. The second is a gateway to the global network, which executes a series of functions (described in detail further below). The third part is the remote device (e.g. a management or visualization application running on a PC), which accesses the Internet via the fieldbus.

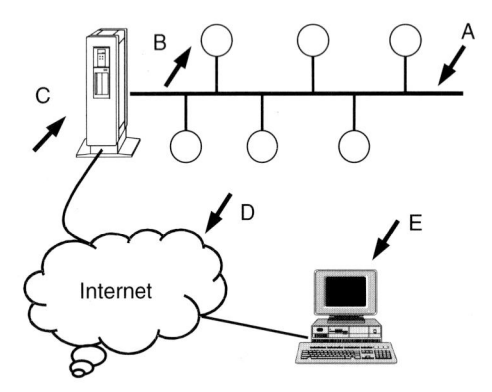

Figure 7.6: Possible Attacks on a Fieldbus

If we now examine the possibilities of attacking such a system, we can define five main areas:

– *Internal attacks* (A) represent the most widespread and dangerous threats. They are aimed at the information flow via the fieldbus. The manipulation or simultaneous reading of data leads to the loss of integrity and confidentiality. That the possibility of such attacks even exists lies in the nature of the fieldbus protocols, which seldom offer any security services (see chapter 3.6). This means that any physical access to the network also offers full access to the flow of information.

- *Network node abuse* (B) is not so common in connection with fieldbuses. It still needs to be considered however, as it represents a great potential for danger. The intruder in this case is the fieldbus node itself, which under certain circumstances contains a so-called "Trojan Horse". The aim of such an application is the unauthorized passing on or manipulation of data.
- *Gateway attacks* (C) are directed at the Internet gateway. Authorized access to the services of this gateway can open up a multitude of possibilities for the intruder for breaking into the network.
- *Attacks via the Internet* (D): It is of course obvious that the Internet represents an unsafe environment. This means that any data to be securely transmitted via the Internet needs to be protected against unauthorized access by third parties. This can be achieved via suitable security mechanisms such as encoding or digital signatures.
- *Indirect attacks* (E) are aimed at the computer that is used to access the fieldbus via the Internet. One possibility is the Trojan Horse (e.g. a Java applet that has been downloaded by another computer). However, the computer can also be attacked from the outside via the Internet. The following example illustrates how dangerous such attacks can be: Encoding mechanisms are used to protect the data exchange. The relevant code can be stored on the local computer. If an intruder were to "steal" this code, the communication end device (in this case the computer) is compromised and the intruder gains full access to the network.

All of the above-mentioned attacks can represent threats to the fieldbus. In the following sections of this chapter we will take a close look at individual elements of secure system architecture, which offer a high level of security for fieldbus applications connected with the Internet.

7.5.3 Secure Architecture Elements

A high level of security can only be achieved with a combination of different security mechanisms. A central issue in the construction of secure fieldbus architecture is the adherence to the respective OSI guidelines [ISO 89], already discussed. In this section we would like to concentrate on the corresponding components such as secure Internet management protocols, smart cards and Java technology. These form the basis of secure system architecture, which is then described at the end of the chapter.

7.5.3.1 Chip Cards and Smart Cards

Chip cards are the prime example of *tamper proof hardware*. This means devices that are "protected against external influences and falsification" [Scha 91]. Once the French Postal Service issued several million such cards (containing memory chips) as telephone cards in 1985/86, the technology that had been developed in the late seventies finally gained widespread acceptance.

The term *chip card* denotes all (plastic) cards that are fitted with an IC. For this reason, they are also sometimes referred to as *integrated circuit card* (ICC). The formats of these cards are standardized in accordance with ISO7816-1, however, it is not possible to make any general statements about the circuitry of the IC. In the past, it was common to use *memory cards* in which the IC essentially comprised memory components (ROM, EEPROM). Since then, optical memory cards have been developed which are initially written and then readable via laser, with memory capacities of several Mbytes [Sonn 97].

Smart cards on the other hand contain a microprocessor, which means they are also intelligent (hence the term smart). The typical layout of a smart card is illustrated in Figure 7.7. In addition to the CPU (central processing unit) it contains a ROM for programs, a RAM for transient and an EEPROM for permanent data as well as an I/O unit for communication via the serial interface. With many ICs there is also a special security block, which may for example contain a cryptographic co-processor.

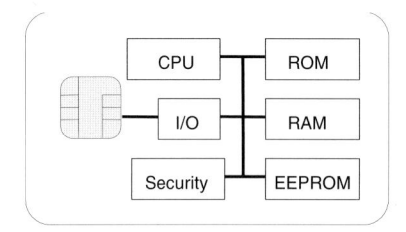

Figure 7.7: Layout of a Smart Card

The CPUs are generally based on 8-bit architecture. Due to the high requirements on the chip surface area, RAM is used economically; the size therefore ranges from 256 bytes to 1 kByte. The ROM size typically lies between 8 and 96 kBytes and depending on the intended use, an EEPROM is integrated with a capacity of up to 32 kBytes.

Smart cards usually only contain a single IC, into which all functions are integrated. The background to this architecture is the view that security can only be guaranteed when each IC integrated into the card can only communicate with the outside world via a permanent program that is immune to modification from the outside. A single-chip solution is therefore the safest.

A differentiation is also made between *contact* (still the most common variant) and *non-contact* chip cards. Because the transmission of data and supply energy between card and card reader does not occur via electrical contact with the non-contact variety, but via capacitive or inductive coupling, they are not susceptible to dirt, wear and tear on the contacts or electrostatic discharge. Non-contact cards need not be placed into a card reader but can communicate with the terminal from distances of 10 cm and more. This makes them interesting for applications such as entry control, where the constant insertion of cards is inconvenient and time-consuming.

As with most of today's computer systems, chip cards also usually have an operating system with overlying application programs. The operating systems however must satisfy special requirements, as they must not only work with very limited memory resources, but in terms of their design there must be no gaps in the security concept and they must be implemented free from any errors. The essential tasks of a chip card operating system are:

– Management of the file system,
– Communication with the card reader,

– Processing the operating system commands.

The operating system must have a secure status in every situation, i.e. it must not falter as a result of voltage or temperature fluctuations in particular.

For the first time, the *Java Card API* offers the possibility of chip card programming that is independent of hardware. As is normal with Java, the software is written in a high level language and interpreted by a *Java Virtual Machine*, which is adapted for the respective platform. In principle, Java cards also allow the reloading of applications in the field, this feature can of course also be deactivated [Kais 98].

7.5.3.2 Security Features of the Third Version of SNMP

The protocols, via which data is exchanged between the Internet gateway and the remote computer, must also satisfy specific security requirements. The SNMP management protocol, its capacity and application possibilities in EIB networks have already been discussed in this chapter. In the latest version, SNMPv3, security mechanisms have been used which appear to allow the successful use of SNMPv3 for linking fieldbuses with the Internet.

The security functions of SNMPv3 can be split into two groups, the security subsystem and the access control subsystem [Sta 98]. The main tasks of the security subsystem are authentication and encoding, in order to guarantee confidentiality and integrity. Access control provides authorization services which are used to control read and write access to the objects contained in the management information base (MIB).

7.5.4 Construction of Secure System Architecture

This section takes a look at secure architecture of fieldbus-Internet links and the corresponding guidelines for implementation. An important issue is the question of software tools. It is practically impossible to find a universal programming language that is suitable for all possible applications. The choice of language and programming environment is based on the requirements of the application. Boundary conditions such as functionality, restrictions and support offered by libraries or special hardware must also be taken into consideration.

With these considerations in mind, the Java programming language appears in a very good light. It has several convincing advantages such as platform independency (although this does involve differences in hardware access) and with that simple portability, security mechanisms for the application code and it is supported by a number of hardware and software manufacturers. Java is rapidly developing in the direction of embedded systems[31], which is interesting for fieldbus applications. A further major advantage of Java is the solid security concept, which has been greatly improved in the recent version (Java 2.0). Despite the advantages, we must not overlook the disadvantages, which can be critical for some applications in the control field. The first is the necessity of a Java Virtual Machine (JVM), which interprets the Java code and places high requirements on the system resources. Performance and reliability are also quite low. On balance however, the advantages outweigh

---

[31] The term "embedded system" generally refers to an application-specific computer unit that is incorporated into an electrical device for control or communication purposes. Slim-line implementation is characteristic, i.e. only the functions (those of the operating system and user interface in particular) that are required for the specific tasks are actually provided.

the disadvantages and Java is still considered the most suitable programming language for fieldbus-Internet links. This is discussed in more detail in the subsequent sections.

7.5.4.1 General Considerations

We will now combine the above-mentioned architecture elements to give an overview of secure system architecture. Figure 7.8 illustrates a recommended layout. In the overall system, a gateway plays the main role in permitting access to the fieldbus from the outside. For this task, two different groups of Internet protocols can be used. SNMP is very well suited to pure data exchange between the client in the Internet and the gateway (with attached fieldbus) as server.

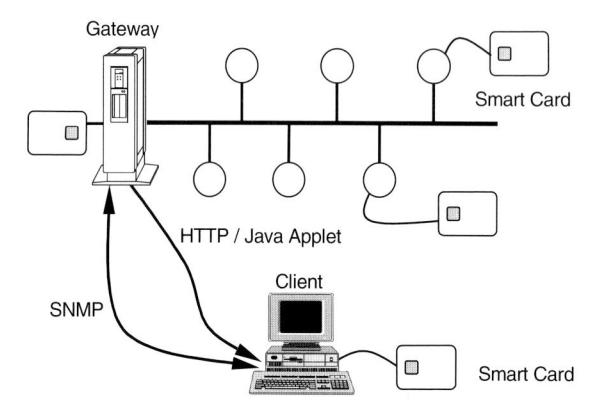

Figure 7.8: Secure System Architecture (General)

In some applications it might also be necessary to make a specific interface available to the user, via which he has full access to the fieldbus from anywhere in the Internet. The logical approach therefore is to download a www-page with Java applet from a www-server installed on the gateway. The applet contains a complete client application that is started in the Internet browser and which communicates with the server. The system architecture is much simpler for applications that do not require such a comfortable interface.

An essential aspect of the architecture is the implementation of security mechanisms, which protect the data flow at the fieldbus level as well as data exchange between gateway and client. Implementation is based on smart cards, which currently offer the best security with good manageability. The following sections take a closer look at implementing the security mechanisms.

7.5.4.2 Data Security at the Fieldbus Level

The most important axiom in the world of secure systems is that no system can offer 100% security. It is therefore essential to define limits between the secure and non-secure parts of a system. Implicit in this, is that the risk of attack in some areas is much lower than in others. The former are termed secure and the latter non-secure. The further development of the system is ultimately based on this fundamental assumption.

The same principle is valid for the development of secure fieldbus applications. It is always possible and necessary to identify secure and non-secure areas. For a typical fieldbus-Internet

7.5 Security Aspects

coupling it is sometimes possible to assume that the threat to data security on the fieldbus is much less than that in the Internet, if for example, there is no physical access to the fieldbus. Under these conditions, it is possible to do without security mechanisms for the fieldbus itself and all efforts can be concentrated on the data flow via the Internet. In other areas however, such as home and building automation where the fieldbus is easily accessible, security at the fieldbus level is of extreme importance. Security mechanisms at this level also represent an additional hurdle to an intruder who has attempted to overcome the barriers on the Internet side.

To illustrate these security mechanisms Figure 7.9 provides a general overview of secure communication between two fieldbus nodes on the basis of smart cards. The principle is easily extended to communication between one or more fieldbus nodes and a management station. We will assume that node A contains specific input data that needs to be processed and then securely transmitted to node B. Node B then produces corresponding output data.

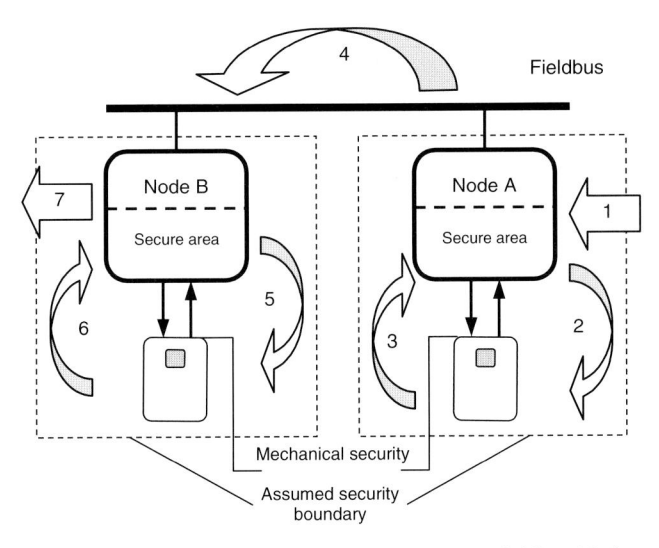

Figure 7.9: Secure Communication between Two Fieldbus Nodes

The first step is to draw the security boundaries. In this it is assumed that the communication between the nodes and the smart cards is secure[32]. The card itself can be considered secure, data exchange between the card and node however could in theory be manipulated or bugged. The most reliable protection against this is of a mechanical nature and involves placing card and node in a common housing, possibly sealed, which cannot be opened without destroying the node. The connection between node and card represented schematically in the diagram can be implemented in practice in a number of different ways.

Another important point is the security of the inputs and outputs of the node. If the inputs can be manipulated, system security at the fieldbus level is totally superfluous because the sys-

---

[32] This of course is only an assumption. Smart cards have already been successfully attacked on a physical level several times [Lemm98]. The effort needed to achieve this along with the associated cost however remains high.

tem could already contain incorrect input data. A possible solution is again offered by mechanical protection. Only when the inputs, the node itself and the smart card together form a secure system component can we talk of a "secure fieldbus node".

The prototype of a secure communication procedure goes something like this: Node A receives a secure input signal (1), which it then processes. If the security regulations of a secure transmission demand, the pre-prepared data is transmitted to the smart card on which the security services run (2). After encoding and/or the formation of a digital signature the card returns the data to the node (3), which then transmits the data via the fieldbus (4). Although this crosses the boundary between secure and non-secure area, it is not a matter for concern as the data has already been protected by the smart card. Node B receives the data and passes it on to its smart card (5), which decodes it and/or checks the signature. The card returns the data to the node (6), which can now generate the corresponding output signal.

The only danger in this system comes from the fieldbus architecture and is ultimately due to the fact that fieldbuses were never designed for security[33]. To the contrary, great effort was made to ensure simple, flexible and powerful access to the data within a fieldbus. From the point of view of the system functionality, this is a major advantage but for the standpoint of security, it is quite the opposite. A large number of fieldbuses permit direct access to the application data of the nodes. This is meaningful for the network management but not for data security.

In the given example, it could be that the memory area in node A or in node B is directly accessible via the fieldbus. In this way, an intruder could easily bypass the security mechanisms and get access to data that should actually be secured. In this situation, neither a smart card nor encoding would provide the necessary protection. A counter measure would involve the introduction of a "secure memory area" for applications and data, which is not accessible from the outside whether directly or indirectly. Although not simple, the implementation of such a concept is important.

To conclude our discussions on safety at the fieldbus level, we should point out one last important fact. According to statistics, around 80% (!) of successful attacks to data security can be attributed to administrative errors within a company. This includes instances such as unauthorized third-party access to computers, passwords, which are stuck onto computer screens with post-its, and much more. If the company policy cannot eliminate such trivial errors, then even the strictest of security mechanisms are of little use in preventing attacks.

7.5.4.3 Security Concepts for a Gateway and a Client

It has already been emphasized that security concepts must be incorporated into the development stage of a fieldbus-Internet gateway. Figure 7.10 shows the general architecture of the gateway. As can be seen, the entire gateway should remain within a secure area. This requirement is not easy to meet in practice, as the hardware and the software often have complex structures. It is essential however that there is a secure connection to the smart card.

---

[33] A good comparison is the first prototype for the Internet, which only involved four universities in the USA and was intended purely for academic purposes. There was no need therefore to consider data security. As the Internet began to grow, the aspect of data security became increasingly important. Today, there is a multitude of secure Internet protocols and services available.

7.5 Security Aspects

This critical requirement can be satisfied in the same way as for a node with the use of mechanical protection.

In the previous sections, we have dealt with SNMP in some detail and shown that it can be a good basis for fieldbus-Internet links. The diagram below illustrates the essential modules that a gateway must have in order to fulfill all the requirements. All can be represented in the world of SNMP. The gateway core comprises a large part of the SNMP agent. The security module uses the security and access control subsystems of SNMP and also allows access to the security functions that run on the smart card. The communication module processes the SNMP messages on the basis of PDUs that are received from or transmitted by the IP ports of the gateway. The fieldbus module represents the connection between the gateway application and the fieldbus.

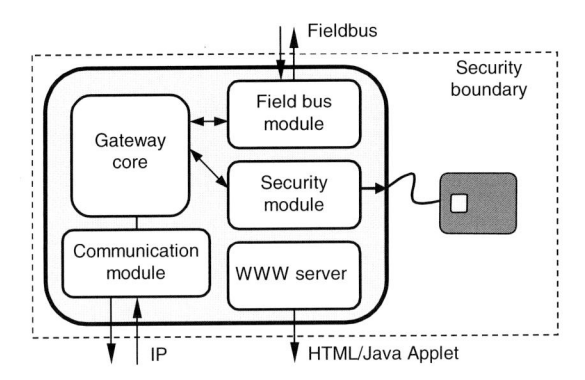

Figure 7.10: General Architecture of a Fieldbus Internet Gateway

And finally, a word on the client. Its implementation is also based on the concepts presented here. A very flexible form is the Java applet which in reality consists of an SNMP manager application with a specific user interface. It also contains essential SNMP modules such as the security subsystem, which uses the security mechanisms of the smart card. Java also provides additional security measures such as for example a digital signature for the applet code. This further increases the security of the overall system.

8 Tools

A complex bus system like EIB would be unthinkable without standardized tools. The success of EIB is partly due to the use of powerful software tools, which are available on common operating system platforms. This chapter outlines the most important of these tools and takes a look at their respective areas of application. There are basically four different groups of tools, each with different requirements as regards application area, user knowledge and user interface. For example, it should not be necessary for an electrical engineer to program, nor should a product developer have to verify each individual bit in the EEPROM of a device prototype.

There are four user groups:

1. The *Product Manufacturer* needs tools to program and test devices. It is important to test not only the static correctness of the application programs in the devices but also the run time behavior and interoperability.

2. The *Electrician* requires an easily operated tool to plan and commission full EIB systems. Uniformity is of the utmost importance here – such tools must be able to work with all EIBA certified devices.

3. The *Application Developer* relies on efficient drivers for bus communication and database access. Such programming interfaces are prerequisites for the development of additional products such as remote control, visualization, links to products in the field of facility management and much more.

4. The *End Customer* in the home (but also sometimes in functional buildings) wants tools that are extremely easy to use and which facilitate not only remote control but also configuration possibilities for lights, household appliances and other EIB linked areas. Here it is important to avoid overloading the user with too many details, comfort takes highest priority.

8.1 Development and Test Tools

Traditionally, development and testing tools used for EIB do not involve large software packages, but are elegant solutions for well-defined tasks. For example, there are separate tools for product development and for the testing of run time behavior. The behavior of a product with regard to interoperability is verified with another tool. It is generally true to say that these specialized applications have become significantly more complex and powerful with the passage of time. Because the user has a large influence on the design of the tools, targeted further development is guaranteed.

8.1.1 Integrated Development Environment

The integrated development environment (EIB-IDE, which runs under Windows 3.1/95/98) facilitates the development of application programs for BCU 1, BCU 2 as well as BIM 112. In principle, this environment represents a front-end for the "icc6811.exe" compiler (version 4.20B) and the "a6801.exe" assembler (version 1.00G). The compiler and assembler (manufacturer: IAR Systems AB in Uppsala, Sweden) must be installed separately to EIB-IDE.

In addition to a mere front-end for the compiler and assembler, EIB-IDE also offers the possibility of creating complete program templates for BCU 1, BCU 2 and BIM 112 at the touch of a button. This saves the developer a great deal of time as all initializations (association tables, address tables, communication objects) are generated automatically. We know this principle of *Application Wizards* from the latest development tools for the Windows operating system. It is easy for example, to establish the number of communication objects or the size of the address and association tables; many other BCU/BIM parameters are just as easily determined.

The EIB-IDE also offers the possibility of elementary debugging. It is possible to define variables in the application program whose values are then shown at run time. For application programs within a BCU 2 it is possible to declare the EIB-IDE as the *polling master*, which significantly speeds up the process of troubleshooting.

Concerning device access, the EIB-IDE not only facilitates the loading of the generated program into the device memory, but for application programs within a BIM 112 above *mask version 0701* it also offers access to *EIB objects*. This form of access to the properties of a device is denoted by the following terms: A device (e.g. a product on the basis of BCU 1, BCU 2 or BIM 112) has various EIB objects, which in turn consist of one or more *EIB properties*. An EIB property is restricted to a value range by its *EIB property type*, comparable with variables and data types in higher order programming languages such as C or Java. Each EIB property also has specific access rights for read and write processes. In addition to the access rights, each EIB property has a value that is displayed in the EIB-IDE and which depending on the access rights can also be modified.

8.1.2 Bus Monitor

The "EIBus Monitor" program that runs under MS-DOS (with Windows 3.x/95/98 in the DOS window), generally referred to as bus monitor, is used to trace all data traffic on the bus. All telegrams are represented bit by bit. It is also possible to poll each device; complete memory excerpts can be shown on screen in hexadecimal format. Such services are necessary in order to be able to verify the correct run time behavior of EIB products. As the bus monitor provides the telegram data with accurate time stamps, any timing errors are easily determined.

8.1.3 EIB Interworking Test Tool

The EIB Interworking Test Tool (EITT) is used to check the EIB conformity of end devices. It is used at the product development stage (when testing the application) and then for certification. In the current version 2.1 (which runs under Windows 3.x/95/98) EITT can be used to test the application layer of end devices as well as network management functions. It is important for end devices to be able to work in accordance with the protocol even at high bus load. EITT 2.1 facilitates detailed tests of the network management at low and high bus loads. EITT 2.1 is also able to send faulty telegrams, which is useful for testing the robustness of end devices[34].

EITT 2.1 can also check the conformity with regard to the EIB Interworking Standard (EIS). It is possible, for example, to easily create any desired EIS types and pack them into a telegram; this also provides a good means of testing the functionality of a product (how are maximum and minimum values handled, etc.). The automatic generation of a PIXIT header that is implemented in EITT (see [EIBA 99] and chapter 10) is used to accurately check not only the end device itself but also the associated documentation. It is then possible to substantiate statements made by the product manufacturer, such as for example, "The blind is raised when the input variable takes on the value 1".

Future versions of EITT will be able to test the OSI layers of an end device individually. The intention is to use EITT when developing new bus coupling units as well as during quality control. A current demo version of EITT can be downloaded from the EIBA web site (www.eiba.com).

8.2   EIB Tool Software

One reason for the success of EIB is the philosophy of the EIB Tool Software (ETS). Right from the beginning in the field of software tools, great importance was attached to the view that practical interoperability of a bus system essentially depends on two factors:

1. End devices which follow the EIB standard must be certified and with that satisfy the EIB Interworking Standards.
2. The software for planning, commissioning and maintaining EIB projects must be recognized as a standard on a world-wide scale. Every EIB device must be able to work with this software.

Point 1 is trivial. Without well-defined protocols and standardized data types the concept of "interoperability" would be unthinkable (see chapter 10). Point 2 on the other hand is again a question of philosophy: If the development of planning and commissioning tools was left to the open market, there would be a rapid crystallization of different software packets for the various areas of application. What could initially be seen as an advantage for the bus system, is ultimately a major disadvantage for the end user. In this context it is vital to consider the universal nature of tools that are developed by profit-making companies. Such tools from manufacturer A are often unable to commission products from manufacturer B, or do not allow the installation of routers or other related products, for example.

In the field of EIB the situation was made clear right from the outset with ETS: There must only be one planning and commissioning tool for EIB and it must come from one source. It was the task of the non profit-making EIBA based in Brussels to develop and maintain the ETS. Of course, the member companies of EIBA also have a say in the specification – those

---

[34] EITT 2.1 facilitates the generation of faulty telegrams that must at least be accepted on the physical layer of a BCU. Thanks to additional hardware, future versions of EITT will also be able to send telegrams that are not compliant with the BCU (telegrams that do not correspond to the physical layer). This is particularly useful if BCUs are supplied from various manufacturers.

who are concerned with new developments in particular (which then result in changes to ETS) make a major contribution to the ETS specification.

8.2.1 Application Areas of ETS

ETS is split into the following application areas:

1. Manufacturer's tool: This is often referred to as ETS+ and allows the product developer to generate an ETS database entry for his planned product. This is necessary in order to be able to certify the product and then to introduce it to the market.

2. Project planning (or simply planning): The ETS allows the user (in this case an electrician or system integrator for example) to create a virtual picture of the EIB project. It involves the definition of individual buildings, rooms, control cabinets and also EIB devices. The connections between them are also specified: This group address switches on the floor lighting, one is responsible for the blinds, etc. The important fact when planning is that the entire EIB network is virtually represented on the PC; all physical addresses, association tables, address tables and such like are created automatically.

3. Commissioning: Here, ETS is used to load all devices. If the project has been clearly defined, then in the ideal situation a mouse click is enough to put a complete building into operation. The virtual representation (the project database) of the EIB project is downloaded and translated into network management calls. After successful commissioning the EIB network is installed and – depending on the quality of the planning – ready to use. Of course, ETS also allows the subsequent manipulation of individual devices, and the EIB project can be modified or extended whenever necessary.

4. Management: The ETS offers various modules for product and project management. When considering the wealth of products and projects an average ETS user is confronted with, it is often necessary to preserve clarity. Work often involves extending old EIB projects. For this purpose, ETS offers the possibility of converting product databases from older ETS versions into the current version.

One of the main reasons for the high acceptance of ETS is the very comfortable user interface, part of the ETS philosophy. Even the very early versions of ETS for example used "drag & drop", providing the users with a comfortable means of defining connections between EIB devices. Equally comfortable is the possibility of opening several views within an EIB project: There is the building view, the topological view, the discipline view – it is up to the user how he divides or represents his project.

In its current version "ETS2 v1.1", the software is available for Windows 3.x, Windows9x as well as Windows NT 4.0. A complete description of ETS would extend the scope of this book, and the user is referred to the relevant literature [ETS 99]. The following sections concentrate on a few very specific subject areas, of particular interest to product developers.

8 Tools

8.2.2 Manufacturer Extensions of ETS

ETS2 v1.1 offers the possibility of integrating manufacturer-specific extensions on the basis of separate screen dialogues and Dynamic Link Libraries[35] (DLL). [CDI 98] outlines the corresponding technical conditions for such extensions. A basic differentiation is made between two variants:

1. An ETS screen dialogue can be replaced by a manufacturer-specific dialogue.

2. The ETS can be supplemented with manufacturer-specific DLLs. Read and write access to the ETS database is permitted.

Point 2 in particular allows comprehensive extensions of the ETS. Whether it be the management of a separate database or a connection to manufacturer-specific software – using the Complex Device Interface allows any extensions to be integrated into ETS.

8.2.3 ETS2 as the DDE Server

The term Dynamic Data Exchange (DDE) stands for a Windows standard for exchanging data within various applications. Data is transmitted from a DDE server to a DDE client via the operating system. ETS2 v1.1 contains a DDE server, which allows clients to retrieve the most important data from the database. Wherever necessary, it is also possible to change a value in the ETS database with the "DDE Poke" command.

The addressing of data with DDE is denoted by three terms: Applications, topics and items. In the case of ETS, the application is called "ETS"; the individual topics and items are described in [DDE 98]. The example below showing how it is possible to display the name of the first building in an ETS project from a spreadsheet application such as Microsoft Excel, illustrates the simplicity of DDE:

Assuming that the ETS DDE server has been started (for details please refer to [DDE 98]), an element can be verified under Excel as follows:

```
=ETS|BuildingId!GetFirst
```

With that, Excel shows the building identifier of the first building in the currently opened ETS project. Using this identifier it is then possible to display the name of the corresponding building (in the following example the building identifier is used as a constant, as an Excel macro would exceed the scope of this chapter):

```
=ETS|Building!'building_name(12345)'
```

A detailed description of all topics and items, as well as an explanation of the commands associated with "DDE Poke" is given in [DDE 98]. The DDE server of the ETS is mainly used to provide access to the ETS database for other applications. DDE is mainly interesting to applications that are not able to interact with ETS via the Complex Device Interface (see

---

[35] A Dynamic Link Library is a dynamically loadable file which contains executable code. This method of calling up functions is common under Windows; by using DLLs functions can be used dynamically from the initiating program. With that, the applications themselves are more compact. The principle of DLLs has existed since the introduction of Windows.

above). In the eteC and ETS next generation infrastructure, the Eagle component is the successor of this DDE API.

8.2.4 ETS via Internet

The iETS ("Internet ETS") is an add-on that allows ETS to be used with the Internet. Instead of addressing the EIB network via the serial interface, iETS permits an Internet connection to an iETS gateway, which is connected to the EIB network. This allows the iETS user to service an installation from afar, i.e. without being on site. The economic advantages for owner and installer are massive.

The combined iETS, consisting of the off-the-shelf ETS completed with an iETS client communicator module, offers all the functionality of the standard ETS. It is ideally suited to the remote parameterization of EIB networks. However, it is not intended for complete commissioning as it is necessary to manually press the programming key when programming the physical addresses – and this is not yet possible via the Internet!

8.2.5 EIBnet/IP and iETS Gateway "Server"

On the gateway side, connected to the actual EIB network, one may use a Windows PC running the EIBA's own iETS Server. Alternatively, the open IP tunneling protocol has been implemented in embedded iETS gateway devices, offered by several EIB manufacturers.

Indeed, iETS uses the EIBnet/IP tunneling protocol on a TCP/IP connection (as described in the previous chapter) which, together with the mechanisms of the Windows NT operating system, offers a high level of security.

In fact, the iETS Server implementation from the EIB Association acts as a PC-based universal EIBnet/IP gateway in its own right – also for other clients than just iETS.

8.3 EIB Tool Environment Component Architecture

ETS2 consists of various applications (e.g. manufacturer's tool, commissioning and others), each of which is strongly monolithic in character. As the requirements on ETS are constantly rising, a division into components is useful for reasons of maintainability alone. For this, there is a multitude of possibilities, depending on the programming language and operating system. With Java for example, the natural thing to do is to break down larger components into so-called JavaBeans [Ecke 99]. Microsoft's COM (Component Object Model) offers a comparable component concept. This concerns a collection of classes, which carry out a well-defined task and with that are independent of other program parts [Vale 99]. The emphasis is on interface contracts, rather than directly on class implementation. The division of an application into components is a very difficult process with the need to consider many different factors. This component forming is a relatively young field of research, with many works dedicated to the underlying description languages [Fowl 98] or diverse architectures [Slam 99]. A good introduction to the fundamentals of diverse component forming technologies is given in [Patt 97]. For a practical overview, [Szy 98] is a useful reference.

ETS is software for Windows that has been developed in C++. With that, COM suggests itself as the underlying model [Rog 98]. It is currently only available for Windows and has become the de-facto standard in this context too. In its network-capable form of DCOM

(Distributed COM) it offers a comfortable means of integrating programs as distributed applications. Every COM component can be easily converted into a DCOM component without recompiling [Rubi 98]. It is equally possible to make COM components capable of scripting, so that languages such as Visual Basic for Applications (VBA) or other macro languages can access the respective components.

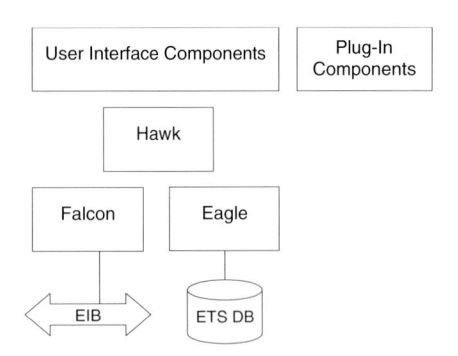

Figure 8.1: EIB Tool Environment Component Architecture

The components which make up the ETS are combined under the term "EIB tool environment components architecture" (eteC), see Figure 8.1. eteC satisfies two primary requirements. Firstly, it provides elementary software components for the ETS developer and secondly it allows developers of visualizations, remote controls or other software packets to program their applications with ETS conformity. This is a very important point, as absolute interoperability must be guaranteed in the software area too. It should be possible to address an EIB project from any EIB visualization, compatibility problems must be avoided from the outset. In general, specialized software is never totally error-free. If every manufacturer of EIB visualizations (there are many solutions on the market) were to write his own bus driver and ETS database driver, the probability of incompatibility would be very high. For this reason, the two most important components of eteC are the bus access component "eteC Falcon" and "eteC Eagle", a component for controlling the ETS database. These two components are examined in more detail below.

8.3.1 An Overview of Previous EIB Components

Within ETS there are certain parts that have emerged as candidates for components. The first worthy of mention is the bus access component whose task it is to encapsulate all bus accesses at the telegram level. Developers of EIB software, for example, should not have to set up a series of telegrams in order to access the memory in a BCU 1. The programmer should be relieved of this work.

A second candidate is database access. The ETS is based on a complex relational database which manages all products, buildings, group addresses, etc. Access to this database should occur via well-defined interfaces, in the same way as access to the bus, direct access via SQL (structured query language) should be avoided at all costs. In a similar way to the previous example involving database access, entry of a new device to a line should not be achieved by an SQL command but by calling up "AddDevice( )" of class "Line". The tabular structure of the database should be encapsulated in the components, similar to how the bus access component conceals the telegram layout from the programmer.

ETS2 provides both these components in the form of Dynamic Link Libraries (DLL). However, as regards bus access a certain degree of knowledge concerning the telegram layout is necessary and the database component does not offer any consistent object hierarchy for a full EIB network.

The reason for the reduced level of componentization in ETS2 lies in the operating system: ETS2 needs to run under Windows95/98 (and NT with additional software), but also under Windows 3.x. The finer aspects of the market have a role to play here[36]. Under Win16 (the name of the operating system platform of Windows 3.x) it has not been common to split applications into components. As the applications became more complex, there was a trend towards more monolithic programs. It was only with the introduction of COM/OLE (Object Linking and Embedding) that componentization became more widespread.

The bus access component of ETS2 is called "EIBlib" and runs under the MS-DOS operating system. The database component DAL (Database Access Library) is based on the 16-bit version of the SQL Sybase database. Both program libraries are described in [Sdk 99]. They form the basis for the development environment for complex devices (see above). Of course, both Eiblib and DAL can be used for applications other than ETS2.

8.3.2 Models Behind the New eteC Components

Whereas in the previous generation approach described above, any models implicitly emanated from the tools themselves, roles have been radically reversed in the development of the eteC infrastructure for the next generation ETS tools: first abstract Object Models were drawn up, to represent the different aspects of the domain, relevant for the design and configuration of EIB network installations. Subsequently, these models were realized in a multi-tier component architecture, as summarized in the following sections. Finally, the actual "tools" form a comparatively lean layer, on top of the Model Components.

Here is a summary of the eteC Models:

- *Network Access & Protocol Model* encapsulates the physical access to the EIB network, and provides an abstract method API to the EIB protocol. Realized in the Falcon component.

- *Product Template Persistence Model* models the data-driven product plug-in, provided as template by the manufacturer for each of his EIB products. Realized as subset of the Eagle component.

- *Project Persistence Model* is the Core Repository Model, which captures all configuration-relevant aspects of an EIB installation. This is also the key to off-line design with ETS. Also realized as subset of the Eagle component.

- *Network & Device Management Object Model* captures the resource and address management on the bus, together with the complexity of the various node implementation models of EIB (such as BCU1, BCU2, ...), which result from the fact that the system is not bound to any specific processor architecture. Realized in the Hawk component.

---

[36] In the most important ETS markets the depreciation periods for PCs are typically 4-5 years. This means that many ETS users are still bound to Windows 3.x.

8 Tools

– *ETS Domain Object Model* unifies all previous, underlying models in a common application framework. When applicable, a domain object, a Device say, is both repository- and network-aware. This means that the application can update the persistent copy of the object in the repository, or write its updated configuration to the bus– always with simple and generic manipulations. The Domain Object Model also carries the know-how for specialized or collective Operations on Domain Objects, involving several other components, or entire collections of objects. The Domain Object Model is realised as a family of eteC components, which are selectively accessible to plug-ins living in the ETS Next Generation framework.

Resulting from this approach, we find a rich set of object API's, evidently matching the components' interfaces, for the entire EIB domain. Several of these API's are published, and available for 3rd party use, as explained below.

8.3.3 eteC Falcon

eteC Falcon (Falcon for short) offers COM based access to EIB and supports the serial interface in accordance with BCU 1 or BCU 2. When drawing up the specification, it was crucial that Falcon was not just to be used for the sending and receiving of telegrams, it was much more about changing to a higher level and providing a true application programming interface (API). For example, if you wanted to send a group telegram with Eiblib it would first be necessary to put it together. With Falcon, this would be achieved by calling up a special function, which relieves the developer from the task of creating the telegram. A welcome side effect of this API approach is that Falcon does not require any detailed EIB knowledge. If we include the scripting capacity of Falcon, then it is possible to access the EIB via Visual Basic applications.

Falcon offers various interfaces such as, for example, GroupDataTransfer or MemoryAccess. Please refer to the corresponding documentation for more details [Falc 00]. As with COM, these interfaces offer various methods, with GroupDataTransfer there are, for example, methods such as "GroupDataWrite" or "GroupDataRead".

Apart from group telegrams and memory access, Falcon also offers read/write access to physical addresses, association tables, address tables as well as system identifiers or domain addresses. The latter are required in connection with open media such as radio or powerline (see explanations in chapters 5 and 6). With Falcon, network management is covered by the PropertyAccess component.

Mapped to SOAP calls, the Falcon API is also a formal part of the ANubis extensions of the EIB system, for IP. This defines a "high-level representation" of "low-level access" to the bus, across any internet link.

Falcon supports Windows9x, NT4 and Windows2000. A version for Windows CE called "eteC PocketFalcon" is also available (this does not offer full Falcon functionality [PocF 00]).

8.3.4 eteC Eagle

The "eteC Eagle" COM component (Eagle for short) has been developed to allow comfortable access to the ETS Repository, realized as a database. The specification made sure that programmers could execute all transactions in the project part of the ETS database without requiring SQL know-how (structured query language). The project part is important to developers as it contains all details concerning the EIB project. This data is used to commission an EIB network. The product part of the ETS database contains the EIB products that are required for a specific project. Eagle must offer read/write access to this area too. In the manufacturer part ETS saves its knowledge of EIBA member companies. And lastly, the certification part represents information about certified EIB products. This part is only used when registering new EIB products.

In total there are four different levels of Eagle:

1. Read access to the project part: This level is typically used by application developers.

2. Read/write access to the project part, read access to product part: This level is necessary to allow the creation of ETS projects in other applications, e.g. for the automatic generation of EIB networks in third-party software.

3. Read/write access to the project part, product part and manufacturer part: This level is used with ETS.

4. Read/write access to the entire database including certification part: This level is only used by EIBA; the "usual" ETS cannot use this level. Only the Certification Department has access – after all, the certification of products can be decided on this level.

Eagle is completely based on OLE DB (Object Linking and Embedding for Databases); Figure 8.2 shows the underlying model. The application that is based on Eagle need only use a special object hierarchy [Eagl 00], as is customary with OLE DB. Collections and items offer a comfortable means of looking through or even modifying the database. By using this de-facto standard of OLE DB, Eagle offers very simple database access, the scripting capacity additionally substantiates the advantages for developers. For example, it would be feasible to use a spreadsheet program to edit the ETS project database.

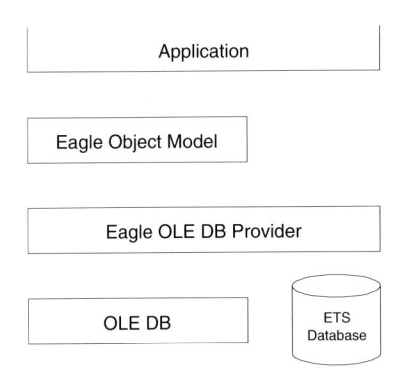

Figure 8.2: Structure of Eagle

8.3.5 Other Components

In addition to the basic components of Falcon and Eagle, eteC also offers "eteC Hawk", which is responsible for the commissioning process of individual EIB devices. With few commands, Hawk facilitates the downloading of complete S19 files (file format for microprocessors) into the correct memory area. This saves the developer a huge amount of work – the different mask versions of the respective devices are encapsulated with all properties in Hawk.

eteC also offers components to implement the look and feel of ETS in other applications. These represent extensions of the familiar Windows Controls. For example, the tree representation of the group view is characterized by particularly detailed list views. ETS drag and drop functions are included in the user interface components. For every EIS data type there are separate input and output functions, and specific screen dialogues from ETS are directly reusable.

A very important topic in connection with ETS is expandability. The idea of being able to incorporate manufacturer-specific extensions into the ETS is derived from the above mentioned Complex Device DLLs. With eteC, ActiveX offers a standardized way of providing the ETS with separate modules. In a similar way to plug-ins for various web browsers, it is possible to incorporate extensions quickly and simply with the ETS components of Falcon, Eagle and Hawk [Sdk 00].

8.4 EIB OPC Server

OPC stands for "Object Linking and Embedding for Process Control" and represents the de-facto standard for communication between Windows applications and drivers for fieldbuses, special devices or other systems outside the PC, in short: SCADA applications. OPC defines how an OPC server (in our case representing the fieldbus for example) communicates with an OPC client (typically a visualization, analysis program, remote control, gateway or even facility management software) [Opc 99].

OPC has enjoyed success on the market. Every visualization now has the functionality of an OPC client. A good many fieldbus gateways are nothing more than standard software, such as LabView for example, with a separate OPC server per fieldbus. The economic advantages of this standardization far outweigh the cost of development of an OPC server.

The EIB OPC server is fully based on eteC Falcon and accepts direct file export from ETS. This guarantees a high level of compatibility, which is being extended to the new XML configuration meta-model, proposed by the ANubis Common Datapoint Model Functionality includes group and physical addressing methods; separate configuration software allows the comfortable editing of all data points within an EIB project.

9 Facility Management

Facility management represents a tool that takes into account the entire life cycle of a building and whose aim it is to reduce the overall costs and to optimize the use of buildings. It goes without saying that this methodology has a huge effect on the organizational layout and processes within a company. We will show that digital building automation and control, and as a result fieldbus systems, are important tools for facility management.

9.1 Definition

In practice, the strategic part of facility management has been christened corporate real estate management, which as the name suggests, deals with the management of real estate functions. Actual facility management, or sometimes object management, on the other hand deals with the optimum usage and management of existing buildings (see [Zech 97]).

Facility management integrates the areas of facilities, employees, costs and processes (Figure 9.1).

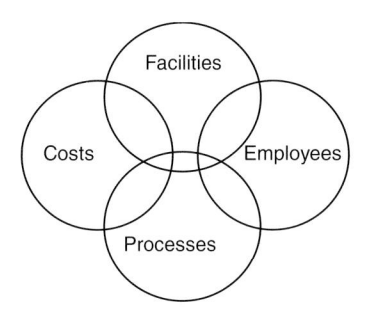

Figure 9.1: Integration of Different Areas (according to [Zech 97])

Facilities are plots of land, systems, buildings, infrastructures, machines and setups – i.e. everything representing the fixed assets of a company.

Strategic approach

In the USA, corporate estate management and facility management form a strategic alliance [Ghah 98]. There, facility management is integrated into the business planning of corporate real estate (CRE) management. In the late seventies, the occupation of "Facility Manager" was established and the area gradually came to be known as "Facility Management".

In 1982 the United States Library of Congress defined facility management as:

9 Facility Management

"The practice of coordinating the actual physical workplace with the people and processes within a company; it combines the business management principles with the architectural principles and those of behavioral and engineering sciences" [Rond 95].

Operative approach

The following definition was issued in Glasgow in 1990 by the national European FM associations:

"Facility management is responsible for the coordination of all efforts related to planning, designing and managing buildings and their systems, equipment and furniture to enhance the organization's ability to compete successfully in a rapidly changing world."

Building management – concentration on usage

In order to clarify the position of fieldbus systems in the context of operative FM (facility management), we will make a further differentiation between facility management and building management. Building management covers all services that are necessary for the maintenance of buildings. In other words, building management involves the management and running of existing buildings. The term building management covers all coordination tasks that are necessary to guarantee the effective use of buildings.

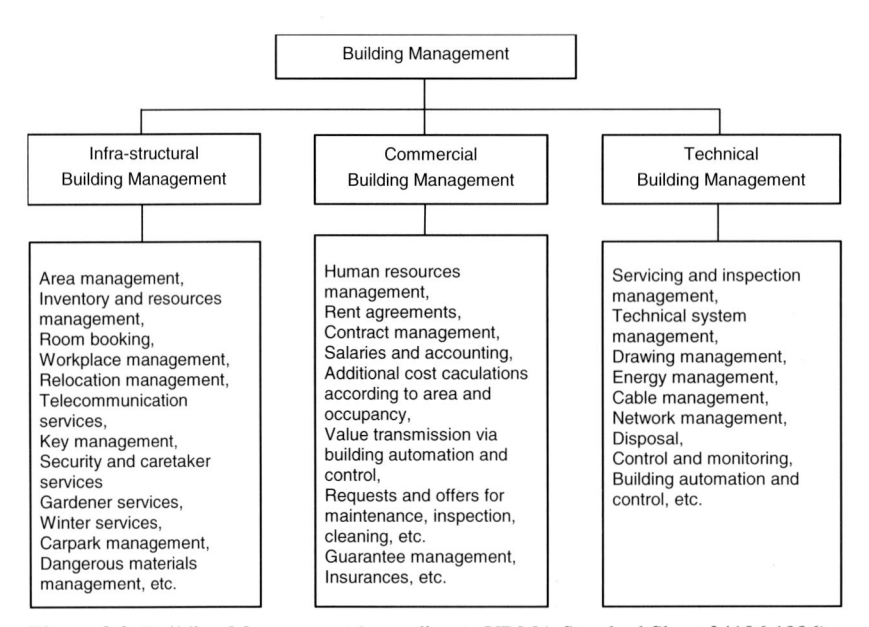

	Building Management	
Infra-structural Building Management	Commercial Building Management	Technical Building Management
Area management, Inventory and resources management, Room booking, Workplace management, Relocation management, Telecommunication services, Key management, Security and caretaker services, Gardener services, Winter services, Carpark management, Dangerous materials management, etc.	Human resources management, Rent agreements, Contract management, Salaries and accounting, Additional cost caculations according to area and occupancy, Value transmission via building automation and control, Requests and offers for maintenance, inspection, cleaning, etc. Guarantee management, Insurances, etc.	Servicing and inspection management, Technical system management, Drawing management, Energy management, Cable management, Network management, Disposal, Control and monitoring, Building automation and control, etc.

Figure 9.2: Building Management (according to VDMA Standard Sheet 24196,1996)

The definition of building management reads [Nävy 98]:

"The totality of technical, infra-structural and commercial services with regard to using buildings/real estate in the context of facility management" [Vdma 96].

This definition divides building management into three areas. We differentiate between infra-structural, commercial and technical facility management (see Figure 9.2). The term building management is used to denote all tasks during the operative use of a building. All material resources outside the building are excluded from building management. Only the actual usage phase is taken into consideration, all others are excluded.

However, this approach is associated with certain problems:

- In practice, these three areas cannot be viewed independently. If, for example, there is a change in the technical area, this often results in a change in the commercial and infra-structural areas.
- Restricting considerations to the usage phase only cannot yield the best results as a large part of the subsequent processes and costs are established during the design and planning stages.

In order to avoid the disadvantages of this approach, an attempt has been made to find a functional definition of FM which takes into account the entire life cycle of a building and which also avoids the classical division into technical, commercial and infra-structural areas.

Total functional approach

This approach is based on the life cycle of a building (Figure 9.3). The various phases are as follows:

- Planning and preparation,
- Construction,
- Usage,
- Renovation,
- Redevelopment,
- Demolition.

It should be noted that the individual phases do not always run sequentially, but in cycles. Examples include renovation and redevelopment which both occur during usage. As soon as they are complete, usage resumes.

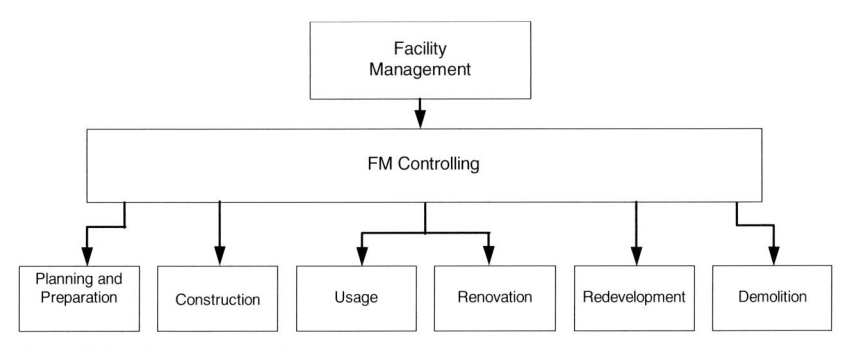

Figure 9.3: Hierarchic Function Tree

This functional definition has been worked out by the International Facility Management Agency (IFMA) in cooperation with the Technical University of Vienna. It represents a collection of individual tasks or functions during the phases of the life cycle. FM controlling forms an intellectual, organizational and business management blanket across all phases. It is

9 Facility Management

ordered above all the other functions and is applicable to all life stages of the building. FM controlling guarantees the integration, coordination and controlling of the individual areas. Although this term also covers a few general activities, FM controlling is comparable with the classic term of controlling more familiar in business management fields.

In each phase, the object is recorded and documented by the facility management. The importance of a common view of the building from its conception right through to its demolition is clear from the hierarchic function tree or the strong emphasis of the FM controlling philosophy.

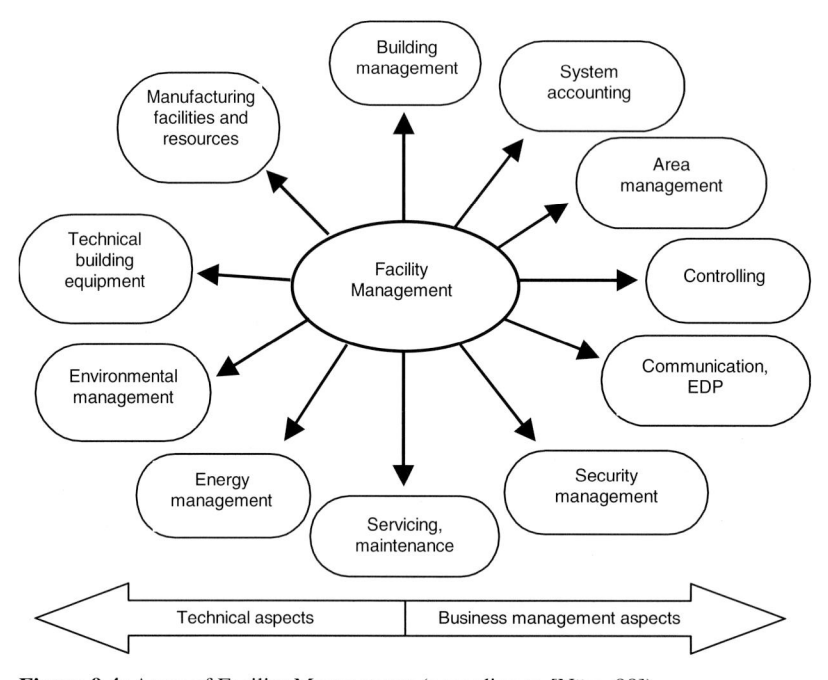

Figure 9.4: Areas of Facility Management (according to [Nävy 98])

To ensure optimum use of the resources, information is used across all areas. All data with regard to the object is collected and made available. It can be seen that there are both technical aspects and business management aspects (Figure 9.4). However, a common view is essential for facility management [Nävy 98].

Integrated computer systems, which can access all building automation and control data, create a link between the technical operation and commercial management of a building [Schn 97].

9.2 Prerequisites

An organization that is planning to introduce FM must be clear about the effects:
- *Organizational changes*: The introduction of facility management will result in organizational changes. All activities must be coordinated within one department or at least be

under the control of this department. This is the only way to manage the many and varied tasks.

- *Data recording*: Before EDP tools can be used effectively, all necessary data must be recorded. The building and room plans together with the information concerning the usage of the rooms and cost centers form the basis for a CAD-based FM program. Depending on the individual application, it may be necessary to include additional data.
- *Maintenance work*: The work is not complete with the one-off recording of the data. All modifications must be incorporated in order to ensure the current status is represented at all times. Planning can only be achieved accurately on the basis of up-to-date information. It is also necessary to have the correct data in the system to be able to calculate the costs accurately.

This additional effort however pays off with a series of advantages:

- *Cost reduction and improved usage*: In most cases, using FM results in a reduction of usage costs with a simultaneous increase in benefits for the individual.
- *Fast, flexible planning*: On the basis of the constantly updated data, which is available to the planning, planning is faster and more flexible. It is possible to envision more scenarios and the quality of the solutions is improved.
- *Access to information*: It is easy to ensure simple, controlled access to information for all those involved in the processes.
- *Relocation management*: On the basis of the master data, information can be derived for other processes. An example is relocation management. Lists for the forwarding department concerning what is to be transported from location A to location B can be generated automatically.
- *Cost certainty*: The correct allocation of costs to users. This yields huge potential for savings and makes the work of the FM department more transparent.
- *Resource optimization*: All these approaches result in the optimum use of the available resources.

The necessary data always represents an essential part of every FM system.

9.2.1 Profitability

Overall, the above-mentioned requirements result in an increase in the construction costs of the building. Until now, this has put a lot of builders off introducing such systems. Usually however, the calculations involve a comparison of the construction costs alone. This approach has changed dramatically in the past few years.

Nowadays, the total cost of ownership is included into any assessment of economic viability. With regard to buildings or other resources, it has been shown that simply reducing the construction costs is almost always associated with increased expenditure during the usage phase. This quickly results in much greater overall costs. With the passage of time, the divide is becoming greater. The cost situation can only be improved slightly, because at this time it is only possible to make minor changes to the structure of the outputs.

Using the example of the fieldbus, we will attempt to clarify the situation. The installation of such systems results in additional costs, costs however that can be recouped in the usage phase. This can be highlighted with the example of converting two separate one-man offices into a single workspace for two employees. In the original status, both rooms have an on/off switch near the door and a light fitting in the middle. Within the course of the conversion when a classic 230V cable has been used, the electrician is called on after the dividing wall has been removed to reconfigure the system. The on/off switches must be exchanged for three-way switches and the lamps need to be connected in series. This means the electrician either has to lay some new cable or re-lay the existing cables to create the necessary circuit. The work is quite involved and messy and prolongs the conversion time by a few hours. The coordination of an additional discipline also needs to be considered. This means increased costs for the conversion work and longer down times for both offices. If a fieldbus had been used on the other hand, there would be no need for any change to the circuitry on site. In the central building systems control center, the system could be reconfigured with a few simple software commands. The fieldbus devices receive the new commands and immediately behave in accordance with the new needs of the user.

In addition to a pure saving in costs, there is also a noticeable increase in the benefits to the user, particularly as regards comfort. The common use of offices represents a simple application example. This model assumes that the employees are never all in the office at the same time. This means there is a massive potential for savings if the number of workplaces can be reduced. However, it also means that the employee no longer has a "fixed" workplace that belongs to him alone, he now has to share his desk. The disadvantage of this situation is the work required to regulate the room in accordance with the requirements of the individual inhabitants, i.e. to offer them the best lighting conditions, heat, etc., even though they vary strongly from person to person. A fieldbus system offers the perfect solution. As soon as an employee switches on his networked PC, he is detected by the system which sends the corresponding control commands to the fieldbus devices to create the desired work environment.

Possibilities are also opened up in the field of preventive maintenance. By integrating suitable sensors it is possible to foresee breakdowns and with that initiate timely maintenance work, thereby avoiding a system failure and the costs that are associated with this. The necessary maintenance and inspection work is also reduced, because it is possible, for example, to exchange bulbs shortly before they blow out instead of the usual periodic changing, and with that to achieve longer average operating times. Damage prevention is also supported. If for example there is a water leak or some sort of malfunctioning, then the device can not only send an error message to the central building systems control center but it can also transmit commands to other devices in order to substitute its own supply or limit the damage, for example by stopping the leak. From the fault, if there is a clear picture of the damage, a maintenance order can be initiated directly, which defines the necessary personnel (internal and external) who are dispatched immediately and afterwards paid accordingly.

There are of course other possible scenarios. In hotels, thoughts are turning to the possibility of monitoring beds using fieldbuses in order to prevent using single rooms as doubles. In such application areas, it is always necessary to weigh up between the benefits for the operator and what is legally permitted or what can be tolerated on a social or ethical level.

As can be seen from the above examples, the use of FM and fieldbuses results in higher costs for construction but there are greater savings during the usage phase and ultimately greater profit levels.

9.2.2 Data Acquisition

Information quickly accumulates when planning and constructing buildings. This information however is not usually concentrated in one place but distributed among the individual planners and companies.

The architect creates a digital model of the building. If the architect and specialist planner use different CAD tools, this model is exchanged with the help of the DXF[37] format. Although there are ISO standards for the exchange of CAD data, e.g. the STEP[38] standard, the DXF format from Autodesk is currently used as the de-facto standard, especially in the construction field. This interface however, only regulates the format of the transfer of individual "dumb" drawing elements.

The general specification of this interface is not sufficient to ensure optimum further processing possibilities in other programs. It is often necessary to establish extra rules and structures. The structuring, i.e. the grouping of individual logically connected drawing elements, is one of the main issues. This structuring is achieved in CAD programs (and in the associated DXF format) partly through "layers". A layer contains all drawing elements that represent a specific type of object (e.g. walls, openings). Other programs use similar structures, but use other terms such as levels and filters.

As there is no standard for structuring on the basis of layers, software manufacturers have taken different approaches that are mostly incompatible with one another. This means that if exchanged without further controls, the data can no longer be correctly interpreted by the programs. In other words, a wall is no longer represented as a wall but as two independent lines.

A further problem is that the entire intelligence of the individual drawing elements (doors, walls, windows, etc.) is implemented using macros. These are basically small programs that control the look of the objects according to scale for example. When exchanging data between different CAD programs using the DXF format, the object intelligence is lost.

The approaches currently being used to overcome these problems are:

– When using the DXF format, there must be a uniform ruling for the layer structure. All those involved in the planning must supply their data in this structure. This guarantees not only that the data can be correctly interpreted by those concerned but also that it can be imported into an FM system without requiring a great deal of extra work.

– An additional simplification can be achieved when importing into the FM system if blocks are also used. Blocks are the second structuring possibility available in DXF format. Using blocks it is possible to group elements such as for example, light switches, ventilation ducts, etc. All drawing elements that denote a light switch are then combined into one object. It is also possible to attach attributes to blocks, which for a light switch could specify the supply circuit for example. Almost all common FM tools allow such blocks to be copied as single objects. Furthermore, with the help of configuration files, it is also possible to insert the attributes into the database tables of the FM tool automatically and to link them with the object.

---

[37] Drawing Interchange (eXchange) Format from the Autodesk company

[38] Standard for the Exchange of Product Model Data, ISO 10303

- A significant improvement of the current situation would be achieved if object-based data exchange between the individual applications were possible. These objects – walls, windows, fieldbus lines, devices, etc. – would have to be detected and correctly processed by all software products. Attempts at such a file format have already been around for some time. The STEP standard, an ISO standard, common in the field of mechanics, does offer such possibilities but cannot be translated for use in construction. The International Alliance for Interoperability (IAI), has been working on a practical building model for some years now. The individual objects correspond to the parts of the building and are denoted as Industry Foundation Classes. This alliance unites research institutes with leading software companies. The underlying objects have now been defined and the first software products supporting this interface are now available. Within the scope of an IAI project, the elements necessary to describe fieldbus lines and devices have been worked out. In version 3.0 of the Industry Foundation Classes, it is now possible to exchange fieldbus information in an object-oriented way with the help of this file format.

When using the DXF format for data exchange, the best approach has proven to be to combine the entire fieldbus device into one block. Its properties (physical address, variables, etc.) correspond to attributes that are added to this block. When using the IAI interface there are separate objects for the fieldbus devices available to the user which also offer the possibility of managing all device-specific characteristics.

In both cases, the user is provided with all information for the installation after the planning stage – in the form of CAD plans and in the FM database. Subsequent changes can be planned with the FM software, whereby the up-to-date information is always available to the facility manager. This data not only facilitates planning and installing, but with the help of an interface to the fieldbus system, the monitoring and controlling of the available devices.

9.2.3 Maintaining and Updating Data

In addition to the collection of data during construction, there is a further problem. Once the building is handed over to the end user, up-to-date information diminishes very rapidly (Figure 9.5). This situation is seen in most companies of today.

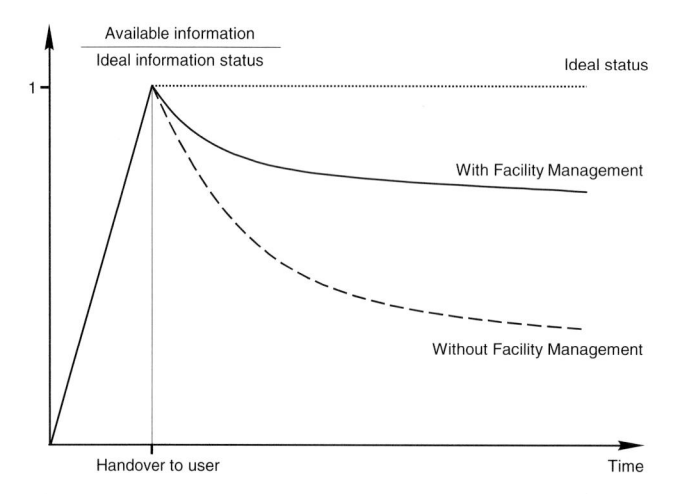

Figure 9.5: Status of Information

In an ideal world, the entire knowledge base would be retained after the building is handed over. The loss of information is scarcely noticed by the users as plans are only compulsory in the case of changes in usage caused by new environmental conditions. The procedure for collecting the data for the new situation is usually both time-consuming and costly.

The theoretical ideal is unattainable. The use of facility management however does make it possible to slow down and reduce the inevitable loss of information. Facility management ensures the availability of correct data, which is essential for strategic long-term planning [Nävy 98]. The rate of data loss can be minimized, important knowledge can be retained, managed and increased.

Of significant importance in this connection is the procedure for new plans or re-planning. Ideally, all necessary information should be stored in the FM database and from there made available to the planner. After the design has been drawn up and the structural changes made on the basis of this data, the updated data should be replayed into the FM system in order to ensure the data stock reflects the current situation. In this context we naturally encounter the same problems as with data acquisition.

It is foreseeable that semantic building product models and object-oriented building information systems will form the basis for long-term facility management systems and with that offer a platform for future integrated building automation, building systems control and fieldbuses (see [Kran 97]).

9.2.4 Facility Manager

The area of conflict in which the facility manager finds himself is considerable. He must comply with the needs of the customer, the management, the core business and the operational services. We will see that the facility manager is a reactive manager, who must operate under the constraints of the economic viability and cost transparency of his organization.

Facility managers operate within an area of conflict, which Cornelius Geertsma and Peter P. Felix [Zech 97] characterize with four main parties; these are (Figure 9.6):

Customer

This denotes the occupants or users of the building or any person within the building. This category includes requirements such as complaints, wishes, information, faults and everything not directly related to the primary process.

Operational services

This includes the facility organization, departments, external suppliers and services. These services include planned and ad-hoc activities. The latter are problematic. The facility manager can maintain an overview of the planned activities such as periodic maintenance, cleaning etc. The ad-hoc activities on the other hand present him with problems. The facility manager does not know how many bulbs will blow in the next six months, how often the doors to the stairwell will be locked or how often the coffee machine will break down. These faults in the regular services make life more difficult for the facility manager and take up most of his time.

However, the facility manager is assessed by the customer on the basis of these ad-hoc activities and this area determines the image of the facility organization. If, for example, the grass remains un-mown for more than a week, complaints are not forthcoming. If however, the coffee machine is out of order during a meeting or a flickering bulb is not immediately replaced, the inhabitants of a building make their displeasure known almost without delay (see [Zech 97]). A fieldbus system with "intelligent" sensors and actuators can successfully diffuse this situation.

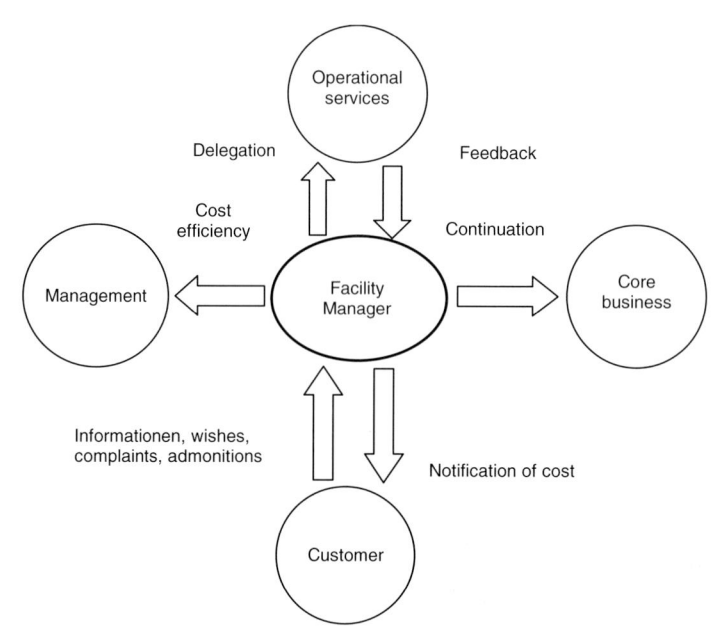

Figure 9.6: Field of Conflict of the Facility Manager (according to [Zech 97])

Management

An important measurement of efficiency is money, and with that the management can be seen as a further influencing area for the facility manager. The management are primarily interested in the costs of the facility services, which means that these expenses must be justifiable.

Core business

The core business of a company is the production process or actual service that is to be provided, i.e. the actual task for which the entire organization has been founded. Facility management serves and manages the organization. The facility manager suffers from a chronic lack of information about the business management aspects (budget, costs, ...). In most cases his simple mission is to save costs. The possibilities of improving productivity by means of efficient facility management therefore remain in the background.

9.3 EDP Tools for Facility Management

In order to maintain an overview of the situation, the facility manager requires practical tools such as automated systems. Only then is he in a position to initiate suitable actions aimed at reducing costs. He can then execute the necessary managerial tasks by simply registering and requesting information based on actual and up-to-date figures. Management on the basis of assumption can finally be assigned to the past.

Special FM software is usually used to support the activities of the facility manager. It allows, for example, costs to be calculated, buildings to be managed and checked and cleaning areas to be established. On the basis of this information other calculations can be made and statistics and reports generated. These products provide excellent support for the static management of buildings.

In larger buildings, building systems controls are increasingly being used to support technical operation. These are used to operate and monitor control systems within buildings. In most cases, fieldbuses facilitate communication between the individual devices and control units. This tool can be used to record the current status of the building.

The use of two different software systems however is associated with certain disadvantages:

- Two systems must be bought and maintained.
- Data redundancy increases.
- The same data must be maintained in two separate systems.
- This increases the number of errors and the costs.

One possible solution would be the integration of the FM system and the building systems control tool. This tool can be used to manage the static data, but it also means the current status of the building is available. The prerequisite for this is a direct connection between the FM system and the devices within the building.

In an initial stage, the functionality is extended with features that allow the status information of the individual fieldbus devices to be made available in the FM system. In a second stage, functions are implemented allowing the individual devices to be controlled.

The polling of the status of a device as well as its control are now commonplace. However, the applications being used are often unwieldy, complex and expensive. It is therefore more advantageous to control the devices directly from the FM tool. It should be possible to click on a symbol representing a lamp, and in this way turn the lights on and off.

9.3.1 CAIFM Tools

Today, there is a wide range of Computer Aided Integrated Facility Management (CAIFM) tools. They consist of a CAD interface and a database with which it is possible to manage object-related information for every drawing element contained therein.

The buildings or rooms are recorded in two-dimensional plans. These plans contain all relevant equipment items, devices and cables. Each object is additionally linked with database elements.

In order to make the above-mentioned concept a reality, the CAIFM tools must have two important properties:

- The software product must be capable of setting up a simple data connection to another computer via the Internet.

- It must be capable of OLE[39] automation in order to allow "remote control" from external programs.

9.3.2 Facility Management and Fieldbus Systems

The interface should not be directly connected to the fieldbus driver software, as there is no common standard for fieldbus systems. It would then be necessary to implement a separate interface in the FM tool for every fieldbus. This means that the interface should use a server as a gateway or interpreter. This server [Schw 98] forms the link between the interface of the FM software and the EIB. This concept does not only facilitate the use of EIB, it is also suitable for the simultaneous operation of several fieldbuses. It also allows control via the Internet.

Java is suitable as the programming language for the server, as it can be used independent of platform. A further advantage of Java is that it is very easy to incorporate extra fieldbus systems. Access to the different buses is encapsulated in a special class. This basis is not changed. This means that nothing changes on the interface when a new fieldbus is attached. The new fieldbus is automatically supported by the FM interface.

The FM tool interface sends a control command via a TCP/IP link. These commands can be used to poll and modify the status of fieldbus devices. After the execution of every command, the server returns a message to the client, in this case the FM interface.

For every device, the FM software stores all variables, their maximum and minimum values as well as the degree of granularity. On the basis of this information, it is then possible to check whether a particular value is permitted by a device even before the command is executed. These check routines significantly reduce the probability of error when executing commands.

9.3.3 System Architecture

If FM software supports the OLE automation interface, i.e. represents an automation server, it is possible to access the data and functions of the FM tool via a different application. However, this communication only functions in one direction. The server cannot send messages to the client if a user request is active.

The main aim of such extensions of FM tools is to use functions that are contained directly in the FM program to control the fieldbus devices. It must not be necessary for the end user to operate two programs. It should be possible to execute all necessary activities by simply selecting one or more objects within the graphic and calling up the desired functions.

In order to achieve this, the following information is stored in the database of the FM product – current status and Internet address as well as server port numbers (Figure 9.7). When the user selects an object with the mouse and wants to display or modify the current status, he simply selects the corresponding function from the menu. This call-up action initiates the

---

[39] Object Linking and Embedding, now also referred to as "ActiveX": Underlying this is the Microsoft Windows standard, which is mainly intended for uniform data exchange as well as the integration of software components in applications. OLE automation is the standard, which allows applications to make their functionality available to other programs.

additional program part – the FM interface – which then sets up contact with the server via a simple TCP/IP connection. This translates the command into the respective fieldbus language and transmits it on the fieldbus. After the device has executed the command, the feedback is returned to the FM where it can be visualized.

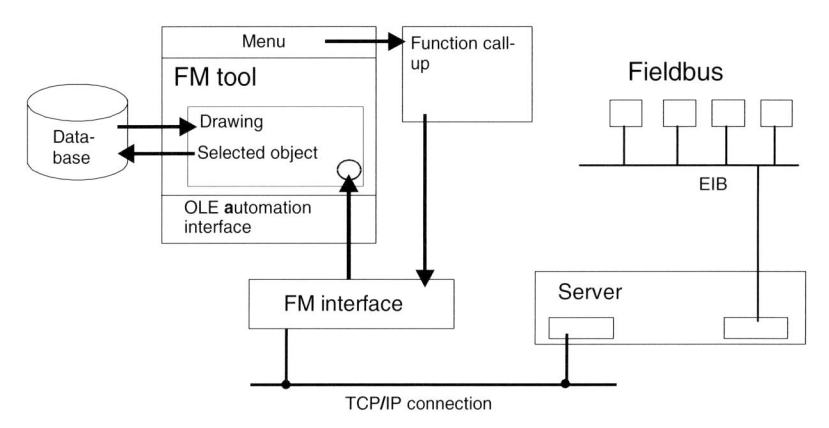

Figure 9.7: Data Flow between FM Software and Fieldbus Devices

The role of the fieldbus system with regard to the requirements on facility management becomes clear. The fieldbus allows the facility manager access down to the lowest level of building systems control. In the case of failure, the location of the defective device is immediately obvious. The facility manager can ascertain the actual status of his system components at any time down to the very last detail.

9.3.4 Facility Management and EIB

An EIB installation does not require a central computer as all monitoring and control functions can be carried out on a decentralized basis by the microprocessors built into the sensors and actuators. However, it is often desirable to use a central control location from where it is possible to determine whether or not any lights remain on within the building, a blind is still raised or a window or door open. Of course, it should also be possible from this center to be able to turn the lights on and off and to remotely control other systems. Technical faults and alarms must not only be reported, they should also be documented with date and time [Kris 97].

If a building automation and control system has been installed in a functional building, then it is advantageous for the building systems control based on EIB to be connected to it via gateways. Take for example the new Commerzbank building in Frankfurt, Germany. Here, more than 3000 data points are passed from EIB to the building automation and control system (see [Kris 97]).

In the Allianz Treptowers building in Berlin, all data-based disciplines, i.e. in-house systems, security systems, DCC and EIB are connected to a governing control room, which is independent of manufacturer, on the basis of the BACnet protocol standard. In order to be able to access all room functions and data from the building management control center, more than 18,000 data points of the approximate 50,000 EIB data points are incorporated via twelve gateways. The gateways are connected via fiber-optic cable in a segmented Ethernet network, to which the computers of the central building management are also connected. Other disciplines are integrated via direct BACnet interfaces. This means that the data points that are provided for all disciplines can be visualized in the neutral control center within 1.5 to 2 seconds. In the Treptowers, the future has already begun: A building management level has been created with any number of access points to functionally structured areas, e.g. maintenance management, visualization, lifts and energy optimization [Prod 98].

9.4 Virtual Facility Management

The virtual facility management world represents an abstract model of the real world for the tasks of planners and users. Here, the various areas of usage can be visualized but also with corresponding interfaces, modeled and programmed. The level of the virtual project world describes the long-term, stable working relationships and is used for the space and time based orientation and coordination of people in the real world. There are a number of interactions between these two levels that are worthy of visualizing [Zech 97]. Visual representations (so-called avatars) are introduced for people, their actions and for objects in these working relationships (see [Benn 97,Step 94]).

The virtual building

High-end visualizations for industry, architecture and construction allow draft documents to be clearly presented. The objects can be inspected before the first cut of the spade. Even at the planning stage the building can be viewed in virtual reality scenes. The viewer walks through a virtual factory complex, can check the spatial arrangement and height of the machine park, sees the lighting effects and can even experience the noise levels. Errors occurring in the planning stage of production processes, whose correction is then expensive, do not even come into the equation (see [Ruha 98]) .

Virtual prototyping

Virtual prototyping represents one of the most important industrial applications of the future. It will replace expensive physical models, and will mainly be used in the air and space industries as well as in shipbuilding and car manufacture. A virtual reality concept for planning and constructing steam power stations, whose link with a database containing all nongraphic data has already been presented [Virt 98].These databases contain material data, prices, delivery times and information on procedures.

Facility management in the future

The consistent development of CAFM software and its symbiosis with virtual reality tools will lead to the option of managing real estate via the PC with Cyberbrille and Trackball (Dataglove). In the foreseeable future, semantic building product models and object-oriented building information systems will form the basis of long-term facility management systems and with that offer a platform for future integrated building control systems, among other things (see [Kran 97]).

Naturally, there is quite a lot of work involved to save a building in the computer with all associated data, before it can be visited in the virtual world. Afterwards however, and depending on the level of detail, there is unrestricted access to all information. If, for example, we walk in the virtual world from the corridor into an office we can clearly identify, and where applicable, operate the lights, carpets, information technology and office furniture. With a simple mouse click, carpet areas, office furniture makes, PC inventory numbers, etc. are evaluated, named and possibly printed out in the form of piece lists. All equipment is easily evaluated simply and quickly, even after changes have been introduced [Faci 98]. This would mean that the dreaded but often necessary moves within a building lose their unnerving quality once visualized. Reams of plans and drawings are reduced to an absolute minimum if all relevant information can be requested on screen in a central and virtual manner.

Fieldbus systems in future virtual facility management

The combination of a correspondingly modified EIB Tool Software (ETS) and the virtual reality software of tomorrow will have far-reaching consequences due to the associated open system architecture:

- The system integrator can walk through the virtual building and is able to configure the network topology in cyberspace. There, he correctly positions the necessary actuators and sensors that are taken from libraries. He presses the switching elements and checks their functions in the virtual room.

- The performance data of the respective bus systems and field devices integrated into cyberspace allow the system integrator to test the network topologies with regard to function and performance by simulations. This yields immediate results and is particularly useful with large numbers of end devices.

- The simulations can apply to the individual disciplines. In the future, a boiler manufacturer will provide its efficiency simulation for different loads via the building model.

- The building work is started once the virtual planning of the building or building complex has been completed and tested. The potential for savings with this method can be considerable.

- Once the building complex exists in reality, the building model is used for virtual facility management. The available data, already extensive in volume, is constantly updated. The management and maintenance of the building is achieved via the virtual facility management right down to the field level.

- All statuses and locations of the respective building components, disciplines and system components can be viewed and diagnosed in cyberspace. Errors can be removed immediately. Any manufacturer from anywhere in the world can check his components at the request of the user or carry out routine servicing or functional tests. Defects and damage to sensitive equipment, e.g. in hospitals, can be predicted in advance or rectified immediately if they have already occurred.

The scenario depicted here is technically feasible and is the aim of current research and technology. In the next few years, economy and practice will show exactly what is feasible and what is truly meaningful.

10 Standardization

10.1 Interoperability

Interoperability is one of the key words for the durability of communication technology on the free market. The rules and connections, dealt with by the field of interoperability, are just as applicable to a conversation between two people as they are to communication between two electronic devices. It is true to say that two individuals capable of interoperable communications (either people or computers) can work together, although they may never have actually "met". This section deals with the rules of interoperability for bus systems in general and EIB in particular.

We will first look at an example from daily life – buying a train ticket from London to Manchester. This requires certain communication skills. If we take an abstract view of the process, one of the first things that both partners must agree on is the choice of communication medium. There would be no communication if the first partner (i.e. the customer) was only capable of sign language and the second partner (i.e. the ticket clerk) used the spoken word. If both partners are using the same medium, let us assume the spoken word, then the next step is the language itself. Even when both parties have selected English, this in itself is still not enough. The language must be used in the correct context and may under certain circumstances be supplemented with specific expressions. For example, "2nd class" can be used to describe things other than ticket class, whilst "sleeper car" and IC are specific terms used in the rail industry. Both communication partners must use these expressions in the same way and have the same understanding of them. In this case, English serves as the medium for a specialist language. The specialist language represents a tool that ensures information can be exchanged without being misunderstood.

It is now necessary to define the sequence in which the relevant data is exchanged. The use of the specialist language is embedded in a protocol that specifies for example that the first step is to introduce oneself and then to ask what the other requires, etc. Furthermore, both partners have specific expectations as regards the functional and temporal behavior of the other. Two ticket clerks, who satisfy all of the above requirements but who can only work day or night respectively, cannot be interchanged. It is the same as regards function. After successful communication, it is the task of the ticket clerk to then sell the ticket. Only when the temporal and functional behavior are the same can we conclude the checklist for the interoperability of a ticket clerk and certify him as interoperable for this application. He can fulfill his tasks, regardless of what school he attended, his nationality or sex (a fact that in the context of bus systems leads us to the expression "multi-vendor system"). With that, the communicative and functional requirements are defined for the "desk clerk" function.

To summarize then, we find (depending on the degree of abstractness and accuracy of our model) a series of properties that build on one another that both communication partners must have in order to be able to carry out a common task:

- Medium,
- Language,
- Specialist language,
- Interaction protocol,
- Temporal and functional behavior.

It is the same for general bus systems. Just using the same technology (e.g. twisted pair) has for some time been no guarantee that the physically connected partners are actually connected on a functional level and with that able to exchange data.

The original motivation for interoperability or interworking standards were so-called "open systems" or "open standards". In the past, conventional bus systems and automation networks were generally the development or invention of a single company. It was then common for this company to develop the corresponding devices for the network and associated software. This usually happened in parallel. The result was a system with specialized tools and components. The system parts could only be acquired from the one company, as other companies did not have the necessary know-how. If available at all, specifications were not published. We were completely at the mercy of the aforementioned company. The mechanisms of the free market could not take effect and automation technology was expensive and inflexible. The advantage of this method was that every new device designed for the bus was physically and functionally compatible with the compound system, as it had been tested extensively by the manufacturer in whose interest it was always to have a functioning system on the market. If a network-based temperature sensor version 1.0 was replaced with its successor version 2.0, then it could basically be assumed that the functionality was the same. It was a closed system; the manufacturer was in possession of the technology and communication.

Modern communication technology follows a different path. An open system is based on a collection of specifications that are made accessible to the public. These specifications describe which communication medium is to be used, how the data is coded, the sequence in which the partners transmit, etc. These specifications allow any number of manufacturers to develop products for this technology. A temperature sensor from company A for example, if it fully complies with the corresponding standards and interchangeability guidelines of the system, can be exchanged for a temperature sensor from company B. This means the customer is free to choose between manufacturers whose products may differ in terms of price, quality and accuracy, but which can all be seamlessly incorporated into his system.

A so-called multi-vendor system is a system whose components have been developed by different manufacturers but work together to achieve a common functionality. There are a number of system integrators, installation companies and operators for the maintenance of the system as the specifications are freely available and anyone can acquire the necessary knowledge. We are no longer dependent on any one company. The technology is offered by a variety of manufacturers, which together market a common network concept. There are advantages for both sides. For the customer, there is a buoyant market, a wide choice of products and numerous suppliers. The manufacturers can market their technology together.

10 Standardization

This means that even small companies can play on a worldwide scale. The competition of the bigger players is balanced out by the support they offer. A manufacturer can use products from another for his own developments, such as for example, configuration software, ICs or drivers. Smaller manufacturers are therefore able to fill niche markets whilst the larger manufacturers can concentrate on the mass markets.

Open systems arise when for example, an existing, closed system is made public. The system specifications are then made accessible and represent the rules that the system components must comply with in order to be able to work with other components. These rules are called interoperability standards. Interoperability is not however the only term that needs to be mentioned in this connection. Within a working paper (TC65 SC65C WG7 of the IEC "Interoperability Definitions") the IEC defined further divisions of the term interoperability (see Figure 10.1).

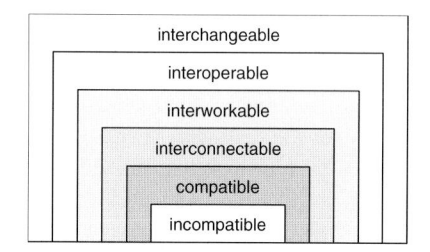

Figure 10.1: Stages of Interoperability

The range extends from *incompatible* to *interchangeable*. The outermost layer of the model in Figure 10.1 represents devices that can be interchanged. Such devices can participate in the functionality of the compound system simply by installing them – there is no need for parameterization or any operation. The other end of the scale is denoted as *incompatible* and offers no functionality. In the case of communication infrastructure based on the ISO/OSI model, the terms of the application functionality can be specified more accurately.

The first line of the matrix in Figure 10.2 shows the application functionality that needs to be achieved. The left-hand column defines the necessary requirements on functions, data, definitions, etc. The gray fields show the connection between the individual stages of the application functionality and the requirements.

Figure 10.2: Application Functionality

266

The most fundamental requirement is that both partners use the same protocol. If this requirement is not satisfied then the partners are *incompatible*. Here, the protocol comprises a group of protocols that can be termed protocol layers. If all seven ISO/OSI layers are defined, this requirement means selection of the same transmission medium, the same addressing possibilities, etc. With the systems dealt with here, the protocol is often implemented as firmware in a micro-controller. An example of this is the EIB BCU. If this microcontroller is used as the bus coupling unit, then the protocol requirements are implicitly satisfied. Such a device is also considered compliant with the corresponding protocol. Conformity tests for ISO protocols are described in the ISO standard ISO 9646. Conformity alone however does not facilitate meaningful cooperation between two partners [Mark 90]. For this, further agreements are necessary.

The way in which data is made available in the network is not defined by layers 1-7 of the protocol. This only offers services which then allow data to be transmitted in a number of different ways. Packets can have a request-response mechanism, there are a number of different ways of addressing, and even the telegram types can have different layouts under certain circumstances. If there are two types of represented data, a shared memory that can be accessed and an event, a message that signals an event, then both partners must be agreed on which data type they want to use for their application. It must also be clear how this data is to be reached, e.g. via an address, an index or a variable name. These requirements are generally embedded in the programming system of the fieldbus and offer no room for incompatibility.

The definition of the data goes a step further. If it is possible to read and write a data area of the partner it must be clear how the data is to be stored there. If the communication part of the partner is arranged as variables, the variable type must be known to both partners. If one partner sees the memory area as a BCD variable, but the other sees it as a 16-bit unsigned integer, meaningful data exchange is not possible. Support is generally offered here by the development tool. If the programming environment is designed for distributed systems, an error at this level is detected in the development phase.

To a certain extent, these semantics already take into consideration the function of the system. A variable takes on physical relevance. A value is not just a 16-bit packet; it also has a physical unit, an accuracy range and other attributes that make a value from the pure data. If a temperature sensor for example has two separate sensors for interior and exterior temperature, we establish here how it is possible to differentiate between the two variables that have been defined in the same way.

The functional behavior of the communication partners describes what happens to the transmitted and received data. The functional description of a network-capable motor, includes for example, the fact that there is a star-delta start-up after the "ON" command and the actual number of revolutions is given in variable number 5.

And lastly, the temporal behavior specifies requirements such as maximum reaction time to a command or event. These requirements are absolutely essential for the global function of a distributed system. The behavior of the network can fundamentally change if one component is exchanged for another which may have the same function and communication but a totally different timing.

The functionalities that can be achieved are as follows:

- *Incompatible*: The two partners have no common ground and their connection, if physically possible, could result in damage. The connectors, voltage levels and modulation methods can be different.

- *Compatible*: The communication partners use the same protocols for layers 1 to 7 in accordance with the ISO/OSI reference model. The devices can be connected, the applications however are not in a position to establish meaningful cooperation as they cannot make sense of the received packets.

- *Interconnectable*: The partners can recognize whether packets were addressed to them and extract or code the data. They use the services of the application layer in the same way.

- *Interworkable*: Applications can correctly exchange data, they use the same coding. A sensor can send its value as a binary number for example and a controller can correctly receive and process this value.

- *Interoperable*: The partners send data with additional information such as physical type of the data, meaning and purpose. An interoperable node can be used immediately without additional description, its network interface is unique. With an interoperable temperature sensor, the following definitions have been made for example – the exterior temperature is the 5th variable, it has an absolute accuracy of 0.1 °C and as a 16-bit value has a range from 0 °C to 6553.5 °C.

- *Interchangeable*: The network partners also correspond to additional temporal conditions. If these conditions satisfy the needs of the application, any node that is interchangeable can be used here.

The above listed requirements are generally written in specifications and profiles. A profile defines which variables carry which data, how they are offered, etc. By virtue of this profile, it is possible to subject devices, which are claimed to satisfy this profile, to a corresponding conformity test. The compatibility of a device is generally also tested when used within a multi-vendor system.

10.2 EIB Interworking Standard

The EIB Interworking Standard (EIS) makes statements on how EIB nodes can achieve the property "interworkable" [EIBA 95]. Requirements are defined and recommendations made. The recommendations are as follows:

- Communication objects should not be bi-directional.

- Communication objects should be used when a device has status information.

- Necessary time delays should occur in the actuators or controllers and not in the sensors.

EIS is available in eleven variations, denoted EIS 1 to EIS 11, as listed in Table 10.1.

As an example we will describe EIS 9 (floating point value) in more detail. EIS 9 is used to transmit physical values such as temperatures or electrical voltages. Every EIS has a code with which it can be uniquely identified. In the case of EIS 9 the code is 90XX. The last two characters specify the physical unit of the value.

10.2 EIB Interworking Standard

Table 10.1: EIS Overview

EIS Number	EIB Function
EIS 1	Switching
EIS 2	Dimming
EIS 3	Time
EIS 4	Date
EIS 5	Value
EIS 6	Scaling
EIS 7	Drive control
EIS 8	Priority
EIS 9	Floating point value
EIS 10	16-bit counter
EIS 11	32-bit counter

Table 10.2 shows a few random examples for the use of EIS 9. The values are transmitted in floating point format in accordance with IEEE 754. The transmitted 32-bit value is illustrated in Figure 10.3. All EIS are described in the same way in the EIBA Handbook for Development [EIBA 95].

Table 10.2: EIS-9 Examples

Quantity	Code	Symbol	Unit
Electric charge	9018	EIB_value_electric Charge	C
Area	9010	EIB_value_area	m²
Frequency	9033	EIB_value_frequency	Hz
Thermal capacity	9071	EIB_value_thermal Capacity	JK^{-1}

If a function is to be implemented for which an EIS exists, then this EIS must be used. If no suitable EIS is available, the used telegrams must be documented by the manufacturer. The used EIS, as well as the functions and data types that cannot be covered by any EIS, are established in a PIXIT (Protocol Implementation Extra Information for Testing, see chapter 10.4).

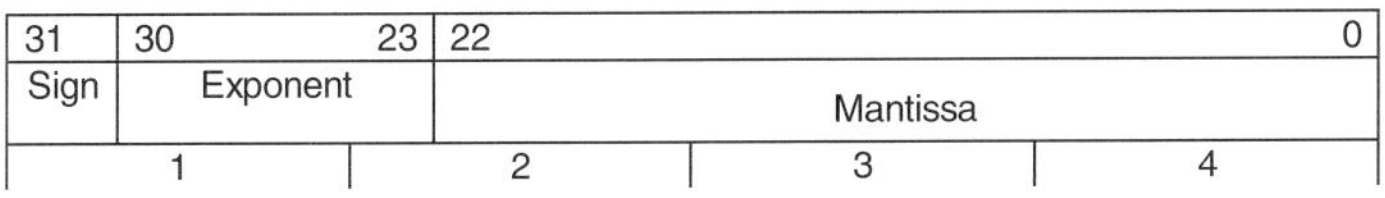

31	30	23	22			0
Sign	Exponent		Mantissa			
1			2	3		4

Figure 10.3: Floating Point Format according to IEEE 754

EIS forms the basis for the "Object Interworking Standard", ObIS. ObIS defines functional profiles that facilitate interoperability. With ObIS, the functions of an EIB node are divided into objects, whose properties represent the communication objects for shared variables.

10.3 The Standardization of EIB

The standardization and proof of conformity of a product through certification with regard to a particular standard, are not only meaningful from a technical point of view, they are also an essential part of successful marketing. It is therefore advantageous to follow the European and international standardization activities in the respective areas. In a similar way, the national EIBA committees have a part to play in the corresponding national technical committees. This section will provide a short summary of the current situation as regards the standardization of EIB, and we will take a look at the most important European and international standardization committees in this context. We will also examine the important de-facto standardization initiatives within industrial consortia.

10.3.1 The Current Situation

EIB is based on a development from the field of installation technology. In November 1992 in Germany, the preliminary standard of DIN V VDE 0829 was published under the title "Home and building electronic systems (HBES)", which describes the technology and functionality of EIB. Every implemented layer of the ISO/OSI reference model is described in a separate part of this preliminary standard (see Table 10.3 and [VDE 92a, VDE 92b, VDE 92c, VDE 92d, VDE 92e, VDE 92f, VDE 92g, VDE 92h]).

Table 10.3: Division of DIN V VDE 0829

Layer in OSI model	Part of the standard
Application layer	DIN V VDE 0829 T320
Presentation layer	-
Session layer	-
Transport layer	DIN V VDE 0829 T410
Network layer	DIN V VDE 0829 T420
Data link layer	DIN V VDE 0829 T521
Physical layer	DIN V VDE 0829 T522

EIB implementations were not possible on the basis of this documentation as it contained too little information on the various protocols. Within the context of the standardization activities it was necessary to draw up and submit a detailed standardization paper. The technical committee TC 247 of CEN published the European preliminary standard ENV 13154-2 in June 1998 under the title "Data communication for HVAC application field net - Part 2: Protocols". This preliminary standard contains four appendices describing the communication systems of BatiBUS, EHS, EIB and LonTalk.

The EIB standard published in this preliminary standard differs on some points when compared with the German preliminary standard and is much more detailed as regards the description of the protocols. Work has already started on the development of EIB in the direction of the automation level; more details on this are given in chapter 12.

10.3.2 Relevant Standardization Committees

CENELEC TC 205

Prompted by activities in other parts of the world, the technical department of CENELEC (Comité Européen de Normalisation Electrotechnique, European committee for electro-technical standardization) decided in 1986 to appoint technical committee 105 (later renamed TC 205). Its task is to draw up standards to guarantee the integration of information processing, monitoring, control and communication in buildings as well as data transmission for the purpose of information processing, monitoring and control in industrial and non-industrial buildings or their environments. Added to this is the standardization of gateways to different transmission media and public networks taking into consideration all aspects of electromagnetic compatibility and electrical safety. The title of this TC was defined as "Home and building electronic systems (HBES)". It is also concerned with the establishment of system-typical properties for hardware and software interfaces as well as testing requirements.

As a result of this standardization work there are now a whole series of documents including technical reports on HBES applications and requirements or network management as well as requirements on the functional security. There are European standards on the HBES system overview and for the important technical requirements on HBES systems. Draft standards are currently being written for the infrared and radio media and for checking the conformity of HBES products. Other standardization work deals with the requirements on the HBES installation and its release.

CENELEC TC 205A

The sub-committee TC 205A has written an important European standard on communication via the 230-Volt mains supply – EN50065/1 "General requirements, frequency bands, electromagnetic interference for data communication on low-voltage networks in the range 3 kHz to 148.5 kHz". Further supplements are currently in progress. A work group has also been appointed for broadband communication via the electrical distribution mains, as this technology could represent an interesting alternative for satisfying the telecommunication needs of the customer.

Other CENELEC/IEC TCs

Standardization activities within CENELEC/IEC SC 23B series 60669 "Electronic switches and remote switches" are also important for the EIB field. The work of CENELEC/IEC SC 34C on digital control interfaces for electronic fluorescent lamp ballast as well as that of CENELEC/IEC TC 72 on automatic controllers for household appliances is also relevant.

CEN TC 247

In 1990, CEN (Comité Européen de Normalisation, European standardization committee) appointed the TC 247 with the title "Mechanical Building Services". Since then, this committee with its four work groups has drawn up a number of important European standards and preliminary standards. The structure of the TC is illustrated in Figure 10.4.

Completed standards include EN 12098-1: 1996 "Outside Temperature Compensated Control Equipment for Hot Water Heating Systems" as well as the preliminary standard ENV 1805-1: 1998, with which the BACnet specification drawn up by ASHRAE (American

Society of Heating, Refrigerating and Air-Conditioning Engineers) has been accepted as the European preliminary standard. Furthermore, "The company-neutral data transmission system for public buildings and real estate" from Germany has been accepted as the European preliminary standard ENV 1805-2: 1995.

Another preliminary standard is ENV 13321-1: 1999 "Data communication for HVAC application field net – automation level" with the protocols BACnet, WorldFIP and Profibus. In the second part of this preliminary standard, prENV 13321-2: 1999 "Data communication for HVAC application field net – automation level", the EIBnet protocol has been specified. There is also the aforementioned ENV 13154-2: 1998 "Data communication for HVAC application field net – field level" with which EIB, Batibus and EHS as well as LONTalk have been established as standardized protocols for the field level.

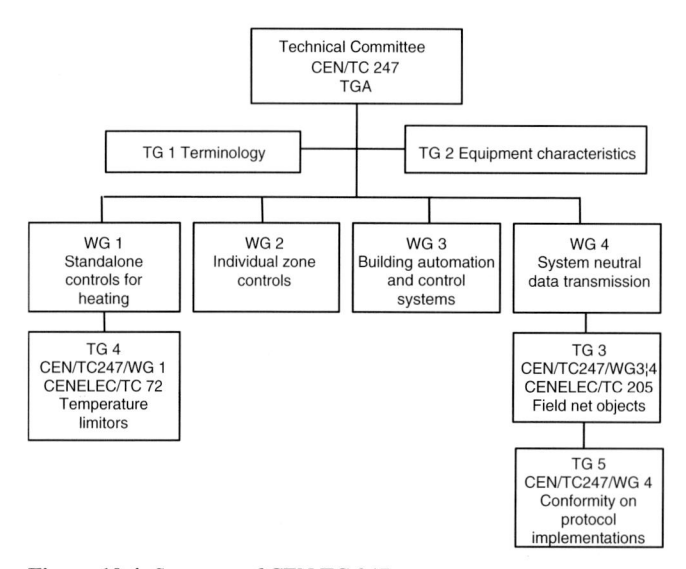

Figure 10.4: Structure of CEN TC 247

Draft prENV 13154-1:1999 is in the final stage which will standardize objects of the field level and services for system-neutral communication for HVAC applications. The EIBnet protocol for the automation level is also near completion.

Within the scope of WG3 of TC 247, concerned with the standardization of systems in building automation and control, work was begun in Europe on the subject of "Overview and definitions" and "System functions for HVAC" in the wake of the European building product guidelines of 1989, before the international activities of ISO TC 205/WG3, which only started in 1996. Figure 10.5 provides an overview of these activities.

The recent decision of the experts of CEN WG3, that the requirements for disciplines other than HVAC, such as for example CENELEC TC 79 or CENELEC TC 205 should be made in close cooperation and in a uniform way, could be forward looking. The description of the functions in part 2 is such that they can also be applied to control, optimization and monitoring functions of other disciplines, which stems from the history of discipline-crossing building automation and control.

The German example shows that the results from the CEN standardization have a direct influence on the VOB/C (contract procedure for building works) and on the standard ratings book (performance range 070/071/072 and 053, i.e. building automation and control with field, automation and management levels and the performance range "low voltage systems – installation units"). With that, these standards have a direct influence on the specifications for planning engineers when planning building automation and control. They also influence the companies carrying out the construction with respect to calculations and accounting.

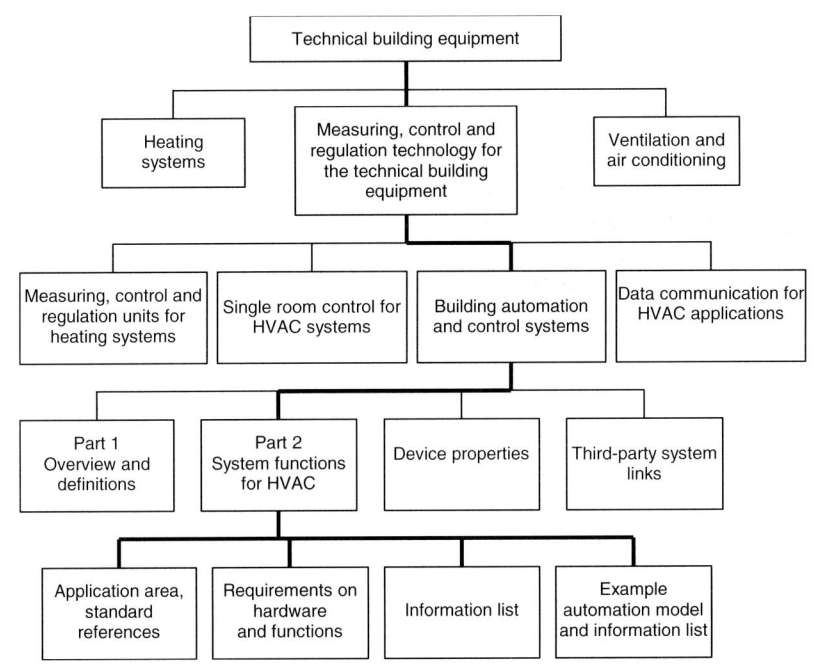

Figure 10.5: Overview of the Activities of TC 247 WG 3

ISO TC 205 WG3

Work group WG3 was founded in 1996 within the scope of ISO TC 205 "Building Environment Design", with the tasks of ASHRAE and ANSI, including translation of the BACnet specification into an international standard.

Apart from the work carried out within CEN TC 247, work items influenced from Japan also emerged from ISO WG3. The main task at ISO is the specification of the ANSI transmission protocol, i.e. the translation of the BACnet specification from a national US standard and European preliminary standard into an ISO standard. The specification should include comparison lists between the BACnet objects, CEN objects or EIB objects and other objects of the field level.

ISO/IEC JTC1/SC 25/WG 1

JTC1 was founded at the end of the eighties with the combination of the standardization activities of ISO and IEC in the fields of information and communication technology. Relatively shortly thereafter, within the scope of SC25 "Connecting information technology

10 Standardization

devices", work group WG1 was formed, under the title "Electronic system technology for the home". Since then, apart from IEC standard 948 "Numeric keyboard for HES", there has only been one draft standard "Introduction to HES" along with a series of technical reports on the architecture, communication levels and models of HES lighting control or energy management systems. The latest papers however, demonstrate significant overlapping with the work of CEN TC 247 and ISO TC 205. The main reason for this is the (non-agreed) extension of the work area from "homes" to "buildings".

10.4 Certification

In order to guarantee consistent good quality and successful interaction of EIB products, every product must pass a conformity test. Only when this test is completed successfully is the EIB certificate awarded, which is then valid for a period of five years. The certification procedure complies with the requirements of EN 45011 [CEN 98], and within EIBA the certification board monitors adherence to this standard. The three partners involved in the certification process – applicant, tester and certification office are represented in Figure 10.6.

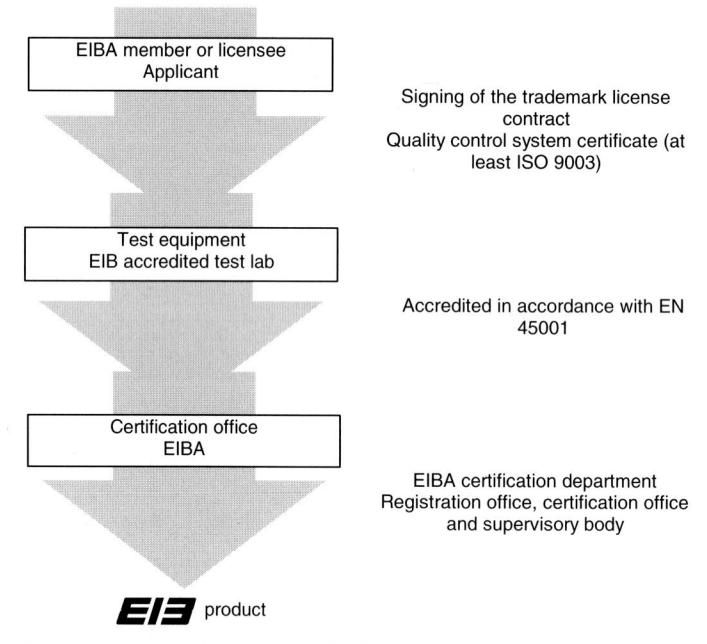

```
┌─────────────────────────┐
│ EIBA member or licensee │
│       Applicant         │
└─────────────────────────┘
           ▼                    Signing of the trademark license
                                             contract
                                Quality control system certificate (at
                                          least ISO 9003)

┌─────────────────────────┐
│     Test equipment      │
│  EIB accredited test lab│
└─────────────────────────┘
           ▼                    Accredited in accordance with EN
                                             45001

┌─────────────────────────┐
│   Certification office  │
│          EIBA           │
└─────────────────────────┘
           ▼                    EIBA certification department
                                Registration office, certification office
                                       and supervisory body

          EIB product
```

Figure 10.6: Participants in the Certification Procedure

The certification itself is divided into:

– Application phase,

– Registration phase,

– Test phase

– Certification phase.

In the application phase, i.e. before an application can be made to register a newly developed device, every applicant must sign the trademark license agreement which establishes the rules for using the EIB logo. A quality control system certificate must also be presented.

In the registration phase, an application for product registration is made. The product data is then transferred into the product database (temporarily for the time being) and released for use. The conformity test of the registered product must be carried out within six months.

During the six-month test phase, the applicant submits the documentation on his product to a testing office that has been accredited in accordance with the European standard EN 45001 [CEN 89]. This includes the PIXIT (Protocol Implementation eXtra Information for Testing) and possibly also the PICS form (Protocol Implementation Conformance Statement) as well as manufacturer declarations with test reports on tests completed for EMC, climate and electrical safety.

In the certification phase, the testing office sends the test report to the applicant, who then forwards a copy to EIBA. If the result is positive, EIBA issues the corresponding EIB certificate for the product and the product is definitively accepted into the product database. If the result is negative, the applicant must withdraw the product from the market and it is removed from the product database.

10.5 European Installation Bus Association – EIBA

10.5.1 Development of EIB

Process computers were used really early on to monitor and control the various technical operating systems in functional buildings, e.g. for the heating, ventilation and air conditioning systems as well as the lighting. This resulted in the development of building automation and control with the 3-level concept still in use today: The control level with its central computers, the automation level with specific controller elements and the field level containing sensors and actuators.

With the advent of local computer networks (LANs) and in particular the enormous success of the PC, it became possible to distribute the necessary computing power throughout an entire network of such PCs. This did demand however, like with process control systems, suitable real-time operating systems.

A corresponding initiative, to use such a computer network to link the resources within a building was developed and presented in Japan in 1984 (Professor Nakamura) within the TRON house. It networked a total of approximately 1000 computers of varying sizes for a variety of automated functions within the building, such as for example, controlling the lighting, heating, air-conditioning ventilation and hot water supply. It even provided a means of monitoring the cleaning, the washing and the ironing as well as the general maintenance.

Inspired by these ideas and supported by the low-cost microprocessors now readily available, similar initiatives sprung up around the world. Within the scope of the EIA the CEBus (ANSI) Organization was founded in the USA. Another consortium started the SMART-HOUSE concept. In Japan, the HBS specification (Home Bus System) was developed, which served as a template for the EHS (European Home System) within the framework of the Eureka and Esprit projects in Europe and in America as the forerunner of the CEBus.

It was the aim of all these initiatives to connect the narrow band control bus for controlling the various applications and resources within a building with the broadband communication requirements. European initiatives were started in 1987 in France with the Batibus, primarily used for energy management and in Germany with the instabus group, which later became one of the cornerstones of EIBA (European Installation Bus Association). Companies in Germany had already been working on the concept of installation bus systems for central management, including ABB with the Sigma i-bus and Busch Jäger with the Timac-X10.

10.5.2 The Aims and Tasks of EIBA

The next milestone in gaining success with the developing EIB system was the foundation of EIBA with its headquarters in Brussels. On May 8th, 1990, 15 leading companies from the field of installation technology in the most important EU countries got together to create this new organization and with that the international establishment of EIB home and building electronic systems, HBES.

The most important aims of this new organization were:

- Promotion of EIB home and building electronic systems by establishing the technical specification with quality guidelines and testing instructions,
- Management and expansion of the EIB trademark to a stamp of quality on the basis of conformity tests for interoperable products by independent testing institutes,
- Creation of a favorable market (international and national initiatives for marketing and communication, cooperation in the standardization, common EIB training system, standard EIB tools and components).

Every technically qualified company can become a member of EIBA, on the condition that it wants to manufacture EIB system components or EIB products and is committed to the corresponding quality requirements. It must also support the trademark. Products that are to carry the EIB logo must undergo certification. Also, any manufactured EIB products must be marketed by technically qualified companies in the field of installation technology. To date, there are more than 110 EIB registered companies [Rose 00].

As with the instabus group, certain aims were pre-specified right from the start of the EIB development:

- Simple planning and installation,
- Easily modifiable, extendable system,
- Practical, free line arrangement; line length must not influence communication (within the system limits),
- Planning, installation, modifications, service and maintenance must not pose problems for the electrician,
- Decentralized system with distributed intelligence, i.e. extension of the principle of building automation and control, that does not allow for information processing at such a field level,
- Adherence to valid safety standards,
- Open system, always expandable for new applications,

– Interoperability between products of different manufacturers.

When working out the system specification, for which the term "home and building electronic systems" was introduced, the following disciplines were taken into consideration right from the start:

– Lighting control

– Blind control

– Heating, ventilation and air-conditioning

– Energy management and load control

– Reporting and monitoring.

Design of the EIB products that was geared towards the electrician was one of the major considerations during development. Close cooperation with the corresponding organizations in Germany, the ZVEH, ensured a great deal of support in this.

A totally new step was taken with the introduction of the neutral, standardized configuration and planning software, ETS (EIB Tool Software). It was one of the keys to the success of EIB. The concept of modularity of the hardware together with software modularity in connection with ETS (the functionality of the hardware is determined by the ETS download of the application software) gave the EIB system and EIB products a great deal of flexibility for the planners and installers but also for the manufacturers. A further consequence of the manufacturer-neutral ETS was the possibility of properly responding to the need for training as a result of all the new systems.

The given objectives demanded the following actions:

– Establishment of the EIBA work structure with work groups and decision-making bodies.

– Rules had to be established and made binding in order to ensure the interworking of products from different manufacturers.

– A certification system in accordance with EN 45011, not only for EIB products but also for training institutes, trainers and qualified installers had to be set up and put into operation. This certification system has been running since 1992.

– A suitable product database had to be defined, which was to contain the EIB-certified products with all data necessary for installation and which was to allow access (only) via the ETS.

– A standard training concept was to be worked out, in order to allow integration of the electrician as the technically predestined service provider.

With that, EIB offered the trade the opportunity of developing into more ambitious service providers. A standard, neutral and international EIB training system was developed on the basis of ISO 9002, through which both the EIB trainers had to be qualified but also the now more than 60 international training centers.

11 Performance Aspects

The tests outlined in this chapter on the transmission behavior of the EIB system provide information on whether, and if so by how much, the values stated in the EIB documentation (e.g. in [EIBA 95]) differ from those measured in a small EIB system. The system uses twisted-pair cabling and comprises one line through two rooms with a total of more than 50 EIB nodes. The environment is not free from interference; in particular, there is a train track underneath the two rooms with regular traffic representing a source of interference that should not be underestimated. The data is accessed directly on the physical medium with the use of a very simple receiver circuit, which converts the EIB signals into the signals of a regular serial V.24 interface. The data is evaluated with the help of a protocol analyzer.

The measurements and evaluations described in this section can be split into two groups. Firstly there are measurements on the physical level, where the signal propagation time is examined in different situations, and secondly there are measurements on the user level, which involve recording the time between the user triggering an action and reaction by the process. An example of the second group would be to measure the time from pressing a switch until the corresponding light comes on.

11.1 Measurements on the Bus Level

The following measurements are made on the bus level:

– Latency time,

– Reaction time,

– Response time,

– Data throughput.

11.1.1 Latency Time Measurement

The latency time is the transmission duration of a bit from one end of a transmission channel to the other [Pete 96]. The signal delay of a line in the EIB bus system is caused by the properties of the transmission medium between the end devices (actuators and sensors).

In an ideal situation, the signals would be transmitted at the speed of light, and with that the latency time would only depend on the separation between end devices. Of course, signals are never transmitted ideally as every line consists of a complex arrangement of parasitic capacitances, inductances and ohmic resistances. A simplified representation consisting of a series of low pass RC filters is generally sufficient for the theoretical model. The line characteristics for loop resistance and line capacity depend on length and for that reason are specified per km. The line capacity also depends on frequency. It is possible to calculate a time constant after which the line capacity, with regard to the voltage, is 63 % charged or 37 % discharged and after which it is possible to assume that the voltage level can be detected by the receiver.

The EIB bus permits two line types – PYCYM 2×2×0.8 and IY(ST)Y 2×2×0.8 (with restricted laying methods). Cable type PYCYM has a line capacity of 0.12 µF/km (at 800 Hz) and a loop resistance of 72 Ω/km, which is easily calculated:

$$\frac{R_{Loop}}{km} = \frac{2 \cdot \sigma_{Copper}}{A} = \frac{8 \cdot 0.018 \frac{\Omega \cdot mm^2}{m}}{\pi \cdot 0.64 mm^2} \approx 72 \frac{\Omega}{km}.$$

The established values for the line capacity and loop resistance represent the maximum and minimum values respectively so that it is possible to work out an approximation of the maximum delay time of the line:

$$t_{V\,real} = R_{Loop} \cdot C_{Line} \Rightarrow t_{V\,real} = 8.64 \frac{\mu s}{km} \approx 9 \frac{\mu s}{km}.$$

The values established by the measurements carried out were not particularly meaningful due to the existing environment and lay below the expected values [Fris 98]. Even if the latency time of an EIB line were equal to 10 µs, it can be confidently neglected when compared to a bit duration of 104 µs. In the case of simultaneous bus access by two end devices on the line, one can be certain that the resulting collision is clearly detected at the first different bit of the two telegrams during this bit duration.

11.1.2 Reaction Time Measurement

The reaction time t_R is the period between the time at which an order is issued and the time at which the decision concerning its acceptance is reported [DINV 92].

This definition is intentionally kept very general so that it can be applied to all layers of the ISO/OSI reference model. The only question concerns exactly when an order is valid as being issued. In this case, the reaction time refers to layer 1 (physical layer) of the ISO/OSI reference model, so that an order is valid as "issued" when the last bit of a telegram is sent from the unit issuing the command (sensor) to the bus. The message concerning acceptance is only given once the last bit of the acknowledgement is sent from the unit receiving the command (actuator) to the bus. In accordance with this definition, the reaction time can be determined as follows:

The last bit of a transmitted telegram is the stop bit of the check character. After this stop bit there is a two-bit pause. Afterwards, the bus rests for a further 13 bits (pause time) and waits until the acknowledgement is sent. The acknowledgement character consists of 13 bits, whereby the last two represent the pause time of the bus. For this reason they are omitted from the calculation of reaction time and with that the acknowledgement character only consists of 11 bits. Furthermore, the reaction time also includes the latency time of the transmission channel, as the telegram needs a certain amount of time to reach the addressee. Therefore:

$$t_{Reaction} = t_{Pause\,bit} + t_{Pause} + t_{Acknowledgement} + t_{Latency}$$

11 Performance Aspects

$$t_{Reaction} = (2 + 13 + 11) \, bit \, * \, 104 \, \mu s/bit + 2.7 \, \mu s = 2.706 \, ms$$

According to this calculation, the theoretically attainable reaction time is equal to 2.706 ms. The measured values are shown in Table 11.1.

Table 11.1: Measured Values of Reaction Time

Measurement	Value / ms
1	2.706
2	2.704
3	2.705
4	2.705
5	2.706
6	2.705
7	2.706
8	2.706
9	2.705
10	2.706

The measured values are constant, difference from the theoretical value is minimal.

11.1.3 Response Time Measurement

The response time is the period between the time at which the issuing of a command from one unit to another ends and the time at which acceptance of the result of carrying out the command (or corresponding message) begins at the issuing unit [DINV 92].

Again, like the definition of reaction time, this definition is kept very general. Because the response time is also determined on the physical layer, the issuing of a command is complete when the last bit of the telegram is sent from the unit issuing the command. Transmission of the result is complete when the last bit of the acknowledgement has arrived at the command issuer. The difference with respect to reaction time lies in the measurement of the last acknowledgement bit: In the case of the response time, the last bit of the acknowledgement is measured at the unit issuing the command and not as in the case of reaction time at the unit receiving the command. This means that reaction time and response time only differ by the latency time.

11.1.4 Data Throughput

The data throughput is the number of bits or bytes per unit of time that can be measured on an interface.

How the time unit is defined is decisive for the data throughput. The smallest possible unit of time is one bit length. The maximum possible data throughput is then worked out by projecting this value to the standard time unit of one second. This value can be equated with the transmission speed.

11.1 Measurements on the Bus Level

For the twisted pair transmission medium, the bit duration is fixed in EIB at 104 µs, which yields a transmission speed of 9615 bit/s. For the user data, this maximum data throughput only represents a theoretical value that can never be attained as every telegram also includes control bits, pause times and acknowledgement. We will first look at a message containing 2 bytes of user data, which is used to switch on a lamp, and then at a message containing 16 bytes of user data. For both messages, the bus idle time, the pause time and the acknowledgement are omitted from the calculation.

The short message consists of a telegram with nine characters, the long one of a telegram with 23 characters of 13 bits (which always includes a 2-bit pause) plus 1 character acknowledgement and 50 bits bus idle time ($t1$) before the telegram and 13 bits pause ($t2$) between telegram and acknowledgement. The bit lengths are as follows:

Length =	Bus idle time	+	Telegram	+	Pause	+	Ack	
L_{short} =	50	+	9 * 13	+	13	+	13	= 193 bits
L_{long} =	50	+	23 * 13	+	13	+	13	= 375 bits

The short message lasts 193 bits at 104 µs which equates to 20.07 ms and the long message lasts 39.00 ms. However, the actual totals sent are 10 characters or 24 characters of 11 bits, as the various pause times (e.g. the 2-bit pause between two characters) have not been included in the calculation of transmitted data. This gives a data throughput d_D of:

$$d_{Dshort} = 110 \ bits \ / \ 20.07 \ ms \approx 5482 \ bit/s$$

$$d_{Dlong} = 264 \ bits \ / \ 39.00 \ ms \approx 6769 \ bit/s \ .$$

When calculating user data throughput, only the bits of the user data are taken into consideration, i.e. without the start, stop and parity bits. With a short message that leaves 16 bits and with a long message 128 bits. Therefore, the user data throughput, d_{UD}, can be calculated as follows:

$$d_{UDshort} = 16 \ bits \ / \ 20.07 \ ms = 797 \ bit/s$$

$$d_{UDlong} = 128 \ bits \ / \ 39.00 \ ms = 3282 \ bit/s \ .$$

With the message throughput d_M a telegram corresponds to a frame. The number of bits is irrelevant here as a frame always includes all bits of a message. The frame of the long message is bigger, which means that fewer long messages pass the interface per unit of time than short messages. The calculation is as follows:

$$d_{Mshort} = 1 \ frame \ / \ 20.07 \ ms \approx 50 \ frames/s$$

$$d_{Mlong} = 1 \ frame \ / \ 39.00 \ ms \approx 25 \ frames/s \ .$$

In this calculation it is assumed that the messages are sent in direct succession.

In order to measure the data throughput on the EIB, the following quantities were examined:

11 Performance Aspects

- Transmission speed and with that the bit length: A value of 104 µs is expected for the bit length.

- The duration of a telegram: A telegram begins with the start bit of the control field and lasts until the end of the second pause bit of the check field. When looking at data throughput, the two pause bits of the check character are added to the telegram in order to achieve a constant character length. With a bit duration of 104 µs, the transmission duration of a telegram works out at 12.17 ms.

- Pause time $t2$: The pause time begins with the end of the second pause bit of the check field and lasts until the beginning of the start bit of the acknowledgement. The pause time is measured between the beginning of the last logical '0' of the block check and the beginning of the start bit of the acknowledgement, from which the bits to be omitted are subtracted. For example, with a telegram, 4+1 bits (which is equal to 520 µs) needs to be subtracted from the measured time. With 13 bits, the pause time is equal to 1.352 ms.

- The duration of the acknowledgement: The acknowledgement has been defined so that the parity bit always has the value of logical '0'. This allows the acknowledgement time to be measured between the beginning of the start bit and the beginning of the parity bit, to which is added the measured values of 1 stop bit, 2 pause bits and 1 parity bit, i.e. 416 µs. This yields an acknowledgement duration of 1.352 ms.

- The bus idle time $t1$: The time at which the bus coupling unit starts measuring the bus idle time cannot be measured. The only possibility is to send two telegrams one after the other, as the time in which no characters are transmitted must then be equal to the bus idle time $t1$. This is achieved using a special bus connection, which is set so that the time between two messages is kept to a minimum. The bus idle time is then measured between the beginning of the parity bit of the acknowledgement of the first message and the start bit of the second message. The 2 stop bits, 1 pause bit and 1 parity bit of the acknowledgement must then be extracted from the measured value in order to get the bus idle time $t1$. The resulting bus idle time is equal to 5.20 ms.

To show how the measurement works, we have used the example of transmission speed. The start bit of the control field triggers the measurement. A logic '0' always begins with a negative impulse of approx. 40 µs. The falling edge represents the start point of the bit. The transmission speed therefore is measured between the falling edge of the first negative impulse (cursor 1 = -1.6 µs) and that of the second negative impulse (cursor 2 = 102.4 µs), see Figure 11.1. A bit duration of 104 µs was measured giving the following transmission speed:

$$V_T = 1 / t_{Measurement} = 1 / 104 \ \mu s = 9615 \ bit/s .$$

11.1 Measurements on the Bus Level

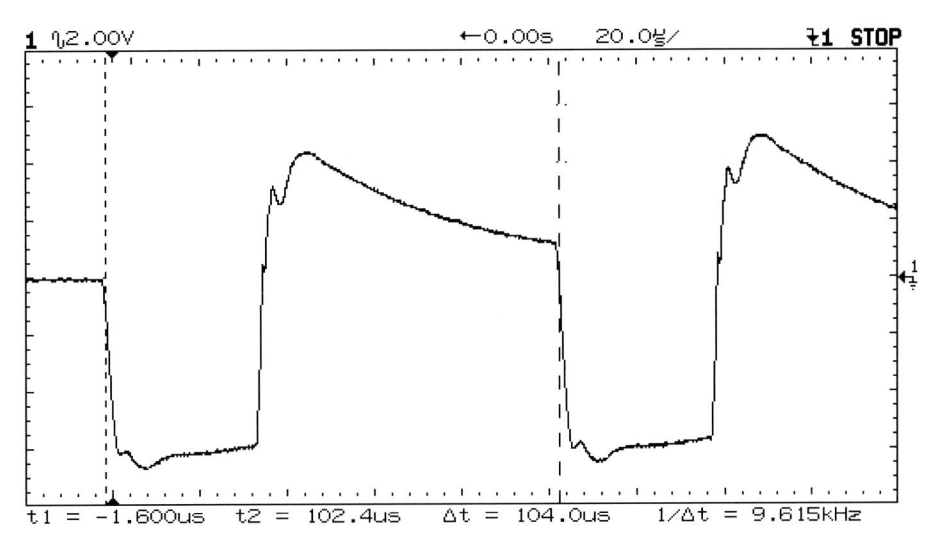

Figure 11.1: Measuring the Transmission Speed

The established values are listed in Table 11.2. The measured values are almost identical to the theoretical values. Possible deviations in the measured values are primarily caused by tolerances in setting the cursors 1 and 2.

Table 11.2: Measured Values

Quantity		Theoretical value	Measured value
Bit duration	µs	104	104
Transmission speed	bit/s	9615	9615
Telegram duration	ms	12.17	12.17
Pause time $t2$	ms	1.352	1.35
Acknowledgement duration	ms	1.352	1.352
Bus idle time $t1$	ms	5.20	5.26

Table 11.3 represents the measured values of data throughput for messages of varying lengths on the premise that they are transmitted in direct succession.

Table 11.3: Data Throughput for Different Telegrams

Data throughput		2 bytes switch/dim	3 bytes set value	4 bytes set value	16 bytes max. size
Data d_D	bit/s	5470	5637	5784	6756
User data d_{UD}	bit/s	796	1118	1402	3276
Messages d_M	frames/s	50	47	44	26

283

11.2 Measurements on the User View

The following measurements are carried out on the user level:

- Waiting time,
- Reliability.

11.2.1 Waiting Time Measurement

The waiting time is the time an order (in multi-program operation in particular) waits for the beginning or the continuation of processing.

Therefore, the waiting time is the period between pressing the button and the existence of the first bit of the telegram on the bus. With regard to the ISO/OSI reference model, this means that for the measurement we are located on layer 7, as the pressing of a button represents an application.

Table 11.4: Waiting Time Measurement Sequence

Measurement	Value/ms
1	49.0
2	47.8
3	52.2
4	47.8
5	47.8
6	48.8
7	50.0
8	51.0
9	48.2
10	48.8

With the CSMA access method, the sender waits a specific length of time before releasing its telegram to the bus. For this reason, the bus coupling unit also waits for the period of the pause time, $t1 = 50$ bits. This time, $t1 = 5.20$ ms, is simultaneously the shortest theoretical waiting time if we assume an ideal bus coupling unit, which does not require any time for the processing of an order. The values obtained by measurement are shown in Table 11.4.

The major difference between the theoretical values and the measured values can be traced back to the processing time of the bus coupling unit. For reasons of interference immunity (bounce suppression), the bus coupling unit requires a specific keying impulse length which is included in the measured result. The measured result also includes the internal processing time of the bus coupling unit.

11.2.2 Reliability Measurement

The reliability of automation systems is an important indication of quality and is determined by the main criteria of availability, malfunctioning and breakdown. The availability is the probability in percent of a particular unit, e.g. a compound system, being in a functional status at a particular point in time. Malfunctioning is true for the unit (system or device or reference point), if an agreed characteristic is not honored. Breakdown represents the change in status of the unit from error-free to faulty. Breakdown is quantified by the failure rate λ. The failure rate states how many breakdowns of a particular unit are expected or established within a given time interval for the given level of usage [DINV 92]. The failure rate is defined as follows:

$$\lambda = \frac{n}{N \cdot t}$$

where n: number of breakdowns, N: number of elements (hardware and software), t: duration of the observed time interval in hours.

The "fit" (*failure in time*) is used as the unit of failure rate. 1 fit is the sum of all breakdowns in the defined time interval. With regard to the physical layer, the failure rate is determined by the ratio of faulty to fault-free bits, the bit error rate (BER), for the duration of the observed time interval.

$$\lambda = \frac{BER}{t}$$

A value of the order of 10^{-5} is expected for the bit error rate, as we are dealing with a twisted pair line.

With the help of a protocol analyzer, which records the data directly on the bus, the received characters, telegrams, repeated telegrams, acknowledgements and parity errors were counted over a time period of 64.5 hours. The recorded values are listed in Table 11.5.

Table 11.5: Measuring the Bit Error Rate

Type of measurement	Value
Number of transmitted characters	96768950
Number of transmitted telegrams	9675881
Number of transmitted acknowledgements	9676895
Number of transmitted repetitions	0
Number of occurring errors	1021

With the values thus obtained, the failure rate and bit error rate can be calculated as follows:

$$\mathbf{BER} = \frac{FaultyBits}{FaultFreeBits} = \frac{NumberOfErrors}{8 \cdot NumberOfCharacters} = \frac{1021}{8 \cdot 96768950} = 1.32 \cdot 10^{-6}$$

11 Performance Aspects

$$\lambda = \frac{n}{N \cdot t} = \frac{BER}{t} = \frac{1.32 \cdot 10^{-6}}{62.5h} = 2.11^{-8} \, fit$$

The bit error rate established with these measurements is better than that which would have been expected theoretically and with a value of 10^{-6} satisfies very high requirements.

The acknowledgement is a component of a message. With that it can be used to check the total number of transmitted characters:

$$9676895 \text{ acknowledgements} \cdot 10 \text{ char per message} = 96768950 \text{ char}$$

This value corresponds exactly to the number of transmitted characters of 96768950.

A further point to be noted with the measured values is that the value of the acknowledgement counter is 1014 more than the number of sent telegrams. However, this effect cannot occur in a system with certified end devices as an acknowledgement is always the confirmation of a telegram. In this case, it is due to the very simple receiver circuit which does not correspond in all points to the EIB standard. A detailed description of the measurements and their results is given in [Fris 99].

12 Outlook

As a network, EIB is among the largest research and development areas of computer networks and with that is subject to rapid progress. This means that many innovations in this field have either a direct or indirect influence on EIB: direct, because as a network EIB must be further developed in order to meet the increasing demands and indirect because the environment in which EIB is used is subject to constant change as a result of other networks.

The previous chapters of this book were naturally dedicated to the current status of EIB and associated network technology and only touched on future aspects. We will now look at developments and subject areas that are likely to have an influence on EIB in the future. It is impossible however to cover all areas. We will present a few of the more important aspects, but this chapter should be viewed as a collection of ideas rather than a comprehensive study.

12.1 Convergence

As can be seen from the various contributions in this book, EIB is constantly being developed. It is not only the communication media that are being expanded – we now have digital transmission on the 230V supply mains, radio and even Ethernet in addition to the familiar twisted pair – but also the system basis. Steps have already been taken to standardize functions and applications to expand the basis of interworking and to promote automatic configuration with ETS.

Other systems began to appear on the European market at the same time as EIB, including Batibus (France) and EHS (from Esprit 2431). It soon became clear that several concurrent systems were not helpful in opening up the market. Talks were begun back in 1994 with the aim of agreeing on a common procedure. This led to the start of the convergence project in 1996, bringing together the Batibus, EHS and EIB systems.

The aims of this process are:

– A joint organization to create a common standard, with the ultimate goal of achieving an EN or even an IEC standard,

– A common system platform (system components),

– An extended multi-vendor market[40],

– A standard logo on the basis of a standard certification of interworking,

– Harmonized communication on the market,

---

[40] A key word for EIB certified products

12 Outlook

– Protection of investments already made in terms of product development, training and projects by means of a binding transition from the current products to the convergence products. EIB takes this into consideration by means of "backward compatibility".

The EIB principles of interworking and the common ETS tool have both been taken on as well as the certification process. Added to this is the possibility of configuration without ETS, and there are new transmission media being planned. A particular matter of concern is the common specification of functions and applications. The 2000+ model, shown in Figure 12.1, provides an overview of the convergence principle (a name is yet to be decided on).

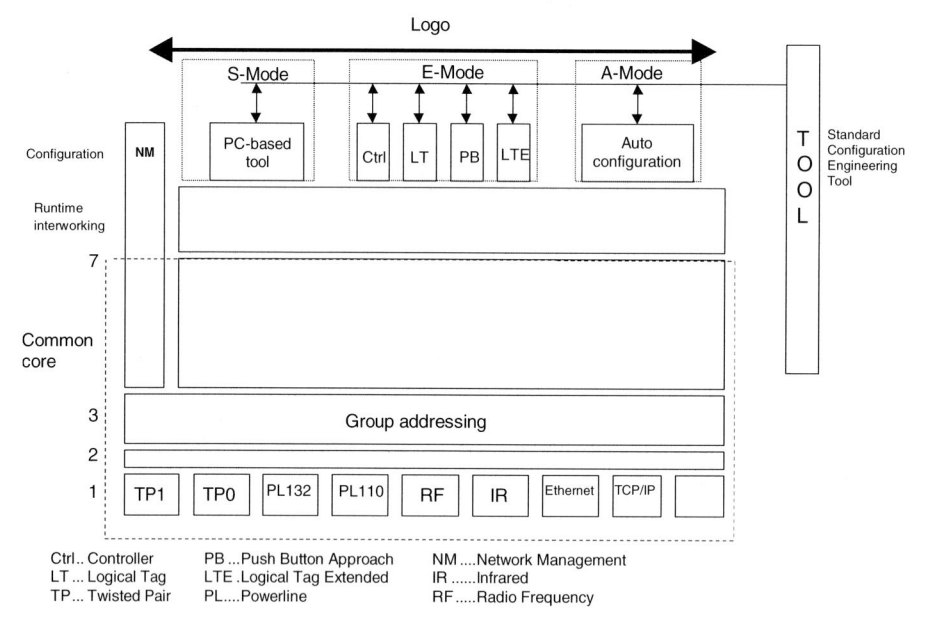

```
Ctrl.. Controller        PB ...Push Button Approach     NM ....Network Management
LT ... Logical Tag       LTE .Logical Tag Extended      IR ......Infrared
TP... Twisted Pair       PL....Powerline                RF .....Radio Frequency
```

Figure 12.1: Convergence System Model

Whilst EIBA is only open to manufacturers of EIB products, the new organization is also open to service providers and other interested companies or even individuals.

Thanks to the synergies that will arise through this cooperation under the heading of convergence, the home and building electronic systems (HBES) market will really open up, particularly in the private homes sector. On the basis of the assumed model 2000+, it will be possible to connect various system designs and their networks either with the ETS or in accordance with the E-mode (easy installation mode), e.g. with a specific installation controller or corresponding white goods device with integrated intelligence, to such HBES networks (CENELEC TC 205) almost automatically.

This convergence involves not only the product manufacturers, but also service providers such as the utility companies (gas, water, electricity), telecommunication companies as well as cable TV operators, etc., all of which are working on the common specification. This cooperation will lead to a greater range of services on offer to the domestic customer (via the Internet, key words being home-gateway, set-top box, etc.). Tailor-made service packages, which will allow occupants to stay within their 4 walls, will result in even more incentives. This is becoming an ever-stronger sales argument even for builders, who are now starting to build houses and apartments with intelligent networks. This is also improving general acceptance.

Convergence however is not only developing in the direction of the home. Thanks to the EIBnet extension (see subsequent sections), the incorporation of BACnet objects and the link to Profibus, acceptance is also increasing in the field of functional buildings too.

All in all, EIB has developed from a simple control bus to a future-safe basis for home and building automation and control.

12.2 BACnet

As the main area of application of the EIB is the field level, an additional communication system must be used for the higher order automation and management levels in larger projects. BACnet suggests itself as a standardized and readily available system.

In 1987, the ASHRAE Standard Project Committee (SPC) 135P began working on a communication standard for applications in the field of building automation and control under the title of "A Data Communication Protocol for Building Automation and Control Networks (BACnet)". An initial draft of BACnet was published in August 1991. 507 objections were received, some of which were incorporated into the second published version of March 1994. On the back of the objections to this version, the third version was issued in the spring of 1995. After expiry of the period for filing objections, the BACnet standard was submitted to the national American standardization committee, ANSI, which officially published the standard on December 19th, 1995 under the reference ANSI/ASHRAE Standard 135-1995 [ANSI 95]. In January 1998, part of this standard was accepted by the technical committee TC 247 "Controls for Mechanical Building Services" in CEN as the European preliminary standard ENV 1805-2 for the management level [EIBA 95]. A further part of the ANSI standard was published in January 1999 by TC 247 as European preliminary standard ENV 13321-1 for the automation level [EIBA 95].

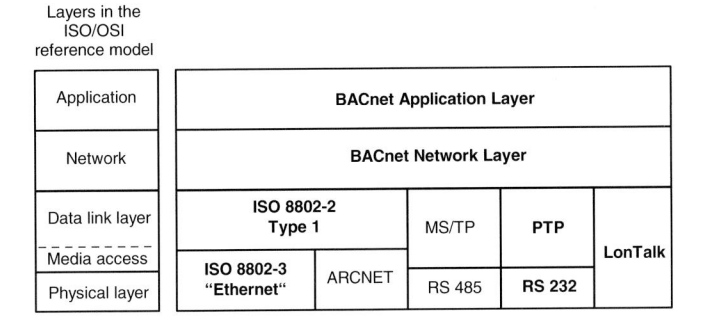

Layers in the ISO/OSI reference model					
Application	BACnet Application Layer				
Network	BACnet Network Layer				
Data link layer	ISO 8802-2 Type 1		MS/TP	PTP	LonTalk
Media access	ISO 8802-3 "Ethernet"	ARCNET	RS 485	RS 232	
Physical layer					

Figure 12.2: BACnet in the ISO/OSI Reference Model

As Figure 12.2 indicates, there are several permitted protocols for data transmission in the BACnet standard. In the European preliminary standard for the management level, standards ISO 8802-3 and ISO 8802-2/Type 1 for communication in local networks were permitted as well as a point-to-point protocol (PTP) on the serial RS 232 interface for communication via a modem or other telecommunications device. The same point-to-point protocol is also

12 Outlook

included in the preliminary standard for the automation level together with the LonTalk protocol, which offers a simple, low-cost connection for automation networks.

The special requirements of building automation and control have been taken into consideration for the specification of the protocols and services of the network and application layers. 18 objects have been defined in the application layer (e.g. analog and digital inputs and outputs, controllers) as well as 13 function classes (e.g. event reporting on the basis of value changes, virtual terminal) and 5 services (e.g. alarm handling, services for virtual terminals).

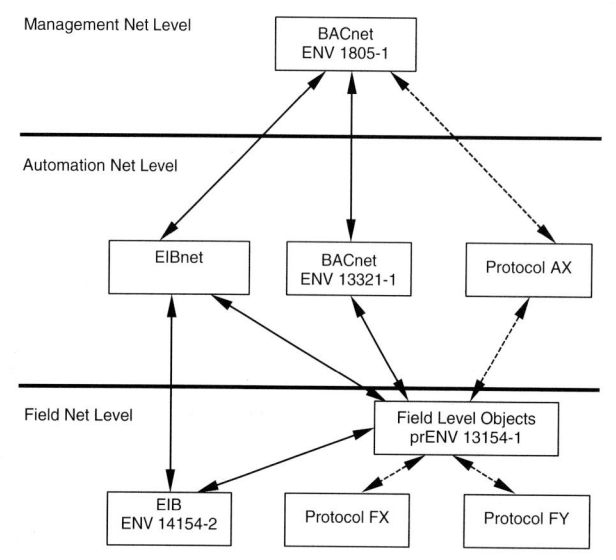

Figure 12.3: Information Flow between BACnet on the Management Level and EIB

Figure 12.3 illustrates the path of this complex information from the management level to the field level. This path can lead to EIB at the field level either via EIBnet or via standardized field objects (see [Fisc 97]). In the first case, the EIB objects and services are mapped onto the BACnet objects and services and vice versa as described in the EIBnet specification appendix. In the latter case, the path either involves another communication system in the automation level or it travels via BACnet to the standardized field objects, which can then be mapped onto the EIB objects and services; the path in the opposite direction follows the same pattern. Other paths are also possible. However, the more one deviates from the two systems BACnet and EIB, the more difficult it is to maintain consistency in the information flow between the systems and levels.

12.3 EIB and Jini

Jini [Arno 99] is a distributed system with that task of making the services and resources of a network more readily accessible to the user. The advantages of a network lie in so-called *resource sharing* and the ability to use non-local data and services from various points within the network. With traditional networks, the problem is always the same – how does a client retrieve the desired services from a server? It is usually necessary to set up and configure this connection manually. Jini attempts to reduce the effects of this restriction. Protocols have been defined that allow the client to find its services automatically and to use them immediately. Of course, it is not always possible for an application to select a suitable service completely automatically. In these situations however, Jini can make a preliminary selection in accordance with certain criteria that is then offered to a human user.

Before we take a closer look at Jini, we would first like to point out that it represents a new technology which opens up new dimensions in the networking of Java-capable devices [Asch 99]. There are now several discussion groups intensively working on Jini and corresponding systems based on it [Jini 99a, Jini 99b].

Jini represents an extension of Java (see [Krüg 99]). All participating components must therefore provide a Java Virtual Machine (JVM) in order to take part in the system (Figure 12.4).

Application	← Service Protocol →	Application
Jini	← Lookup →	Jini
Java	← Java RMI →	Java
Java Virtual Machine		Java Virtual Machine
Operating System		Operating System
	Network	

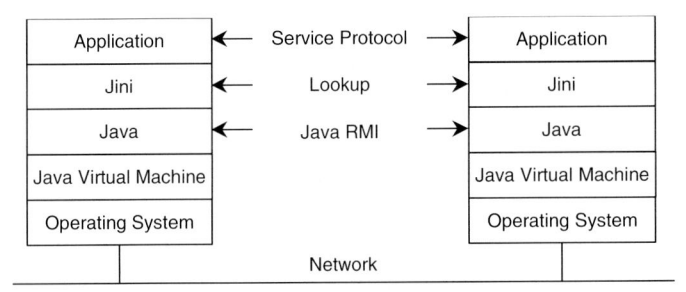

Figure 12.4: Jini Model

The advantage of using the object-oriented programming language of Java is that it supports the creation of reliable applications. This involves for example, strict type checking and automatic memory management (garbage collection). It also permits the development of distributed applications using special network APIs and the mechanism of *remote method invocation* (RMI), which facilitates access to network resources and distributed objects.

Jini lies above the individual JVMs as a logical level. The JVMs are connected together via a network. A logical group of computers on which Jini is used is termed a *Djinn*. There are basically three classes of elements in such a group, and they are *services*, *service users* and so-called *lookup services*.

A service is an abstract term for a resource that can be used in some way via the network. This could be specific hardware for example (sensors or actuators), software or even a communication channel. A service user represents a unit that uses one or more services. In this, the service protocol is not pre-specified, it consists more of a collection of interfaces written in Java. The actual implementation of the network protocol and the communication between service and service user can be selected by the service itself. The service user loads the necessary code dynamically. And finally, a lookup service is any unit that mediates between a service and service user. If a service is new to the Djinn, the discovery protocol first looks for the available lookup services. Afterwards, the service reports its functionality with the associated attributes via the join protocol.

If a service user is interested in a particular service, a connection to an available lookup service is set up via the discovery protocol. The suitable services are then found via the lookup protocol and in accordance with the search criteria (attributes). The service user can now call up the desired resources directly, or refine the selection by specifying further attributes and using the lookup service again.

In addition to this basic exchange functionality, which supports the addition and removal of services within a Djinn, Jini also deals with network problems such as the breakdown of individual computers and network sections as well as security aspects such as the transmission of packets via non-secure lines (details on this are given in [Edwa 99]).

Before we take a look at the possibilities that are opened up by the integration of Jini into an EIB system, we should first name the areas of home automation, which are already provided with communication interfaces or will be in the future and with that are interesting for home and building electronic systems:

- Classic household goods such as freezers, dishwashers as well as "mobile" equipment (e.g. vacuum cleaners and irons),
- Entertainment, including electronic items that are either concentrated in a few rooms (video recorders, hi-fis, games consoles) or distributed throughout the building (portable CD players),
- Information and communication field, consisting of classic end devices such as telephones and answer machines and more modern information processing machines such as PCs and workstations,
- Utilities (gas, water, etc.) with associated devices,
- Security, as already mentioned, as well as
- Other areas that cannot be assigned to any of the above fields (e.g. lifts).

We will now highlight a few examples of possible scenarios that could arise from integrating these application areas with the EIB system.

- Using the TV remote, the user controls the heating or lowers the blinds. The user is guided through the process via corresponding instructions on the TV and any feedback from the EIB devices is also shown there.
- When the TV is switched on, the lights in the room are dimmed and in day light switched off altogether.
- As soon as a person enters the house (or part of a larger building), the heating control reacts accordingly (depending on the values that have been recorded by the EIB weather sensors of course).
- If an electric hob is about to overheat, the lights in the house flash on and off in short intervals to warn the inhabitant of the danger. If the inhabitant does not react, a fire alarm is set off after a certain period of time.
- In an empty house, EIB movement detectors activate entertainment systems in random rooms for short periods of time, in order to give the impression of occupancy.
- The triggering of an alarm in the lift of a building triggers a corresponding alarm in rooms assumed inhabited, ensuring help is on hand.
- All EIB functions that are adjusted locally via corresponding operating units can also be controlled by authorized persons via the Internet (and with that from remote locations).

It would now be desirable if EIB components could automatically initialize themselves as soon as they were connected to the bus system, look for Jini-capable devices and if necessary set up a connection to them, or in the opposite direction if Djinn devices could gain access to EIB components via Jini.

We would now like to present a system architecture that allows an EIB system to be made compatible with Jini [Kast 99a, Kast 99b]. Our aim is to combine both systems so that they can co-exist without problems and if necessary use each other's services. The user should have the impression of a homogenous range of services. Devices that support Jini can use EIB sensors and influence EIB actuators; EIB components on the other hand should be able to profit from the services available in the Djinn. The underlying idea is that parts of EIB that are logically combined into a group (e.g. lights, blinds, etc.) are controlled via an EIB agent. This EIB agent makes the services for which it has been designed available via Jini. Every EIB agent consists of three parts (Figure 12.5):

– Djinn communication module (DC),

– EIB communication module (EC)

– Jini control module (JC).

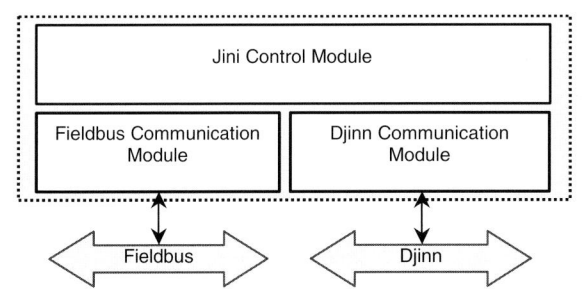

Figure 12.5: EIB Agent

The communication module to the Djinn serves to register the actual services at the lookup service and to look for or call up the desired services. The purpose of the communication module to the EIB is to control the fieldbus logics (polling sensors, controlling actuators). After the installation of one or more new components in the EIB system, it is the task of a control circuit (Jini control module) to configure the new devices via their EIB communication module and to register at the Djinn via the Jini interface. In an initial stage, the control circuit will be developed as a stand-alone module, which among other things has access to the EIB product / project database or can poll current product data from the Internet. In a further stage, integration into the ETS software (e.g. as a plug-in) is both conceivable and advantageous.

In active operation, all requests coming from the Djinn are converted into control commands for the EIB via the Jini control module; requests coming from EIB are executed as service calls in the Djinn via the respective control module. Because EIB requests are handled via the Djinn, it is also feasible that an EIB agent uses the services of another and with that the configuration (and possibly also the communication) of devices on the same bus or across EIB bus limits are transparently converted into Jini service calls. This means that EIB

12 Outlook

devices of different technology can come into contact, so that an EIB agent also acts as a cost-effective EIB gateway (e.g. twisted pair to power net).

An EIB/Jini manager consists of a series of EIB agents (Figure 12.6). Because EIB agents have a modular construction and act independently of one another, it is easy to include new EIB devices (or when new application groups arise within an existing EIB system, these areas) in a separate EIB agent, which can then be dynamically incorporated into the system.

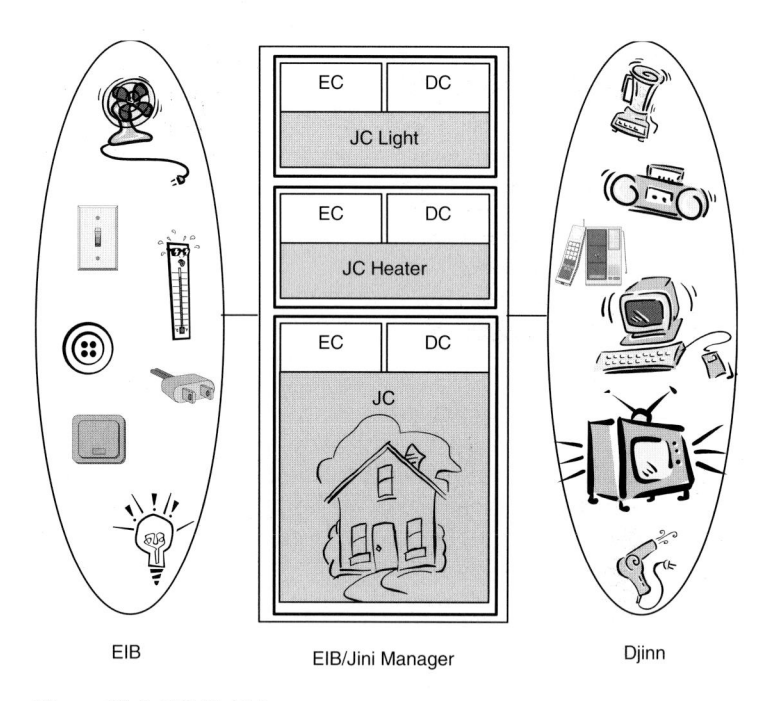

Figure 12.6: EIB/Jini Manager

The addition of a new EIB component to the Djinn occurs in two stages. The first step is the physical installation of the device in the EIB system. Then the EIB agent is installed, which means the EIB communication module can take care of further configuration in the EIB, and the Djinn communication module registers with the lookup service.

An important point is the modeling of the EIB agent as a Java class and the definition of a suitable class hierarchy. As the various manufacturers and devices want to use the services of an EIB agent and at the same time Jini services are found most efficiently by specifying the desired type, it is necessary to define a standard interface for EIB (and other fieldbus systems of the home and building electronic systems area).

With the current concept, every group of EIB services is represented by a basic class split into sub-classes which offer specific functions. All classes stem from a common, abstract superclass which offers certain basic functions. This could include a textual description of the components (manufacturer, installation site, functionality), an applet that displays status information or another applet that permits manual configuration of the EIB device. This also guarantees remote maintenance. These services have to be implemented in a sub-class (e.g. a class that provides access to the lighting of a building). At the same time, this new class can be extended with specific services.

Jini is without doubt a promising, future-oriented mechanism that in a simple and elegant way allows different devices to be connected together to form flexible combinations. This connection occurs automatically and is transparent to the user. The only prerequisite is that all participating components have a Java-based design.

In our opinion, it is not necessary (nor meaningful) to make every individual device Jini-capable in the specific application field of home and building electronic systems. It is more important to have controlling instances that access the underlying fieldbus system and make entire groups of devices accessible to the Djinn. This means it is possible to retain existing network structures, to exploit their advantages and simultaneously open them up for larger application areas.

12.4 IEEE 1394

The serial high-speed bus, IEEE 1394, was originally developed by Apple under the name FireWire [Ande 99] and in 1995 elevated to the status of an international standard [Fire 95]. IEEE 1394, also known as iLink, supports the networking of multimedia devices such as cameras, satellite receivers, video recorders, TVs and PCs. An area in which digitization is already quite advanced within the devices themselves, but as far as the respective interfaces are concerned are still largely analog in nature. This is mainly due to the fact that the bandwidth requirement, or more accurately the guaranteed access to bandwidth in the multimedia field, lies far beyond the possibilities of fieldbuses [Stam 99] or other networks. Guaranteed bandwidth is an important criterion, not least due to the fact that without it even minimal time delays, such as those involved with the transmission of a TV program, would need to be compensated by memory-intensive buffering; a property that is not supported by Ethernet for example, due to its bus arbitration method.

IEEE 1394 is characterized by the following properties:

- Low-cost, serial high-speed bus
- Data rates of 100, 200 and 400 Mbit/s (3.2 Gbit/s being planned)
- Point-to-point wiring, free topology
- Guaranteed access to reserved bandwidth
- Addressing of up to 63 nodes in each of up to 1024 segments
- Supports plug & play
- Relative short distances between nodes
- Optional supply via the bus cable

Due to the extensions and improvements to IEEE 1394 already in progress, the above properties relate to the situation at the beginning of the year 2000. In particular, greater distances between nodes should be possible. The standard [Fire 95] defines a connecting cable of a maximum of 4.5 m and a maximum distance of 15 hops between 2 nodes. The as yet incomplete standard of IEEE 1394b supports distances of up to 100 m between nodes, via copper as well as optical media.

12 Outlook

A particularly noteworthy property of IEEE 1394 is the possibility of providing guaranteed bandwidth. A node that requires guaranteed access to the medium, must request bandwidth and a logical channel from the resource manager before the first transmit process. The resource manager ensures that it is not possible to reserve more bandwidth and channels than are actually available. Several nodes in the network can offer themselves for this task, whereby the allocation is reassigned after every change to the topology. This guarantees that if the resource manager fails, its function can be taken on by another node.

With IEEE 1394 a principal differentiation is made between isochronous and asynchronous data transmission, whereby only the isochronous data can be transmitted in guaranteed time intervals. For this, a bus cycle is started every 125 µs (Figure 12.7) with the isochronous transactions occurring first. Only after all nodes have transmitted their packets can the remaining time to the start of the next cycle be used for asynchronous transmission. As no more than 80 % of the total cycle time can be reserved, guaranteed isochronous transmission is ensured. The advantage when compared with a strict time slot method is that reserved but unused bandwidth can be used for asynchronous transmission.

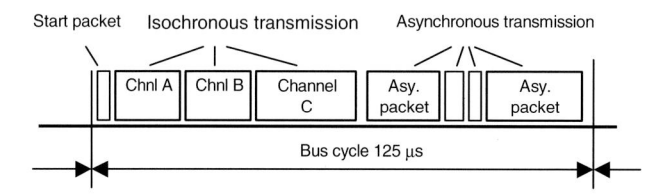

Figure 12.7: Bus Cycle with IEEE 1394

An essential criterion for the success of a network in the field of multimedia devices and consumer electronics is the usability. The configuration and programming of devices should be avoided as far as possible, as the example of the video recorder has highlighted. With regard to a network, this translates to installation involving simple plugging in and out. IEEE 1394 essentially supports this through the following three mechanisms: Firstly, the free choice of topology (the as yet incomplete standard of IEEE 1394a also permits loops) allows the user to connect an additional device to any available IEEE 1394 connection, and secondly every addition or removal of a device is reported to all devices by a global bus reset. As a further support, IEEE 1394 defines directory and data structures as well as their content, which allows devices to identify one another. This means that every device can look for a suitable communication partner within the network and also load the corresponding drivers. IEEE 1394 however only represents the basis for the simple installation and usability of devices. It is also necessary of course for all layers above IEEE 1394, the applications in particular, to use and support these possibilities.

If we classify IEEE 1394 in the sense of the OSI model, then essentially, the lower two layers are supported by the physical layer, the link layer and the transaction layer (Figure 12.8). In the higher levels a number of other protocols come into play of which most are still being standardized. RFC2734 [Joha 99] for example supports the use of the Internet protocols via IEEE 1394, SBP-2 (serial bus protocol) particularly the connection of memory media and AV/C (Audio-Video Command Set) defines commands and status messages for multimedia devices.

12.4 IEEE 1394

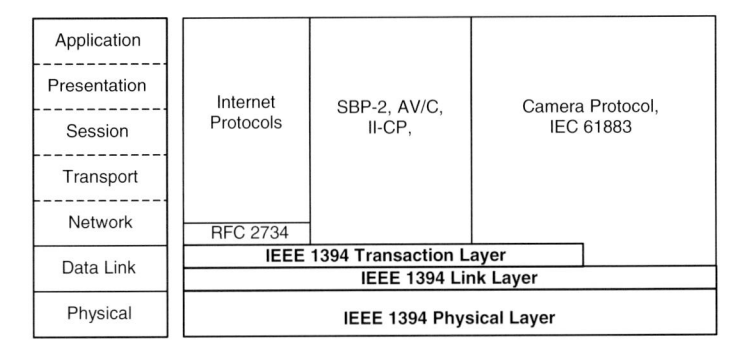

Application			
Presentation	Internet Protocols	SBP-2, AV/C, II-CP,	Camera Protocol, IEC 61883
Session			
Transport			
Network	RFC 2734		
Data Link	IEEE 1394 Transaction Layer		
	IEEE 1394 Link Layer		
Physical	IEEE 1394 Physical Layer		

Figure 12.8: IEEE 1394 and Higher Protocols in the ISO/OSI Reference Model

IEEE 1394 and EIB

As already mentioned at the beginning of the book, the house of the future will contain one or more gateways, which will feed all incoming data into a backbone, which makes the information available everywhere within the house. IEEE 1394 is ideal for this backbone (Figure 12.9), which must guarantee high bandwidths, be user friendly and above all suitable for transmitting multimedia data.

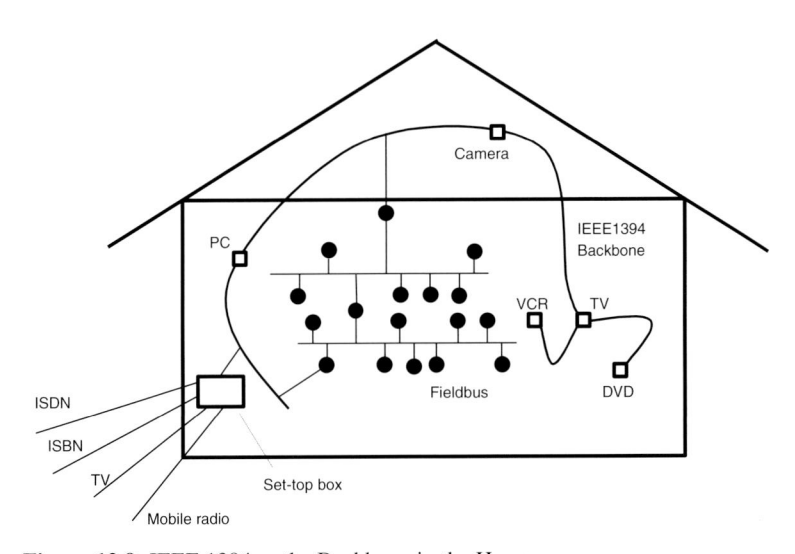

Figure 12.9: IEEE 1394 as the Backbone in the Home

Fieldbuses also allow, in principle, the transmission of voice and picture data [Stam 97]. However, due to their restricted bandwidths, it is better to restrict their usage to the interlinking of sensors and actuators for services such as lighting, heating and ventilation and to provide a link to the multimedia network or backbone. This facilitates the representation of user interfaces for controlling the lights or the downloading of sensor data via the standard TV of the future for example, or the connection of different fieldbus segments via the IEEE 1394 backbone.

12.5 Intelligent Software Agents

What is an agent?

Intelligent software agents (or agents for short) represent a subsection of distributed artificial intelligence. They are used to save and manage knowledge on a distributed basis in order to solve complex problems and to model distributed problem situations. In principle, an agent is a software program. Certain properties however distinguish them from conventional programs.

Table 12.1: Typical Properties of Agents according to [Fra 97]

Property	Meaning
Reactive	Reacts to sensor observations using actuators or communication equipment
Autonomous	Executes its tasks independently without continuous operation
Proactive	Initiates independent actions and is not only reactive
Communicative	Can communicate with other agents and people
Cooperative	Can cooperate with other agents and people

Table 12.1 lists a few typical properties of agents, although it makes no claims on completeness. Other important capabilities of agents are adaptability, mobility and the capacity to learn. An agent therefore is an autonomous, intelligent program that is entrusted with a task which it independently executes using the resources made available to it. These resources may include sensors, actuators and cooperation with other agents. Sensors, actuators and communication are denoted as inputs and outputs of the agent (see Figure 12.10).

Figure 12.10: Agent with Input and Output

At the moment, agent technology is most widely used within the Internet. Agents are used to find information, filter and process it. Mobile agents can deal with electronic purchases and assistant agents help the user to manage his flow of e-mails, look for web sites that may be of interest to him and help him to organize his work more efficiently.

In general we differentiate between two types of agent. The deliberative agent is intrinsically intelligent. It forms internal models of its environment, makes predictions and reflects on its actions. Deliberative agents develop strategies and plans of how to achieve their goals and they also deal with other agents. The various architectures of deliberative agents have a common denominator – they require a great deal of memory and computing power. The second type of agent is called a reactive agent. It has a primitive nature and often only has a reflex-type algorithm. Reactive agents are assembled quickly and easily but are considered non-intelligent. They only gain intelligence within an association. A multi-agent system (MAS) of reactive agents can behave intelligently. Even cooperation and learning are possi-

12.5 Intelligent Software Agents

ble within a reactive MAS. The advantage is that the agents have a very simple construction and with that are more economically priced, robust and easier to maintain [Davi 97].

Agents and EIB

The use of agents in fieldbus systems such as EIB represents the second generation of distributed systems; distributed algorithms are used [Pale 99a]. It is one way of bringing intelligent behavior to the peripherals. Examples of possible applications of multi-agent systems are

- Demand side management (DSM)[41]
- Climate control (heating, ventilation, etc.)
- Lighting technology

Compared with centrally organized applications, agents are flexible, error tolerant and simpler in design.

There are a series of problems associated with implementing agents on the nodes of an EIB network (Figure 12.11).

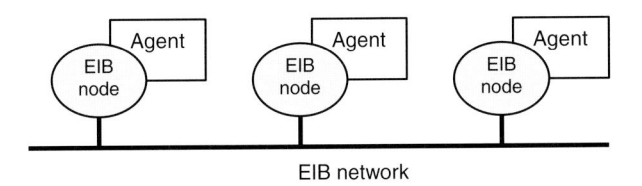

Figure 12.11: Agents on the Nodes of an EIB Network

On one hand, the bandwidth of the network is relatively narrow when compared with the Internet, the computing power in the nodes is limited and they do not have much memory. For this reason, the agents must be kept very simple, which also contributes to the stability and maintainability of the system.

Communication within a multi-agent system is generally message-based. Shared variables (EIB communication objects) can easily be used for this, it is only necessary to keep the length of the messages short (EIB group telegrams can contain a maximum of 14 bytes [EIBA 95]). It is therefore necessary to work out an EIB data format for agent communication. A popular ACL (Agent Communication Language) is KQML (Knowledge Query and Manipulation Language [ARPA 93, Labr 97]). In order to transport the contents of a KQML message via an EIB network in an interoperable way, it is necessary to define an EIS for KQML [Pale 99b]. A separate data type and a separate profile for agents would mean that agents could cooperate and communicate within an EIB network.

Due to the low level of computing power and restricted memory resources, EIB agents are reactive [Pale 99c]. A reactive agent consists of a number of modules (see Figure 12.12) that each carry out a specific and simple task.

---

[41] DSM is the control of electric consumers to avoid peak loads

12 Outlook

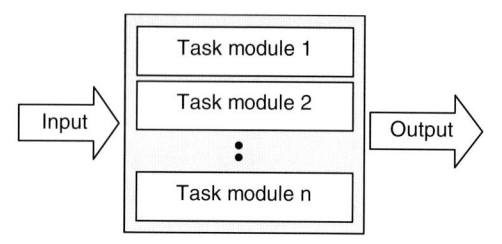

Figure 12.12: Structure of a Reactive Agent

The individual modules carry out specialized tasks, such as for example, controlling temperature or navigating a mobile robot. These tasks are carried out by means of short, simple programs.

An EIB node can house such simple agents for modern, distributed applications. It is expected that like the Internet and other IT areas, agent technology will revolutionize the field of home and building electronic systems.

Abbreviations

ACL	Agent Communication Language	CEPT	Conference Européenne des Administration des Postes et des Télécommunications
AGC	Automatic Gain Control	CGI	Common Gateway Interface
AI	Artificial Intelligence	COM	Component Object Model
ANSI	American National Standards Institute	CRC	Cyclic Redundancy Check
ANubis	Advanced Network for Unified Building Integration Services	CSMA/CA	Carrier Sense Multiple Access with Collision Avoidance
APDU	Application PDU	CSMA/CD	Carrier Sense Multiple Access with Collision Detection
API	Application Programming Interface	DAL	Database Access Library
ARQ	Automatic Repeat Request	DCOM	Distributed COM
ASHRAE	American Society of Heating, Refrigerating and Air-Conditioning Engineers	DDE	Dynamic Data Exchange
		DLL	Dynamic Link Library
		EHS	European Home System
AV/C	Audio-Video Command Set	EIA	Electronic Industries Alliance
BAU	Bus Access Unit	EIB	European Installation Bus
BCD	Binary Coded Digit	EIBA	European Installation Bus Association
BCU	Bus Coupling Unit		
BIM	Bus Interface Module	EIS	EIB Interworking Standard
BPSU	Bus Power Supply Unit	EITT	EIB Interworking Test Tool
BSC	Building Systems Control	EMC	Electromagnetic Compatibility
CAIFM	Computer Aided Integrated Facility Management	EMI	External Message Interface
		ETS	EIB Tool Software
CCITT	Comité Consultatif International Téléphonique et Télégraphique	FAN	Field Area Network
		FDDI	Fiber Distributed Data Interface
CEN	Comité Européen de Normalisation	FM	Facility Management
		FSK	Frequency Shift Keying
CENELEC	Comité Européen de Normalisation Electrotechnique	GAN	Global Area Network

Abbreviations

GPS	Global Positioning System	MAC	Medium Access Control
HBES	Home and Building Electronic Systems	MAS	Multi Agent System
		MAU	Medium Attachment Unit
HDLC	High-Level Data Link Control	MIB	Management Information Base
HES	Home Electronic System	NPDU	Network PDU
HTML	Hypertext Markup Language	NRZI	Non Return to Zero Inverted
HTTP	Hypertext Transfer Protocol	OLE	Object Linking and Embedding
HVAC	Heating, Ventilation, Air Conditioning	OPC	OLE for Process Control
		OSI	Open Systems Interconnections
IC	Integrated Circuit	PAR	Positive Acknowledgement with Retransmission
ICI	Interface Control Information		
IDU	Interface Data Unit	PCI	Protocol Control Information
IEEE	Institute of Electrical and Electronic Engineers	PDU	Protocol Data Unit
		PEI	Physical External Interface
IETF	Internet Engineering Task Force	PIXIT	Protocol Implementation Extra Information for Testing
IFMA	International Facility Management Agency		
		PL	Powerline
IP	Internet Protocol	PLC	Programmable Logic Control
ISBN	Integrated Services Broadband Network	RF	Radio Frequency
		RMI	Remote Method Invocation
ISDN	Integrated Services Digital Network	SAP	Service Access Point
		SCADA	Supervisory Control and Data Acquisition
ISM	Industrial, Scientific, Medical		
ISO	International Organization for Standardization	SDL/GR	Specification and Description Language / Graphical Representation
ITU	International Telecommunication Union		
		SDL/PR	Specification and Description Language / Phrase Representation
JVM	Java Virtual Machine		
KQML	Knowledge Query and Manipulation Language		
		SDLC	Synchronous Data Link Control
LAN	Local Area Network	SDU	Service Data Unit
LDC	Low Duty Cycle	SELV	Safety Extra Low Voltage
LLC	Logical Link Control	SNMP	Simple Network Management Protocol
Lotos	Language of Temporal Ordering Specification		
		SOAP	Simple Object Access Protocol
LPDU	Logical Layer PDU	SQL	Structured Query Language
LU	Logical Unit		

SRD	Short Range Device	UART	Universal Asynchronous Receiver Transmitter
SSDP	Simple Service Discovery Protocol	UPnP	Universal Plug and Play
STEP	Standard for the Exchange of Product Model Data	UTP	Unshielded Twisted Pair
		VDE	Verband Deutscher Elektrotechniker
STP	Shielded Twisted Pair		
TC	Technical Committee	VLDC	Very Low Duty Cycle
TCP	Transmission Control Protocol	WAN	Wide Area Network
TP	Twisted Pair	XML	Extensible Markup Language
TPDU	Transport PDU		

Bibliography

[Ande 99]   Anderson, D.: FireWire System Architecture. Second Edition. MindShare, Inc 1999

[ANSI 95]   ANSI/ASHRAE Standard 135/1995: A Data Communication Protocol for Building Automation and Control Networks. American Society of Heating, Refrigerating and Air-Conditioning Engineers, Atlanta 1995

[Anwe 97]   Handbuch Gebäudesystemtechnik Anwendungen. ZVEI/ZVEH Frankfurt 1997

[Arno 99]   Arnold, K. et al.: The Jini Specification. Addison-Wesley 1999

[ARPA 93]   ARPA Knowledge Sharing Initiative: Specification of the KQML agent-communication language. ARPA Knowledge Sharing Initiative, External Interface Working Group, July 1993

[Asch 99]   Aschemann, G., Kehr, R., Zeidler, A.: A Jini-based Gateway Architecture for Mobile Devices. Proceedings JIT'99, Springer-Verlag 1999, S. 203-213.

[Beck 98]   Becker, M.: Nutzen und Trends der Gebäudeautomation. Universität Kaiserslautern: Tagung Gebäudesystemtechnik und Gebäudeautomation, 4. Juni 1998

[Ben 92]   Bender, K.: Profibus - Der Feldbus für die Automation. 2. Aufl. Carl Hanser Verlag 1992

[Benn 97]   Bennahum, D.: Avatare laden zum Tanz. DIE ZEIT, Hamburg, Nr.9, 21. Februar 1997, S.70

[Caro 97]   Carotta, M.: Home Net Control – Internetbasierende Fernsteuerung von Heimnetzen unter Berücksichtigung von Sicherheitsaspekten. Diplomarbeit, Institut für Automation, Technische Universität Wien 1997

[CDI 98]   EIBA: Documentation of ETE Functions and Interfaces: Enhanced Device Handling / Complex Device Interface. Version 1.1, EIBA s.c., Brussels 1998

[CEN 89]   CEN EN 45001:1989, General criteria for the operation of testing laboratories

[CEN 98]   CEN EN 45011:1998, General requirements for bodies operating product certification systems (ISO/IEC Guide 65:1996)

[CLC 95]	EN50173, Information Technology - Generic Cabling Systems. CLC TC 215, July 1995
[CLC 99]	Draft prEN 50090-9-1, Installation requirements - Generic Cabling for Twisted Pair Class (TP1). CLC/TC 205 (Sec) 171, 11. March 1999
[Davi 97]	Davis, D. N.: Reactive and Motivational Agents: Towards a Collective Minder. In: Müller, J. P., Wooldridge M. J., Jennings N. R. (editors): Lecture Notes in Artificial Intelligence 1193, Intelligent Agents III, Proceedings of ECAI'96 Workshop. Springer 1997
[Dawk 99]	Dawkins, R.: Gipfel des Unwahrscheinlichen – Wunder der Evolution. Rowohlt, 1999
[DDE 98]	EIBA: Documentation of ETE Functions and Interfaces: DDE Server. Version 1.1, EIBA s.c., Brussels 1998
[Diet 95]	Dietrich, D., Neumann, P.: Feldbustechnologie 1995, Schriftenreihe Nr. 9 des ÖVE (Österreichischer Verein für Elektrotechnik), Wien 1995
[Diet 97]	Dietrich, D., Schweinzer, H.: Feldbustechnik in Forschung, Entwicklung und Anwendung. Springer Wien, New York 1997
[Diet 99]	Dietrich, D., Loy, D., Schweinzer, H.-J.: LON-Technologie, Verteilte Systeme in der Anwendung. 2. Aufl. Hüthig 1999
[DINV 92]	DIN V 32734 Digitale Automation für die Technische Gebäudeausrüstung - Allgemeine Anforderungen für die Planung und Ausführung. Beuth, April 1992
[Eagl 00]	EIBA: eteC Eagle V1.0 Documentation. EIBA s.c., Brussels 2000
[Ecke 99]	Eckel, B.: Thinking in Java. Prentice Hall, New Jersey 1999
[Edwa 99]	Edwards, K.: Core Jini. ISBN 0-13-0114469-X Prentice Hall 1999
[EIA 568]	EIA/TIA 568 Commercial Building Wiring Standard
[EIBP 97]	EIBA: EIB-Proceedings Fachbeiträge. Band 1 Brüssel 1997
[EIBA 95]	EIBA: Handbook for Development, Issue EIB 2.21, 1995
[EIBA 98]	EIBA: Handbook for Development, Pre-Release Issue EIB 3.0, 1998
[EIBA 99]	EIBA: Handbook Release 3.0. EIBA s.c., Brussels 1999
[ETS 99]	EIBA: ETS2 V1.1 Manual. EIBA s.c., Brussels 1999
[Falc 00]	EIBA: eteC Falcon V1.0 Documentation. EIBA s.c., Brussels 2000
[Fire 95]	IEEE Standard for a High Performance Serial Bus Std. 1394,. ISBN 1-55937-583-3, 1995
[Fisc 97]	Fischer, P.: Ein Feldbus-Profil auf dem Weg zur europäischen Norm. In: Dietrich, D., Schweinzer, H. (Hrsg.): Feldbustechnik in Forschung, Entwicklung und Anwendung. Springer Wien, New York 1997
[Faci 98]	Florian, V.: Virtual Reality für FM. FACILITY Management Magazin Nr.11a, Wien 1998, S. 22

[Fowl 98]	Fowler, M.: UML Distilled. Addison-Wesley, Reading 1998
[Fra 97]	Franklin, S., Graesser, A.: Is it an agent, or just a program? A taxonomy for autonomous agents. In: Müller, J. P., Wooldridge M. J., Jennings N. R. (editors): Lecture Notes in Artificial Intelligence 1193, Intelligent Agents III, Proceedings of ECAI'96 Workshop. Springer 1997
[Fran 97]	Frank, K.: EIB – ein neues Geschäftsfeld für den Elektroinstallateur. Verlag Technik 1997
[Fris 98]	Frisch, M., Kullmann, C.: Leistungsaspekte des EIB. Ingenieurmäßiges Arbeiten, Fachhochschule Dortmund, Fachbereich Nachrichtentechnik, 1998
[Fris 99]	Frisch, M., Kullmann, C.: Leistungsmessungen an einem EIB-System auf Anwenderebene. Diplomarbeit, Fachhochschule Dortmund, Fachbereich Nachrichtentechnik, 1999
[Patt 97]	Gamma et al.: Design Patterns: Elements of Reuseable Software. 2nd Edition, Addison Wesley, New York 1997
[Ghah 98]	Ghahremani, A.: Integrale Infrastrukturplanung, Facility Management und Prozeßmanagement in Unternehmensinfrastrukturen. Springer 1998
[Gois 98]	Goiser, A.: Handbuch der Spread-Spectrum-Technik. Springer 1998
[Grun 97]	Handbuch Gebäudesystemtechnik Grundlagen. 4. Aufl. ZVEI/ZVEH Frankfurt 1997
[HaWi 82]	Hasse P., Wiesinger, J.: Handbuch für Blitzschutz und Erdung. Richard Pflaum Verlag München und VDE Verlag Berlin, Offenbach 1982
[IEC 95]	ISO/IEC 11801, Information Technology - Generic Cabling for Customer Premises. July 1995
[ISO 89]	ISO 7498-2, Information processing systems - Open system interconnection – Basic Reference Model – Part 2: Security architecture. ISO 1989
[ITSEC 91]	Information Security Evaluation Criteria (ITSEC). Provisional Harmonized Criteria, Commission of the European Communities, 1991
[Jahr 97]	Grütz, A.: Jahrbuch Elektrotechnik. VDE Verlag Berlin, Offenbach 1997
[Jini 99a]	The Community Resource for Jini Technology: http://www.jini.org
[Jini 99b]	Jini-Forum: http://www.informatik.tu-darmstadt.de/de-jini-users/
[Joha 99]	Johanson, P.: IPv4 over IEEE 1394. RFC 2734, December 1999
[Kais 98]	Kaiserswerth, M., Posegga J.: Java auf Chipkarten. Informatik Spektrum 21 (1998), S. 27-28
[Kast 99a]	Kastner W., Krügel C., Reiter H.: Jini ein guter Geist für die Gebäudesystemtechnik. Proceedings JIT'99, Springer-Verlag 1999, S. 213-223
[Kast 99b]	Kastner W., Krügel C.: Jini Connectivity for Home and Building Networks, http://www.auto.tuwien.ac.at/jini

[Kni 97]   Knizak M., Manninger M., Sauter T.: Einbindung von Feldbussen in das Internet via SNMP. e&i 114 (1997) S. 258-262

[Konh 98]   Konhäuser, W.: Industrielle Steuerungstechnik. Hanser 1998

[Kran 97]   Kranz, H.: Building Control, Technische Gebäudesysteme. Automation u. Bewirtschaftung. 2. Aufl. expert-Verlag Renningen-Malmsheim 1997

[Krie 00]   Kriesel, W. Heimbold, T., Telschow, D.: Bustechnologien für die Automation. 2. Aufl. Hüthig 2000

[Kris 97]   Kristen, W.: Anwendung des EIB-Feldbussystems im HLK-Bereich. EIB-Proceedings, Fachbeiträge Band 1, Brüssel 1997, S. 107

[Krüg 99]   Krüger G.: Go To Java 2. Addison-Wesley 1999

[Labr 97]   Labrou, Y., Finin, T.: Semantics for an Agent Communication Language. In Singh, M. P., Rao, A., Wooldridge M. J. (editors): Lecture Notes in Artificial Intelligence 1365: Intelligent Agents IV, Agent Theories, Architectures and Languages. Proceedings of ATAL'97, 1997, S. 209-214

[Lind 67]   Lindner, H.: Biologie. J. B. Metzlersche Verlagsbuchhandlung. Stuttgart 1967

[Lemm 98]   Lemme, H.: Wie sicher sind Chipkarten? Elektronik 16 (1998)

[LonM 96]   LonMark Layers 1-6 Interoperability Guidelines. LonMark Interoperability Association, Version 3.0, 1996

[LonM 97]   The SNVT Master List and Programmer's Guide, LonMark Interoperability Association, Version 8, 1997

[LonM 98]   LonMark Application Layer Interoperability Guidelines, LonMark Interoperability Association, Version 3.1, 1998

[LonW 93]   Echelon: LonWorks Engineering Bulletin: LonTalk Protocol. Echelon Corporation, Palo Alto, CA, USA, 005-0017-01C April 1993

[LNS 97]   LNS Host API. Echelon Corporation, Palo Alto 1997

[LCA 97]   LCA Object and Data Server. Echelon Corporation, Palo Alto 1997

[Mann 98]   Manninger, M.: Netzwerksicherheit durch Chipkarten als Schlüsseltechnologie des Zahlungsverkehrs im Internet. Dissertation, Technische Universität Wien 1998

[Mark 90]   Markley, R. W.: Data communications and interoperability. Prentice-Hall 1990

[Nävy 98]   Nävy, J.: Facility Management, Grundlagen, Computerunterstützung, Einführungsstrategie, Praxisbeispiel. Springer 1998

[Opc 99]   Website: http://www.opc-foundation.org

[Pale 99a]   Palensky, P., Gordeev, M.: Demand Side Management by using Distributed Artificial Intelligence and Fieldbus Technology. Proceedings International Conference on Intelligent and Responsive Buildings, Bruges, March 1999

[Pale 99b] Palensky, P.: On Interoperability and Intelligent Software Agents for Field Area Networks. In Dietrich, D., Schweinzer, H. (Hrsg.): Feldbustechnik in Forschung, Entwicklung und Anwendung. Springer Wien, New York 1997

[Pale 99c] Palensky, P.: Intelligent Software Agents for EIB Networks. Proceedings EIB Scientific Conference, Munich, 28th October 1999

[Patt 94] Patterson, D., Hennessy, J. L.: Computer Organization & Design, the Hardware/Software Interface. Morgan Kaufmann Publishers, San Mateo, California 1994

[Pete 96] Peterson, L. L., Davie, B. S.: Computer Networks – A Systems Approach. Morgan Kaufmann Publishers, San Francisco 1996

[Pigl 90] Pigler, F.: EMV und Blitzschutz leittechnischer Anlagen. Siemens Aktiengesellschaft Berlin, München 1990

[PocF 00] EIBA: eteC PocketFalcon V1.0 Documentation. EIBA s.c., Brussels 2000

[Popp 00] Popp, M.: Profibus-DP/DPV1. 2. Aufl. Hüthig 2000

[Prod 98] Presinell, R., Seemeyer P.: Die Zukunft hat begonnen, Intelligente Gebäudesystemtechnik in den Treptowers der Allianz. Roman Presinell, Produktprofile Nr.2, Berlin 1998

[Raab 98] Raab, V.: Überspannungsschutz in Verbraucheranlagen. Verlag Technik, Berlin 1998

[Rank 99] Rankl, W., Effing, W.: Handbuch der Chipkarten. 3. Aufl. Hanser, Munich 1999

[Redl 97] Redlein, A.: IFC R3.0 Domain Project Documentation, Visualization and Control within FM-Systems. International Alliance for Interoperability Vienna 1997

[Reiß 98] Reißenweber, B.: Feldbussysteme. Oldenbourg, München Wien 1998

[Reit 98] Reiter, H.: Internet Connectivity for Residential Field Area Networks. Dissertation, Institut für Automation, Technische Universität Wien 1998

[Rei 99] Reiter, H.: The Internet Challenge: Establishing Global Connectivity for EIB Networks. EIB-Proceedings 2/99, EIBA 1999, S. 113-124

[Rog 98] Rogerson, D.: Inside COM. Microsoft Press, Redmond 1998

[Rond 95] Rondeau, E. P., Brown, R. K., Lapides P. D.: Facility Management. John Wiley & Sons, New York 1995

[Rose 00] Rose, M., Kriesel, W., Rennefahrt, J.: EIB für die Gebäudesystemtechnik in Wohn- und Zweckbau. 3. Aufl. Hüthig 2000

[Rubi 98] Rubin, W. et al.: Understanding DCOM. Prentice Hall, New Jersey 1998

[Ruha 98] Ruhaltinger, J.: Made in Austria. Industriemagazin Spezial Nr.10, Wien, Oktober 1998, S. 12

[Scha 91] Schaumüller-Bichl, I.: Sicherheitsmanagement. Habilitationsschrift, Universität für Bildungswissenschaften, Klagenfurt 1991

[Schm 98]	Schmit, R.: Kommunikationssysteme - Optimale Methoden des Systemdesigns. Dissertation, Technische Universität Wien 1998
[Schn 97]	Schneider, W.: Praxiswissen digitale Gebäudeautomation. Vieweg 1997
[Schw 98]	Schwaiger, C.: Home Net Control Server – A Gateway between Java and Home Systems. Diplomarbeit, Institut für Automation, Technische Universität Wien 1998
[Schu 97]	Schulz, T.: ISDN am Computer. Springer 1997
[Sdk 99]	EIBA: ETS2 V1.2 SDK Documentation. EIBA s.c., Brussels 2000
[Sdk 00]	EIBA: eteC Plug-In SDK V1.0 Documentation. EIBA s.c., Brussels 2000
[Seip 93]	Seip, G. G.: Elektrische Installationstechnik (Teil 1 und 2). 3. Aufl. Siemens 1993
[Slam 99]	Slama, D. et al.: Enterprise Corba. Prentice Hall, New Jersey 1999
[Sonn 97]	Sonntag, P.: Chipkarten - Stand der Normung. it+ti 5, Oldenbourg, München 1997
[Sta 93]	Stallings, W.: SNMP, SNMPv2 and CMIP. Addison-Wesley 1993
[Sta 98]	Stallings, W.: SNMPv3: A Security Enhancement for SNMP. IEEE Communications Surveys, Fourth Quarter 1998, Vol. 1, No. 1
[Stam 97]	Stampfl, N.: Standbildübertragung über LON. Diplomarbeit, Institut für Computertechnik, Technische Universität Wien 1997
[Stam 99]	Stampfl, N.: IEEE1394 in Comparison with Other Bus Systems. In Dietrich, D., Schweinzer, H. (Hrsg.): Feldbustechnik in Forschung, Entwicklung und Anwendung. Springer Wien, New York 1997
[Stein 99]	Steininger, A.: The Testing of Fault-Tolerant Computers. Habilitationsschrift, Technische Universität Wien 1998
[Step 94]	Stephanson, N.: Snow Crash. Berlin 1994
[Str 97]	Strobel, S.: Firewalls für das Netz der Netze. dpunkt-Verlag Heidelberg 1997
[Szy 98]	Szyperski, C.: Component Software – Beyond Object-Oriented Programming. Addison-Wesley 1998
[Tane 90]	Tanenbaum, A.: Computer-Netzwerke. Wolfram's Fachverlag 1990
[Tane 95]	Tanenbaum, A.: Verteilte Betriebssysteme. Prentice Hall 1995
[Tane 96]	Tanenbaum, A.: Computer Networks. 3rd ed. Prentice-Hall 1996
[Tis 00]	Tiska, R.: Integrated Management Model for Building and Home Automation Systems. Dissertation, Institut für Computertechnik, Technische Universität Wien 2000
[Vale 99]	Valesky, T. C.: Enterprise Javabeans: Developing Component-based Distributed Applications. Addison-Wesley, Reading 1998

[VDE 92a]	DIN V VDE 0829 Teil 100, Elektrische Systeme für Heim und Gebäude - Aufbau der Norm; Begriffe. Beuth, Berlin November 1992
[VDE 92b]	DIN V VDE 0829 Teil 210, Elektrische Systeme für Heim und Gebäude – Systemübersicht; Architektur. Beuth, Berlin November 1992
[VDE 92c]	DIN V VDE 0829 Teil 220, Elektrische Systeme für Heim und Gebäude - Systemübersicht; Allgemeine technische Anforderungen. Beuth, Berlin November 1992
[VDE 92d]	DIN V VDE 0829 Teil 320, Elektrische Systeme für Heim und Gebäude - Anwendungsschicht Klasse 1. Beuth, Berlin November 1992
[VDE 92e]	DIN V VDE 0829 Teil 410, Elektrische Systeme für Heim und Gebäude - Transportschicht Klasse 1. Beuth, Berlin November 1992
[VDE 92f]	DIN V VDE 0829 Teil 420, Elektrische Systeme für Heim und Gebäude - Vermittlungsschicht Klasse 1. Beuth, Berlin November 1992
[VDE 92g]	DIN V VDE 0829 Teil 521, Elektrische Systeme für Heim und Gebäude - Twisted Pair Klasse 1; Sicherungsschicht. Beuth, Berlin November 1992
[VDE 92h]	DIN V VDE 0829 Teil 522, Elektrische Systeme für Heim und Gebäude - Twisted Pair Klasse 1; Bitübertragungsschicht. Beuth, Berlin November 1992
[Vdma 96]	VDMA-Einheitsblatt 24196, Gebäudemanagement Begriffe und Leistungen. Beuth, Berlin August 1996
[Virt 98]	Die Cyberwelt von Arcitec. Virtual Reality, a3 Bau & a3 Byte Magazin Nr.7a, Wien 1998, S. 62
[Zech 97]	Zech, P. R.: Facility Management in der Praxis, Herausforderung in Gegenwart und Zukunft. ExpertVerlag 1997

Index

A/D converter 136
A_Authorize 139, 140
A_Read-PDUs 127, 132, 133, 134, 136, 137, 138, 141, 144, 145
A_Set_PhysAddr 130
A_SetKey 139
A_Write-PDUs 127, 134, 137, 141, 146
Access class 74, 87
Access conflict 72
Access control 229
Access level 139, 148
Access rights 148
Ack 125
Ack_PDU 115
Acknowledgement 114
   Negative 84, 85, 87, 110, 122
   Positive 84
Acknowledgement frame 79, 85
Acknowledgement timer 110, 114, 116
ACL 299
Activation logics 176
ActiveX 248
Address See physical address
Address field 68
Address table 151, 156, 207
Address table object 137, 148, 150
Addressing 49, 81, 155
Agent 222, 293, 298
   Reactive 299
ANubis 226, 246, 248
APDU 141, 146
API 147
Applet 222, 294
Application 143, 145
Application functionality 267
Application layer 28, 51, 109, 125, 126, 143, 145, 150
Application layer protocol data unit see APDU

Application module 145, 169
Application program 156
Application program object 137, 148, 150
Application programming 246
Application programming interface 147
Arbitration 34, 50, 52, 68, 69, 73, 87
ARQ 111
Artificial intelligence 298
ASDU 141
Association table 126, 153, 156, 208
Association table object 137, 148, 150
Authentication 139, 155, 229
Authorize 140
Automatic repeat request 111
AV/C 296
Availability 229

Backbone 54, 71, 102, 218, 219, 297
Backbone coupler 51, 68, 71, 101, 102, 104, 141, 156, 187, 217
BACnet 220, 228, 261, 289, 290
Bandwidth 16, 29, 57, 299
   Guaranteed 295
Batibus 287
BC See backbone coupler
BCU 52, 169
BIM 179, 185
Bit error 95
Bit error rate 285
Bit monitoring 95
Bit time 73, 74, 78, 87, 89
Bridge 96, 101, 106, 150, 216
Broadcast 99, 109, 122, 126, 141, 151
Buffering time 162, 196
Building automation and control 23, 51, 257
Building control system
   Integrated 262

311

Index

Building discipline 23, 61
Building information system 262
Building management 250, 252, 262
Building model 256, 263
Building systems control 42, 51, 257, 259, 261
Building systems engineering 59
Bus 34
Bus coupling unit 52, 145, 169
Bus interface module 185
Bus monitor 239
Bus power supply unit 71
Busy acknowledgement frame 84
Busy frame 85, 87

Cabling 56
CAIFM 259
CAN 49
Carrier Sense Multiple Access 35, 52, 68, 73
Carrier Sense Multiple Access with Collision Avoidance 50
CCITT 27, 37
CEN TC 247 273
CENELEC 288
Certification 51, 185, 239, 247, 270, 274
Character coding 95
Check character 68, 83, 84, 87, 89, 94
Chip card 231
Choke 163
Client 144, 290
Client/server model 144, 150
Collision 35, 68, 69, 99, 279
  Avoidance 69, 72, 73
  Detection 35, 69
  Resolution 50, 160
COM 243, 245, 246
Commissioning 39, 104, 167, 186, 192, 205, 241
Communication
  Connectionless 109, 125, 139, 141, 148
  Connection-oriented 109, 125, 134, 136, 141, 148
  Secure 235
Communication field 292
Communication flag 143
Communication module 168, 185, 293

Communication object 126, 127, 143, 152, 153, 207
Communication object server 147
Communication object table 152
Communication phase 109
Communication relationship 150
Compatible 268
Complex device 245
Complex device interface 242
Confidentiality 229
Configuration 151, 155
  Channel 150
Configuration data 68, 151
Confirmation 65, 66
Conformity 267, 270
Connection setup 111, 121
Connection timer 110, 117, 119
Connect-PDU 111
Control field 68, 90, 91, 125
Control information 125, 141
Controller area network 49
Convergence 287
Costs 19, 22, 36, 41, 57, 59, 218, 253, 259
Coupler 157, 185
Coupling networks 216
CRC 30
CSMA/CA 50, 52, 68, 73, 76, 80, 87, 91, 94
CSMA/CD 35
Cyclic redundancy check 30

DAL 245
Data acquisition 255, 257
Data confidentiality 229
Data frame 68, 79, 81, 82, 125
  Formation 31
  Ranking 92
Data integrity 229
Data interface 59, 158, 164, 169, 191, 196
Data link layer 30, 69, 79, 150
Data point 248, 262
Data rail 163
Data rate 44, 48, 50, 53, 57, 70, 160, 167, 170, 175, 176, 180, 184, 218, 295
Data throughput 278, 280
Data transmission 113, 160
Data type 53, 150, 269

Index

Data_Unack-PDU 122, 123
Datagram 30, 80, 82, 87, 95
Data-PDU 114, 116, 122, 125
DCOM 243
DDE server 242
Delay 155
Democratic concept 28
Destination address 82, 100
Destination address flag 82, 90, 93, 98, 100, 101, 104, 125
Development environment 239
Device address 96, 154
Device addressing See physical addressing
Device configuration 126, 151, 155
Device number 81, 90, 92
Device object 137, 148, 150
Discipline 277
Disconnection 121
Disconnect-PDU 116, 119
Discovery protocol 291
Djinn 291
DM-PDUs 155
Domain address 154, 158
Domain ID 154, 155
Dominant 50, 73, 74, 76, 94
Duty cycle 166, 176

Eagle 243, 244
Easy installation mode 288
Economic viability 36, 257
Economics 59
EHS 272, 275, 287
EIA/TIA 568 56
EIB Interworking Standard 53, 169
EIBA 51, 169, 247, 274, 275, 288
EIBlib 228, 245
EIBnet 243, 289, 290
EIS 53, 169, 268
EITT 239
EMC 41
EMI 151, 227
E-mode 288
EN 50082 42
EN 50090 2-2 42
EN 50170 33, 44, 47
EN 50254 43
Energy management 274, 276
Error control 80

Error detection 95
eteC 227, 243, 244, 245
Ethernet 16, 30, 57, 58, 216, 219, 220, 226, 287, 295
ETS 197, 240, 241, 246, 277, 288, 293
Extension 24, 52, 217, 218
External message interface 151
External Message Interface 227

Facility management 24, 60, 238, 249
    Virtual 262
Facility manager 257
Failure rate 285
Falcon 227, 244, 246
FAN 25
Fiber-optic cable 47, 53, 56, 217
Field area network 25
Fieldbus 25, 42
Firewall 217
FireWire 16, 58, 226, 295
Flow control 80, 84
Frame 67, 79
Frame check sequence 44
Frame format 95
Frequency range 165, 167, 175, 176, 187
FSK 167, 176
FT1.2 199, 214
Fully interconnected network 36

GAN 25
Garbage collection 291
Gateway 36, 217, 219, 221, 227, 231, 234, 261, 297
Global area network 25
Group 93, 122, 151, 293
Group address 68, 93, 104, 126, 152, 153, 228
Group addressing 52, 81, 141, 207
GroupDataTransfer 246

Handler 67
Hawk 245, 247
HBES 57, 271, 288
HDLC protocol 30
Heating, ventilation, air-conditioning See HVAC
Hierarchy 24, 158, 186, 189
    Automation 43
    Network 26, 36

Object 245, 247
Home and building electronic systems 60, 68, 276, 288, 292, 294
Home automation 16, 292
House systems automation 59, 62
HTML 221
HVAC 23, 55, 270, 272, 275, 277, 299

ICI 32, 64
IDE 239
Identification number 148, 210
IDU 32
IEEE 1394 58, 295
IEEE 802.2 30
iETS 228, 243
IFMA 251
iLink 295
Indication 65
Infrared 53, 68, 69, 271
instabus 276
Integrity 229
Interaction time diagram 36
Interbus 43
Interface 64, 143, 291
Interface control information 32, 64
Interface data unit 32
International Facility Management Agency 251
International Organization for Standardization 24, 63
Internet 220, 231, 243, 260, 292, 298
Internet gateway 231, 236
Interoperability 21, 150, 238, 256, 264, 266
Interoperable 268
Interworking standard 268
Interworking test tool 239
ISBN 16
ISDN 16, 25, 220, 227
ISO 24, 63
ISO 8473 30
ITU 37

Java 221, 231, 239, 243, 260, 291
Jini 290
JVM 233, 291

KQML 299

L_Data 82
LAN 25, 217
Latency time 278, 279
Layer design 63
Layer layout 31
LC See line coupler
Lightning protection 41
Line 34, 68, 71, 92
   Attenuation 160
   Connection 163
   Dimensioning 161
   Length 70, 159
   Material 70, 159
Line address 154
Line coupler 51, 68, 71, 92, 101, 102, 104, 141, 156, 186, 187, 217, 219
Line number 81
Link layer protocol data unit 82, 90
Link layer service data unit 82
LLC 30, 79, 86, 94
Local area network 25
Logical address 93, 106, 151
Logical addressing 81
Logical link control 30, 79
Logical unit 69, 71
LonWorks 53, 220
Lookup 291, 293
Low duty cycle 158
LPDU 82, 90
LSDU 82

MAC 30, 80
Main group number 81, 93
Main line 71, 92, 102
Main zone 92
Maintenance 15, 20, 27, 36, 59, 220, 240, 252, 253, 254, 257, 263, 276
Management information base 223
Mask version 141
Master 34, 44, 69, 82
Matched filter receiver 174
MAU 69, 76
Media 157, 159, 189
   Open 205, 246
Media coupler 189
Medium access control layer 80
Medium attachment unit 69, 76
Memory access 155

MemoryAccess 246
MIB 46, 223
Modulation method 176
Multi-agent system 298
Multicast 99, 109, 122, 126, 148
Multimedia 58
   Device 295
Multi-vendor system 264

N_Broadcast 99
N_Data 98, 111
N_Group 99
Nack-PDU 116, 122, 125
Negative acknowledgement frame 84
Network configuration 151, 154
Network layer 29, 79, 150
Network management 126, 143, 149
Network management tool 150, 155, 156
Network protocol data unit 98
Network scan 154
Network service data unit 98
NM-PDUs 154
Non Return to Zero Inverted 31
NPDU 98
NRZI 31
NSDU 98

ObIS 269
Object 136, 143, 148, 155, 210, 239
Object Interworking Standard 269
Object Linking and Embedding 245
Octet 71
OLE 46, 245, 247
OPC 228, 248
Open system 63, 265
Open Systems Interconnection 27, 63
OSI model 27, 63, 64, 65, 106, 150, 266
Overvoltage protection 41

P_Data 74
Packet 67
PAR 111
Parity bit 72, 76, 95
Parity check 72, 95
PCI 32, 64
PDU 32, 64, 65
Peer-to-peer 32, 52
PEI 52, 151, 155, 168, 181, 196
PEI program 156

Physical address 68, 90, 92, 93, 98, 105, 106, 131, 151, 154, 207
Physical addressing 52
Physical external interface 151, 155, 168, 181
Physical layer 31, 69, 79, 84, 88
PIXIT 240, 269, 275
Plug & play 295
Plug-in 248, 293
P-NET 34, 44
Point-to-point connection 19, 33, 219
Polling group 82
Polling master 239
Positive acknowledgement 87, 114, 123
Positive acknowledgement with retransmission 111
Post 22, 109
Power net 294
Power supply 71, 160, 177
Powerline 51, 53, 69, 154, 157, 166, 190, 204, 246
   Band stop 187
   Communication module 181
   Phase coupler and repeater 189
   Transceiver 174
Presentation layer 29, 126, 150
Primary cabling 56
Priority 50, 68, 80, 82, 85, 91, 98, 114
Processing unit 178
Product database 275, 277, 293
Profibus 47, 220, 289
Profile 268
Profitability 253
Programming mode 68, 130, 154, 155
Property 137, 139, 148, 155, 211, 239
Protocol 64, 66, 290, 291
Protocol control information 32, 64
Protocol data unit 32, 64
Proxy 217

Radio 53, 69, 154, 165, 271, 287
   Implementation 181
   Topology 158
   Transceiver 176
Radio frequency 157
RAM flag 143, 152, 192, 203, 207
Range 159, 166, 189, 218
Reaction time 278, 279
Read_Group 128

Index

Read_Mask-PDUs 141
Read_Memory-PDUs 134
Read_PhysAddr-PDUs 132
Read_Property_Value-PDUs 137
Receive counter 110, 114, 125
Receive object 153
Recessive 50, 73, 74, 76, 94
Redundancy 20, 21
Reliability 20, 35, 41, 163, 166, 284, 285
Remote control 219, 238, 248
Remote maintenance 218, 220, 294
Remote method invocation 291
Repeat flag 91
Repeated data frame 91
Repeater 51, 150, 187, 216, 219
Request 65
Reset 155, 195, 204
Resource optimization 253
Resource sharing 290
Response 65, 67
Response time 278, 280
Restart 141
Retransmitter protocol 166
Ring 34, 43, 45
RMI 291
Router 52, 101, 150, 159, 216, 219
Routing counter 98, 99, 105, 108
Routing table 156
RS485 43, 45, 47, 51, 219

SBP-2 296
SDL 36
SDLC 30
SDS 51
SDU 32, 64
Secondary cabling 56
Security 23, 229, 230, 233, 243, 292
Security management 252
Semantics 153
Send counter 110, 125
Sensor/actuator bus 43
Sequence number 110, 114
Serial communication 184
Serial interface 151
Serial number 137, 154
Serial protocol 213
Server 144, 148, 260, 290
Service 63, 290, 291, 293
   Answered 66

Confirmed 66
Connectionless 30
Connection-oriented 30
Locally confirmed 65
Service data unit 32, 64
Service primitive 65
Service provider 32, 63
Service user 32, 63, 291
Session layer 29, 126, 150
Set_Key-PDUs 139
Set_PhysAddr_Req-PDU 130
Set-top box 16, 288
Shielding 68, 69
Short range device 165, 176
Signal coding 69, 76
Signal parameter 77
Signal propagation time 50, 77, 160
Simple Network Management Protocol 222
Simple Object Access Protocol 224
Simulation 263
Slave 82
Smart card 231
Smart distributed system 51
SNMP 222
SNMPv3 233
SOAP 224, 227
Software agent 298
Software device model 193
Source address 82, 92
SSDP 225, 228
Standardization 51, 264
Start bit 72, 78
State machine 156
Stop bit 72, 78
Sub-group number 81, 93
Sub-line 71, 102
Sub-network 159
Summation frame 44
Supply mains 68
Synchronization 29, 72
Synchronous bus system 35
System integrator 23, 241, 263, 265
System object 148, 155, 210
System software 168, 185, 192, 194

T_Broadcast 124, 134
T_Connect 111, 134, 136
T_Data 113, 134

T_Data_Unack 122, 137
T_Disconnect 117, 134
T_Group 123
TC 205 57, 271, 272, 273, 288
TC 247 270, 271, 289
TCP 29
TCP/IP 217, 221, 243, 260, 261
Telegram 67
Tertiary cabling 56
Time monitoring 95
Time slot method 35
Timer 37
Token 35
Token ring 47
Tool 26, 36, 39, 238
Tool environment component architecture 243
Topology 33, 51, 52, 68, 157, 263
TPUART 173
Transceiver 168, 169, 176
Transformer 171
Transmission
    Asynchronous 184, 296
    Isochronous 296
    Synchronous 184
Transmission behavior 278
Transmission Control Protocol 29
Transmission errors 30
Transmit object 153
Transport layer 29, 109, 150
Tree structure 36
Tunnel protocol 217, 227
Twisted pair 42, 45, 51, 57, 68, 69, 76, 87, 157, 167, 178, 278, 281, 285, 294

Twisted pair transceiver 170

UART character 72, 74, 78, 87, 95
Unicast 98, 109, 122, 126
U-PDUs 147
User data 83, 89, 93, 98, 106, 125, 141
User layer 143, 148, 150

Very low duty cycle 158, 166
Virtual building 262
Virtual circuit service 30
Virtual prototyping 262
Virtual token passing 35, 45
Visualization 39, 220, 238, 244, 248, 262

Waiting time 284
WAN 25
Wide area network 25
Windows CE 246
Work environment 254
Write_Group_Req-PDU 129
Write_Memory-PDUs 134
Write_Property-PDUs 138

X.21 31
X.25 30
XML 224, 228

Z.100 38
Z.120 37
Zero 76
Zone 68, 71, 92
Zone address 154, 159

Seip, Günter G. (Ed.)

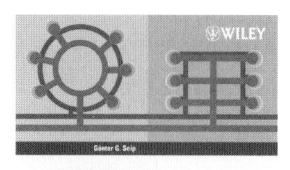

Electrical Installations Handbook

Power Supply and Distribution
Protective Measures
Electromagnetic Compatibility
Electrical Installation Equipment and Systems
Application Examples for Electrical Installation Systems
Building Management

3rd revised edition, 2000, 728 pages, 619 illustrations, 180 tables, 18 cm x 25 cm, hardcover
ISBN 3-89578-061-8
DM 235,00 / € 120,15 / sFr 209,00

This book gives planners, installation engineers, owners, students or trainees, all the information they need for planning and setting up electrical installations. It provides a comprehensive overview of the installation equipment and systems to be implemented, and gives details of latest developments in systems engineering. Special emphasis is placed on installation equipment with communication capability and, in particular, on the way in which this equipment is networked with the instabus EIB bus system for a wide range of applications in residential and functional buildings. Reference has been made, when necessary, to current international, European and German norms, regulations and standards.

Heinhold, Lothar; Stubbe, Reimer; Glaubitz, Wilfried (Eds.)

Kabel und Leitungen für Starkstrom / Power Cables

Auswahl und Projektierung / Selection and Configuration

Version 1.0, 2002
ISBN 3-89578-111-8
DM ca. 213,19 / € 109,00 / sFr ca. 189,00

The CD ROM provides the user with highly effective support in selection of power cables for the low and medium-voltage ranges. In addition to detailed descriptions of the products and of how they can be used, the CD ROM includes tables with up-to-date project planning data (design, electrical and thermal characteristics). For each cable type selected, it also allows the calculation of current carrying capacity and voltage drop. For the technician this CD is an excellent supplement to the standard volume "Power Cables and their Application".

Bezner, Heinrich

Fachwörterbuch industrielle Elektrotechnik, Energie- und Automatisierungstechnik

German-English
4th revised and enlarged edition, 1998, 608 pages,
14,3 cm x 22,5 cm, hardcover
ISBN 3-89578-077-4
DM 134,95 / € 69,00 / sFr 120,00

Dictionary of Electrical Engineering, Power Engineering and Automation

English-German
4th revised and enlarged edition, 1998, 532 pages,
14,3 cm x 22,5 cm, hardcover
ISBN 3-89578-079-0
DM 134,95 / € 69,00 / sFr 120,00

Together with the great number of basic electrotechnical terms, it comprehensively covers large fields of industrially applied electrical engineering with its more than 67,000 entries in Volume 1 (German/English) and 54,000 entries in Volume 2 (English/German). This universal character is also clearly shown by the new title.

Current technical literature, national and international regulations and standards (DIN VDE, IEC, EN, BS, ANSI, CEE, ISO) and the database of the Language Service of Siemens AG were used as an important source for generating the dictionary. Regulations and standards are indicated with the appropriate terms and specialized fields in those cases where a term or translation is closely connected with these. For the 4th edition, the complete database was updated and supplemented by about 2,500 new terms in both language directions.

CD-ROM
Edition 1999
ISBN 3-89578-137-1
DM 310,98 / € 159,00 / sFr 276,00

The CD-ROM contains all entries of both parts of the printed edition, Part 1 (German-English) with over 67,000 and Part 2 (English-German) with over 54,000 entries. All entries have been revised and some 2500 terms added.

With the new designed software one can find quick and easy correct translations of special terms. Additionally users can append their own remarks to the entries. On demand there is a network version of the dictionary available.